Studies in Computational Intelligence

Volume 630

Series editor

Janusz Kacprzyk, Polish Academy of Sciences, Warsaw, Poland
e-mail: kacprzyk@ibspan.waw.pl

About this Series

The series "Studies in Computational Intelligence" (SCI) publishes new developments and advances in the various areas of computational intelligence—quickly and with a high quality. The intent is to cover the theory, applications, and design methods of computational intelligence, as embedded in the fields of engineering, computer science, physics and life sciences, as well as the methodologies behind them. The series contains monographs, lecture notes and edited volumes in computational intelligence spanning the areas of neural networks, connectionist systems, genetic algorithms, evolutionary computation, artificial intelligence, cellular automata, self-organizing systems, soft computing, fuzzy systems, and hybrid intelligent systems. Of particular value to both the contributors and the readership are the short publication timeframe and the worldwide distribution, which enable both wide and rapid dissemination of research output.

More information about this series at http://www.springer.com/series/7092

Ali Ismail Awad · Mahmoud Hassaballah
Editors

Image Feature Detectors and Descriptors

Foundations and Applications

 Springer

Editors
Ali Ismail Awad
Department of Computer Science,
 Electrical and Space Engineering
Luleå University of Technology
Luleå
Sweden

and

Faculty of Engineering
Al Azhar University
Qena
Egypt

Mahmoud Hassaballah
Image and Video Processing Lab, Faculty
 of Science, Department of Mathematics
South Valley University
Qena
Egypt

ISSN 1860-949X   ISSN 1860-9503   (electronic)
Studies in Computational Intelligence
ISBN 978-3-319-28852-9   ISBN 978-3-319-28854-3   (eBook)
DOI 10.1007/978-3-319-28854-3

Library of Congress Control Number: 2015960414

Printed on acid-free paper

This Springer imprint is published by SpringerNature
The registered company is Springer International Publishing AG Switzerland

Preface

Extracting image features has become a major player in many image pertaining applications. Feature detectors and descriptors have been investigated and applied in various domains such as computer vision, pattern recognition, image processing, biometrics technology, and medical image analysis. Driven by the need for a better understanding of the feature detector foundations and application, this book volume presents up-to-date research findings in the direction of image feature detectors and descriptors.

This book includes 16 chapters that are divided into two parts. Part I details the "Foundations of Image Feature Detectors and Descriptors" by four chapters. The rest of the 16 chapters, 11 chapters, are grouped in Part II for covering the "Applications of Image Feature Detectors and Descriptors." Additionally, "Detection and Description of Image Features: An Introduction" is placed in the beginning of the volume for offering an introduction for all the chapters in the two parts of the volume.

This book has attracted authors from many countries from all over the world such as Egypt, Canada, India, Mexico, and Romania. The authors of accepted chapters are thanked by the editors for revising their chapters according to the suggestions and comments of the book reviewers/editors.

The auditors are very grateful to Dr. Janusz Kacprzyk, the editor of the Studies in Computational Intelligence (SCI) series by Springer. The editors are indebted to the efforts of Dr. Thomas Ditzinger, the senior editor of the SCI series, and Holger Schäpe, the editorial assistant of the SCI series. Finally, the editors and the authors acknowledge the efforts of the Studies in Computational Intelligence team at Springer for their support and cooperation in publishing the book as a volume in the SCI series.

November 2015

Ali Ismail Awad
Mahmoud Hassaballah

Contents

About the Editors

Dr. Ali Ismail Awad is currently a Senior Lecturer (Assistant Professor) at Department of Computer Science, Electrical and Space Engineering, Luleå University of Technology, Luleå, Sweden. He also holds a permanent position as an Assistant Professor at Electrical Engineering Department, Faculty of Engineering, Al Azhar University, Qena, Egypt. Dr. Awad received his B.Sc. from Al Azhar University, Egypt, 2001, M.Sc. degree from Minia University, Egypt, 2007, and Ph.D. degree from Kyushu University, Japan, 2012. He has been awarded his second Ph.D. degree from Minia University, Egypt, May 2013. Dr. Awad serves as an external reviewer in many international journals including COMNET (Elsevier), Computers & Security (Elsevier), and SpringerPlus (Springer). His research interests include information security, biometrics, image processing, pattern recognition, and networking.

Dr. Mahmoud Hassaballah was born in 1974, Qena, Egypt. He received a B.Sc. degree in Mathematics in 1997, then M.Sc. degree in Computer Science in 2003, all from South Valley University, Egypt. In April 2008, he joined the Lab of Intelligence Communication, Department of Electrical and Electronic Engineering and Computer Science, Ehime University, Japan as a Ph.D. student, where he received a Doctor of Engineering (D. Eng.) in Computer Science on September 2011 for his work on facial features detection. He is currently an Assistant Professor of Computer Science at Faculty of Science, South Valley University, Egypt. His research interests include feature extraction, face detection/recognition, object detection, content-based image retrieval, similarity measures, image processing, computer vision, and machine learning.

Detection and Description of Image Features: An Introduction

M. Hassaballah and Ali Ismail Awad

Abstract Detection and description of image features play a vital role in various application domains such as image processing, computer vision, pattern recognition, and machine learning. There are two type of features that can be extracted from an image content; namely global and local features. Global features describe the image as a whole and can be interpreted as a particular property of the image involving all pixels; while, the local features aim to detect keypoints within the image and describe regions around these keypoints. After extracting the features and their descriptors from images, matching of common structures between images (i.e., features matching) is the next step for these applications. This chapter presents a general and brief introduction to topics of feature extraction for a variety of application domains. Its main aim is to provide short descriptions of the chapters included in this book volume.

Keywords Feature detection · Feature description · Feature matching · Image processing · Pattern recognition · Computer vision · Applications

1 Introduction

Nowadays, we live in the era of technological revolution sparked by the rapid progress in computer technology generally, and computer vision especially. Where, the last few decades can be termed as an epoch of computer revolution, in which develop-

M. Hassaballah (✉)
Department of Mathematics, Faculty of Science,
South Valley University, Qena 83523, Egypt
e-mail: m.hassaballah@svu.edu.eg

A.I. Awad (✉)
Department of Computer Science, Electrical and Space Engineering,
Luleå University of Technology, Luleå, Sweden
e-mail: ali.awad@ltu.se

A.I. Awad
Faculty of Engineering, Al Azhar University, Qena, Egypt
e-mail: aawad@ieee.org

© Springer International Publishing Switzerland 2016   1
A.I. Awad and M. Hassaballah (eds.), *Image Feature Detectors and Descriptors*,
Studies in Computational Intelligence 630, DOI 10.1007/978-3-319-28854-3_1

ments in one domain frequently entail breakthroughs in other domains. Scarcely, a month passes where one does not hear an announcement of new technological breakthroughs in the areas of digital computation. Computers and computational workstations have become powerful enough to process big data. Additionally, the technology is now available to every one all over the world. As a result, hardware and multimedia software are becoming standard tools for the handling of images, video sequence, and 3D visualization.

In particular, computer vision, the art of processing digital images stored within the computer, became a key technology in several fields and is utilized as a core part in a large number of industrial vision applications [1]. For instance, computer vision systems are an important part of autonomous intelligent vehicle parking systems, adaptive cruise control, driver exhaustion detection, obstacle or traffic sign detection [2, 3], and driver assistance systems [4]. In industrial automation, computer vision is routinely used for quality or process control such as food quality evaluation systems [5]. Even the images used in astronomy and biometric systems or those captured by intelligent robots as well as medical Computer Assisted Diagnosis (CAD) systems benefit from computer vision techniques [6]. A basic computer vision system contains a camera for capturing images, a camera interface, and a PC to achieve some tasks such as scene reconstruction, object recognition/tracking, 3D reconstruction, image restoration, and image classification [7, 8]. These tasks rely basically on the detection and extraction of image features.

Generally, feature extraction involves detecting and isolating desired features of the image or pattern for identifying or interpreting meaningful information from the image data. Thus, extracting image features has been considered one of the most active topics for image representation in computer vision community [9]. Feature extraction is also an essential pre-processing step in pattern recognition [10]. In fact, image features can represent the content of either the whole image (i.e., global features) or small patches of the image (i.e., local features) [11]. Since the global features aim to represent the image as a whole, only a single feature vector is produced per image and thus the content of two images can be compared via comparing their feature vectors. On the contrary, for representing the image with local features, a set of several local features extracted from different image's patches is used. For local features, feature extraction can often be divided into two independent steps: feature detection and description. The main objective of a feature detector is to find a set of stable (invariant) distinctive interest points or regions, while the descriptor encodes information in spatial neighborhoods of the determined regions mathematically. That is, the descriptor is a vector characterizing local visual appearance or local structure of image's patches [12].

In this respect, the number of extracted features is usually smaller than the actual number of pixels in the image. For instance, a 256×256 image contains 65536 pixels, yet the essence of this image may be captured using only few features (e.g., 30 features). There are many types of image features which can be extracted such as edges, blobs, corners, interest points, texture, and color [13–16]. A large number of feature extraction algorithms have been proposed in the literature to provide reliable feature matching [17–19]. Many feature extraction algorithms are proposed for a specific applications, where they prove significant success and fail otherwise because of the different nature of the other applications. A thorough comparison and a detailed analysis of many extraction algorithms based on different application scenarios are reported in [11, 20–22]. On the other hand, several trails have been done to make these algorithms robust to various image artifacts such as illumination variation, blur, rotation, noise, scale and affine transformation as well as to improve their execution time performance to be applicable in real time applications [23, 24].

The use of local feature detection and description algorithms in some applications such as large volume, low-cost, low-power embedded systems, visual odometry, and photogrammetric applications is still limited or negligible to date due to the lack of a worldwide industry standard [22]. Further, most of the aforementioned applications have real-time constraints and would benefit immensely from being able to match images in a real time, thus developing fast feature extraction algorithms is a must. With all these factors and avenues to explore, it is not surprising that the problem of image feature extraction, with various meanings of this expression, is actively pursued in research by highly qualified people and the volume of research will increase in the near future, which has given us the motivation for dedicating this book to exemplify the tremendous progress achieved recently in the topic.

2   Chapters of the Book

This volume contains 15 chapters in total which are divided into two categories. The following are brief summaries for the content of each chapter.

Part I: Foundations of Image Feature Detectors and Descriptors

Chapter "Image Features Detection, Description and Matching" presents a comprehensive review on the available image feature detectors and descriptors such as Moravec's detector [25], Harris detector [26], Smallest Univalue Segment Assimilating Nucleus (SUSAN) detector [27], Features from Accelerated Segment Test (FAST) detector [28], Difference of Gaussian (DoG) detector [29], Scale invariant feature transform (SIFT) descriptor [29], and Speeded-Up Robust Features (SURF) descriptor [30]. The mathematical foundations of the presented detectors and descriptors have been highlighted. In general, the chapter serves as a good foundation for the rest of the volume.

Chapter "A Review of Image Interest Point Detectors: From Algorithms to FPGA Hardware Implementations" studies some image interest point detectors from the hardware implementation viewpoint [31]. The chapter offers a review on the hardware implementation, particularity, using Field Programmable Gate Array (FPGA), for image interest point detectors [32]. The chapter emphasizes the real-time performance of FPGA as a hardware accelerator. However, further researches are demanded for improving the accelerator portability across different platforms.

Chapter "Image Features Extraction, Selection and Fusion for Computer Vision" addresses various research problems pertaining to image segmentation, feature extraction and selection, feature fusion and classification, with applications in intelligent vehicle, biometrics [33–35], and medical image processing. The chapter describes different features for different applications from a holistic computer vision perspective.

Chapter "Image Feature Extraction Acceleration" focuses on accelerating image feature extraction process using hardware platforms [36]. It presents two focal-plane accelerators chips, Application-specific Integrated Circuits (ASICs), that aim at the acceleration of two flagship algorithms in computer vision. The chapter offers the fundamental concepts driving the design and the implementation of two focal-plane accelerator chips for the Viola-Jones face detection algorithm [37] and for the Scale Invariant Feature Transform (SIFT) algorithm [29, 38].

Part II: Applications of Image Feature Detectors and Descriptors

Chapter "Satellite Image Matching and Registration: A Comparative Study Using Invariant Local Features" is devoted for a comparative study for satellite image registration using invariant local features. In this chapter, various local feature detectors and descriptors, such as Features from Accelerated Segment Test (FAST) [28], Binary Robust Invariant Scalable Keypoints (BRISK) [39], Maximally Stable Extremal Regions (MSER) [40], and Good Features to Track (GTT) [41], have been evaluated on different optical and satellite image data sets in terms of feature extraction, features matching, and geometric transformation. The chapter documents the performance of the selected feature detectors for the comparison purpose.

Chapter "Redundancy Elimination in Video Summarization" addresses the redundancy elimination from video summarization using feature point descriptors such as Binary Robust Independent Elementary Features (BRIEF) [42] and Oriented FAST and Rotated BRIEF (ORB) [43]. A method for intra-shot and inter-shot redundancy removal using similarity metric computed from feature descriptors has been presented. Several feature descriptors have been tested and evaluated for redundancy removal with a focus on precision and recall performance parameters.

Chapter "A Real Time Dactylology Based Feature Extractrion for Selective Image Encryption and Artificial Neural Network" combines artificial neural network with Speeded-Up Robust Features Descriptor (SURF) [30] for selective image encryption in real time dactylology or finger spelling. Finer spelling is used in different sign languages and for different purposes [44]. The integrity and the effectiveness of the proposed scheme have been judged using different factors like histogram, correlation coefficients, entropy, MSE, and PSNR.

Chapter "Spectral Reflectance Images and Applications" illustrates the use of spectral invariant for obtaining reliable spectral reflectance images. Spectral imaging can be deployed, for example, in remote sensing, computer vision, industrial applications, material identification, natural scene rendering, and colorimetric analysis [45]. The chapter introduces a material classification method based on the invariant representation that results in reliable segmentation of natural scenes and raw circuit board spectral images.

Chapter "Image Segmentation Using an Evolutionary Method Based on Allostatic Mechanisms" proposes a multi-thresholding segmentation algorithm that is based on an evolutionary algorithm called Allostatic Optimization (AO). Threshold-based segmentation is considered as a simple technique due to the assumption that the object and the background have different grey level distribution [46]. The experimental work shows the high performance of the proposed segmentation algorithm with respect to accuracy and robustness.

Chapter "Image Analysis and Coding Based on Ordinal Data Representation" utilizes the Ordinal Measures (OM) [47] for image analysis and coding with an application on iris image as a biometric identifier. Biometrics is a mechanism for assigning an identity to an individual based on some physiological or behavioral characteristics. Biometric identifiers include fingerprints, face image, iris patterns, retinal scan, voice, and signature with broad deployments in forensic and civilian applications [48].

Chapter "Intelligent Detection of Foveal Zone from Colored Fundus Images of Human Retina Through a Robust Combination of Fuzzy-Logic and Active Contour Model" proposes a robust fuzzy-rule based image segmentation algorithm for extracting the Foveal Avascular Zone (FAZ) from retinal images [49]. The proposed algorithm offers a good contribution toward improving the deployment of retinal images in biometrics-based human identification and verification.

Chapter "Registration of Digital Terrain Images Using Nondegenerate Singular Points" presents a registration algorithm for digital terrain images using nondegenerate singular points. The proposed algorithm is a graph-theoretic technique that uses Morse singularities [50] and an entopic dissimilarity measure [51]. The experimental outcomes prove the reliability and the accuracy in addition to the high computational speed of the proposed algorithm.

Chapter "Visual Speech Recognition with Selected Boundary Descriptors" is devoted for visual speech recognition using some selected boundary descriptors. Lipreading can be used for speech-to-text for the benefit of hearing impaired individuals. In the chapter, the Point Distribution Model (PDM) [52] is used to obtain the lip contour, and the Minimum Redundancy Maximum Relevance (mRMR) [53] approach is used as a following stage for feature selection.

Chapter "Application of Texture Features for Classification of Primary Benign and Primary Malignant Focal Liver Lesions" focuses on the classification of the primary benign and primary malignant local liver lesions. Statistical texture features, spectral texture features, and spatial filtering based texture feature have been used. In addition, Support Vector Machine (SVM) [54, 55] and Smooth Support Vector Machine (SSVM) [56] have been evaluated as two classification algorithms.

Chapter "Application of Statistical Texture Features for Breast Tissue Density Classification" aims to classify the density of the breast tissues using statistical features extracted from mammographic images. It presents a CAD system that is formed from feature extraction module, feature space dimensionality reduction module, and feature classification module. Different algorithms have been used in the classification module such as k-Nearest Neighbor (kNN) [57, 58], Probabilistic Neural Network (PNN) [59], and Support Vector Machine (SVM) classifiers.

3 Concluding Remarks

Detection and description of image features play a vital role in various application domains such as image processing, computer vision, pattern recognition, machine learning, biometrics, and automation. In this book volume, cutting-edge research contributions on image feature extraction, feature detectors, and feature extractors have been introduced. The presented contributions support the vitality of image feature detectors and descriptors, and discover new research gaps in the theoretical foundations and the practical implementations of image detectors and descriptors. Due to the rapid growth in representing image using local and global features, further contributions and findings are anticipated in this research domain.

References

1. Klette, R.: Concise Computer Vision: An introduction into Theory and Algorithms. Springer, USA (2014)
2. Raxle, C.-C.: Automatic vehicle detection using local features-a statistical approach. IEEE Trans. Intell. Transp. Syst. **9**(1), 83–96 (2008)
3. Mukhtar, A., Likun, X.: Vehicle detection techniques for collision avoidance systems: A review. IEEE Trans. Intell. Transp. Syst. **16**(5), 2318–2338 (2015)
4. Geronimo, D., Lopez, A., Sappa, A., Graf, T.: Survey of pedestrian detection for advanced driver assistance systems. IEEE Trans. Pattern Anal. Mach. Intell. **32**(7), 1239–1258 (2010)
5. Da-Wen, S.: Computer Vision Technology for Food Quality Evaluation. Academic Press, Elsevier (2008)
6. Menze, B.H., Langs, G., Lu, L., Montillo, A., Tu, Z., Criminisi, A.: Medical computer vision: recognition techniques and applications in medical imaging. LNCS 7766 (2013)
7. Koen, E., Gevers, T., Snoek, G.: Evaluating color descriptors for object and scene recognition. IEEE Trans. Pattern Anal. Mach. Intell. **32**(9), 1582–1596 (2010)
8. Szeliski, R.: Computer Vision: Algorithms and Applications. Springer, USA (2011)
9. Chen, Z., Sun, S.K.: A Zernike moment phase-based descriptor for local image representation and matching. IEEE Trans. Image Process. **19**(1), 205–219 (2010)
10. Andreopoulos, A., Tsotsos, J.: 50 years of object recognition: directions forward. Comput. Vis. Image Underst. **117**(8), 827–891 (2013)
11. Mikolajczyk, K., Tuytelaars, T., Schmid, C., Zisserman, A., Matas, J., Schaffalitzky, F., Kadir, T., Gool, L.: A comparison of affine region detectors. Int. J. Comput. Vis. **65**(1/2), 43–72 (2005)
12. Moreels, P., Perona, P.: Evaluation of features detectors and descriptors based on 3D objects. Int. J. Comput. Vis. **73**(3), 263–284 (2007)
13. Gonzalez, R.C., Woods, R.E.: Digital Image Processing, 3rd edn. Prentice-Hall Inc, USA (2007)
14. John, C.R.: The Image Processing Handbook, 6 edn. CRC Press, Taylor & Francis Group, USA (2011)
15. Li, J., Allinson, N.: A comprehensive review of current local features for computer vision. Neurocomputing **71**(10–12), 1771–1787 (2008)

16. Liu, S., Bai, X.: Discriminative features for image classification and retrieval. Pattern Recogn. Lett. **33**(6), 744–751 (2012)
17. Tuytelaars, T., Mikolajczyk, K.: Local invariant feature detectors: a survey. Found. Trends Comput. Graph. Vis. **3**(3), 177–280 (2007)
18. Mainali, P., Lafruit, G., Yang, Q., Geelen, B., Gool, L.V., Lauwereins, R.: SIFER: Scale-invariant feature detector with error resilience. Int. J. Comput. Vis. **104**(2), 172–197 (2013)
19. Zhang, Y., Tian, T., Tian, J., Gong, J., Ming, D.: A novel biologically inspired local feature descriptor. Biol. Cybern. **108**(3), 275–290 (2014)
20. Hugo, P.: Performance evaluation of keypoint detection and matching techniques on grayscale data. SIViP **9**(5), 1009–1019 (2015)
21. Bouchiha, R., Besbes, K.: Comparison of local descriptors for automatic remote sensing image registration. SIViP **9**(2), 463–469 (2015)
22. Bianco, S., Mazzini, D., Pau, D., Schettini, R.: Local detectors and compact descriptors for visual search: a quantitative comparison. Digital Sig. Process **44**, 1–13 (2015)
23. Takacs, G., Chandrasekhar, V., Tsai, S., Chen, D., Grzeszczuk, R., Girod, B.: Rotation-invariant fast features for large-scale recognition and real-time tracking. Sig. Process: Image Commun. **28**(4), 334–344 (2013)
24. Seidenari, L., Serra, G., Bagdanov, A., Del Bimbo, A.: Local pyramidal descriptors for image recognition. IEEE Trans. Pattern Anal. Mach. Intell. **36**(5), 1033–1040 (2014)
25. Morevec, H.P.: Towards automatic visual obstacle avoidance. In: Proceedings of the 5th International Joint Conference on Artificial Intelligence, vol. 2, pp. 584–584. IJCAI'77. Morgan Kaufmann Publishers Inc., San Francisco (1977)
26. Harris, C., Stephens, M.: A combined corner and edge detection. In: Proceedings of the Fourth Alvey Vision Conference, pp. 147–151 (1988)
27. Smith, S., Brady, J.: Susan-a new approach to low level image processing. Int. J. Comput. Vis. **23**(1), 45–78 (1997)
28. Rosten, E., Drummond, T.: Fusing points and lines for high performance tracking. In: Proceedings of the Tenth IEEE International Conference on Computer Vision, vol. 2, pp. 1508–1515. ICCV'05, IEEE Computer Society, Washington, DC (2005)
29. Lowe, D.G.: Distinctive image features from scale-invariant keypoints. Int. J. Comput. Vis. **60**(2), 91–110 (2004)
30. Bay, H., Ess, A., Tuytelaars, T., Gool, L.V.: Speeded-up robust features (SURF). Comput. Vis. Image Underst. **110**(3), 346–359 (2008)
31. Schmid, C., Mohr, R., Bauckhage, C.: Evaluation of interest point detectors. Int. J. Comput. Vis. **37**(2), 151–172 (2000)
32. Possa, P., Mahmoudi, S., Harb, N., Valderrama, C., Manneback, P.: A multi-resolution FPGA-Based architecture for real-time edge and corner detection. IEEE Trans. Comput. **63**(10), 2376–2388 (2014)
33. Jain, A.K., Ross, A.A., Nandakumar, K.: Introduction to Biometrics, 1st edn. Springer (2011)
34. Egawa, S., Awad, A.I., Baba, K.: Evaluation of acceleration algorithm for biometric identification. In: Benlamri, R. (ed.) Networked Digital Technologies, Communications in Computer and Information Science, vol. 294, pp. 231–242. Springer, Heidelberg (2012)
35. Awad, A.I.: Fingerprint local invariant feature extraction on GPU with CUDA. Informatica (Slovenia) **37**(3), 279–284 (2013)
36. Awad, A.I.: Fast fingerprint orientation field estimation incorporating general purpose GPU. In: Balas, V.E., Jain, L.C., Kovaevi, B. (eds.) Soft Computing Applications, Advances in Intelligent Systems and Computing, vol. 357, pp. 891–902. Springer International Publishing (2016)
37. Viola, P., Jones, M.: Rapid object detection using a boosted cascade of simple features. In: Proceedings of the 2001 IEEE Computer Society Conference on Computer Vision and Pattern Recognition (CVPR 2001), vol. 1, pp. 511–518 (2001)
38. Awad, A.I., Baba, K.: Evaluation of a fingerprint identification algorithm with SIFT features. In: Proceedings of the 3rd 2012 IIAI International Conference on Advanced Applied Informatics, pp. 129–132. IEEE, Fukuoka, Japan (2012)

39. Leutenegger, S., Chli, M., Siegwart, R.: BRISK: Binary robust invariant scalable keypoints. In: IEEE International Conference on Computer Vision (ICCV), pp. 2548–2555 (2011)
40. Donoser, M., Bischof, H.: Efficient maximally stable extremal region (MSER) tracking. IEEE Comput. Soc. Conf. Comput. Vis. Pattern Recogn. **1**, 553–560 (2006)
41. Shi, J., Tomasi, C.: Good features to track. In: Proceedings of IEEE Computer Society Conference on Computer Vision and Pattern Recognition (CVPR'94), pp. 593–600 (1994)
42. Calonder, M., Lepetit, V., Strecha, C., Fua, P.: BRIEF: Binary robust independent elementary features. In: Proceedings of the 11th European Conference on Computer Vision: Part IV, pp. 778–792. ECCV'10. Springer, Heidelberg (2010)
43. Rublee, E., Rabaud, V., Konolige, K., Bradski, G.: ORB: an efficient alternative to sift or surf. In: Proceedings of the 2011 International Conference on Computer Vision, pp. 2564–2571. ICCV '11, IEEE Computer Society, Washington, DC (2011)
44. Pfau, R., Steinbach, M., Woll, B. (eds.): Sign Language: An International Handbook. Series of Handbooks of Linguistics and Communication Science (HSK), De Gruyter, Berlin (2012)
45. Trémeau, A., Tominaga, S., Plataniotis, K.N.: Color in image and video processing: Most recent trends and future research directions. J. Image Video Process. Color Image Video Process. 7:1–7:26 (2008)
46. Sezgin, M., Sankur, B.: Survey over image thresholding techniques and quantitative performance evaluation. J. Electron. Imaging **13**(1), 146–168 (2004)
47. Sun, Z., Tan, T.: Ordinal measures for iris recognition. IEEE Trans. Pattern Anal. Mach. Intell. **31**(12), 2211–2226 (2009)
48. Awad, A.I., Hassanien, A.E.: Impact of some biometric modalities on forensic science. In: Muda, A.K., Choo, Y.H., Abraham, A., N. Srihari, S. (eds.) Computational Intelligence in Digital Forensics: Forensic Investigation and Applications, Studies in Computational Intelligence, vol. 555, pp. 47–62. Springer International Publishing (2014)
49. Gutierrez, J., Epifanio, I., de Ves, E., Ferri, F.: An active contour model for the automatic detection of the fovea in fluorescein angiographies. In: Proceedings of the 15th International Conference on Pattern Recognition. vol. 4, pp. 312–315. (2000)
50. Fomenko, A., Kunii, T.: Topological Modeling for Visualization. Springer, Japan (2013)
51. Hero, A., Ma, B., Michel, O., Gorman, J.: Applications of entropic spanning graphs. IEEE Signal Process. Mag. **19**(5), 85–95 (2002)
52. Al-Shaher, A.A., Hancock, E.R.: Learning mixtures of point distribution models with the EM algorithm. Pattern Recogn. **36**(12), 2805–2818 (2003)
53. Peng, H., Long, F., Ding, C.: Feature selection based on mutual information criteria of max-dependency, max-relevance, and min-redundancy. IEEE Trans. Pattern Anal. Mach. Intell. **27**(8), 1226–1238 (2005)
54. Lee, C.C., Chen, S.H.: Gabor wavelets and SVM classifier for liver diseases classification from CT images. In: IEEE International Conference on Systems, Man and Cybernetics (SMC'06), vol. 1, pp. 548–552 (2006)
55. Christmann, A., Steinwart, I.: Support vector machines for classification. In: Support Vector Machines. Information Science and Statistics, pp. 285–329, Springer, New York (2008)
56. Lee, Y.J., Mangasarian, O.: SSVM: A smooth support vector machine for classification. Comput. Optim. Appl. **20**(1), 5–22 (2001)
57. Cover, T., Hart, P.: Nearest neighbor pattern classification. IEEE Trans. Inf. Theory **13**(1), 21–27 (2006)
58. Wu, Y., Ianakiev, K., Govindaraju, V.: Improved k-nearest neighbor classification. Pattern Recogn. **35**(10), 2311–2318 (2002)
59. Specht, D.F.: Probabilistic neural networks. Neural Networks **3**(1), 109–118 (1990)

Part I
Foundations of Image Feature
Detectors and Descriptors

Image Features Detection, Description and Matching

M. Hassaballah, Aly Amin Abdelmgeid and Hammam A. Alshazly

Abstract Feature detection, description and matching are essential components of various computer vision applications, thus they have received a considerable attention in the last decades. Several feature detectors and descriptors have been proposed in the literature with a variety of definitions for what kind of points in an image is potentially interesting (i.e., a distinctive attribute). This chapter introduces basic notation and mathematical concepts for detecting and describing image features. Then, it discusses properties of perfect features and gives an overview of various existing detection and description methods. Furthermore, it explains some approaches to feature matching. Finally, the chapter discusses the most used techniques for performance evaluation of detection and description algorithms.

Keywords Interest points · Feature detector · Feature descriptor · Feature extraction · Feature matching

1 Introduction

Over the last decades, image feature detectors and descriptors have become popular tools in the computer vision community and they are being applied widely in a large number of applications. Image representation [1], image classification and retrieval [2–5], object recognition and matching [6–10], 3D scene reconstruction [11], motion tracking [12–14], texture classification [15, 16], robot localization [17–19],

M. Hassaballah (✉) · H.A. Alshazly
Faculty of Science, Department of Mathematics, South Valley University,
Qena 83523, Egypt
e-mail: m.hassaballah@svu.edu.eg

H.A. Alshazly
e-mail: hammam.alshazly@sci.svu.edu.eg

A.A. Abdelmgeid
Faculty of Science, Department of Computer Science,
Minia University, El Minia, Egypt
e-mail: abdelmgeid@yahoo.com

© Springer International Publishing Switzerland 2016   11
A.I. Awad and M. Hassaballah (eds.), *Image Feature Detectors and Descriptors*,
Studies in Computational Intelligence 630, DOI 10.1007/978-3-319-28854-3_2

and biometrics systems [20–22], all rely on the presence of stable and representative features in the image. Thus, detecting and extracting the image features are vital steps for these applications.

In order to establish correspondences among a collection of images, where feature correspondences between two or more images are needed, it is necessary to identify a set of salient points in each image [8, 23]. In a classification task, feature descriptors of a query image are matched with all trained image features and the trained image giving maximum correspondence is considered the best match. In that case, feature descriptor matching can be based on distance measures such as Euclidean or Mahalanobis. In image registration, it is necessary to spatially align two or more images of a scene captured by different sensors at different times. The main steps involved in performing image registration or alignment tasks are: feature detection, feature matching, derivation of transformation functions based on corresponding features in images, and reconstruction of images based on the derived transformation functions [24]. In the context of matching and recognition, the first step of any matching/recognition system is to detect interest locations in the images and describe them. Once the descriptors are computed, they can be compared to find a relationship between images for performing matching/recognition tasks. Also, for online street-level virtual navigation application, we need a feature detector and a feature descriptor to extract features from planar images (panoramic images) [25].

The basic idea is to first detect interest regions (keypoints) that are covariant to a class of transformations. Then, for each detected regions, an invariant feature vector representation (i.e., descriptor) for image data around the detected keypoint is built. Feature descriptors extracted from the image can be based on second-order statistics, parametric models, coefficients obtained from an image transform, or even a combination of these measures. Two types of image features can be extracted form image content representation; namely global features and local features. Global features (e.g., color and texture) aim to describe an image as a whole and can be interpreted as a particular property of the image involving all pixels. While, local features aim to detect keypoints or interest regions in an image and describe them. In this context, if the local feature algorithm detects n keypoints in the image, there are n vectors describing each one's shape, color, orientation, texture and more. The use of global colour and texture features are proven surprisingly successful for finding similar images in a database, while the local structure oriented features are considered adequate for object classification or finding other occurrences of the same object or scene [26]. Meanwhile, the global features can not distinguish foreground from background of an image, and mix information from both parts together [27].

On the other hand, as the real time applications have to handle ever more data or to run on mobile devices with limited computational capabilities, there is a growing need for local descriptors that are fast to compute, fast to match, memory efficient, and yet exhibiting good accuracy. Additionally, local feature descriptors are proven to be a good choice for image matching tasks on a mobile platform, where occlusions and missing objects can be handled [18]. For certain applications, such as camera calibration, image classification, image retrieval, and object tracking/recognition, it is very important for the feature detectors and descriptors to be robust to changes in brightness or viewpoint and to image distortions (e.g., noise, blur, or illumination) [28]. While, other specific visual recognition tasks, such as face detection or recognition, requires the use of specific detectors and descriptors [29].

In the literature, a large variety of feature extraction methods have been proposed to compute reliable descriptors. Some of these feature descriptors were exclusively designed for a specific application scenario such as shape matching [29, 30]. Among these descriptors, the scale invariant feature transform (SIFT) descriptor [31] utilizing local extrema in a series of difference of Gaussian (DoG) functions for extracting robust features and the speeded-up robust features (SURF) descriptor [32] partly inspired by the SIFT descriptor for computing distinctive invariant local features quickly are the most popular and widely used in several applications. These descriptors represent salient image regions by using a set of hand-crafted filters and non-linear operations. In the rest of the chapter, we give an overview for these methods and algorithms as well as their improvements proposed by developers. Furthermore, the basic notations and mathematical concepts for detecting and describing image features are introduced. We also explore in detail what is the quantitative relation between the detectors and descriptors as well as how to evaluate their performance.

2 Definitions and Principles

2.1 Global and Local Features

In image processing and computer vision tasks, we need to represent the image by features extracted therefrom. The raw image is perfect for the human eye to extract all information from; however that is not the case with computer algorithms. There are generally two methods to represent images, namely, global features and local features. In the global feature representation, the image is represented by one multi-dimensional feature vector, describing the information in the whole image. In other words, the global representation method produces a single vector with values that measure various aspects of the image such as color, texture or shape. Practically, a single vector from each image is extracted and then two images can be compared by comparing their feature vectors. For example, when one wants to distinguish images of a sea (blue) and a forest (green), a global descriptor of color would produce quite different vectors for each category. In this context, global features can be interpreted as a particular property of image involving all pixels. This property can be color histograms, texture, edges or even a specific descriptor extracted from some filters applied to the image [33]. On the other hand, the main goal of local feature representation is to distinctively represent the image based on some salient regions while remaining invariant to viewpoint and illumination changes. Thus, the image is represented based on its local structures by a set of local feature descriptors extracted from a set of image regions called interest regions (i.e., keypoints) as illustrated in Fig. 1. Most local features represent texture within the image patch.

Global feature representation   Local feature representation

Fig. 1 Global and local image features representation

Generally, using what kind of features might greatly depend on the applications on hand. Developers prefer the most discriminative ones. For example, a person with a bigger nose and smaller eyes, and a person with a smaller nose and bigger eyes may have similar mug shot in terms of histogram or intensity distribution. Then, local features or the global pattern distilled from local feature clusters seem to be more discriminative. Whereas, for very large datasets in the web-scale image indexing application, it is appropriate to consider global features. Also, global features are useful in applications where a rough segmentation of the object of interest is available. The advantages of global features are that they are much faster and compact while easy to compute and generally require small amounts of memory. Nevertheless, the global representation suffers from well-known limitations, in particular they are not invariant to significant transformations and sensitive to clutter and occlusion. In some applications, such as copy detection, most of the illegal copies are very similar to the original; they have only suffered from compression, scaling or limited cropping. In contrast, the advantage of local features is their superior performance [34]. Meanwhile, using local features for large-scale image search have much higher performance than global features provide [35]. Besides, as the local structures are more distinctive and stable than other structures in smooth regions, it is expected to be more useful for image matching and object recognition. However, they usually require a significant amount of memory because the image may have hundreds of local features. As a solution for this problem, researchers suggest aggregating local image descriptors into a very compact vector representation and optimizing the dimensionality reduction of these vectors [35].

2.2 Characteristics of Feature Detectors

Tuytelaars and Mikolajczyk [27] define a local feature as "*it is an image pattern which differs from its immediate neighborhood*". Thus, they consider the purpose of local invariant features is to provide a representation that allows to efficiently match local structures between images. That is, we want to obtain a sparse set of local measurements that capture the essence of the underlying input images and encode their interesting structures. To meet this goal, the feature detectors and extractors must have certain properties keeping in mind that the importance of these properties depends on the actual application settings and compromises need to be made. The following properties are important for utilizing a feature detector in computer vision applications:

- Robustness, the feature detection algorithm should be able to detect the same feature locations independent of scaling, rotation, shifting, photometric deformations, compression artifacts, and noise.
- Repeatability, the feature detection algorithm should be able to detect the same features of the same scene or object repeatedly under variety of viewing conditions.
- Accuracy, the feature detection algorithm should accurately localize the image features (same pixel locations), especially for image matching tasks, where precise correspondences are needed to estimate the epipolar geometry.
- Generality, the feature detection algorithm should be able to detect features that can be used in different applications.
- Efficiency, the feature detection algorithm should be able to detect features in new images quickly to support real-time applications.
- Quantity, the feature detection algorithm should be able to detect all or most of the features in the image. Where, the density of detected features should reflect the information content of the image for providing a compact image representation.

2.3 Scale and Affine Invariance

Actually, finding correspondences based on comparing regions of fixed shape like rectangles or circles are not reliable in the presence of some geometric and photometric deformations as they affect the regions' shapes. Also, objects in digital images appear in different ways depending on the scale of observation. Consequently, scale changes are of important implications when analyzing image contents. Different techniques have been proposed to address the problem of detecting and extracting invariant image features under these conditions. Some are designed to handle scale changes, while others go further to handle affine transformations. In order to address the scale changes, these techniques assume that the change in scale is same in all directions (i.e., uniform) and they search for stable features across all possible scales using a continuous kernel function of scale known as scale space. Where, the image size is varied and a filter (e.g., Gaussian filter) is repeatedly applied to smooth subsequent layers, or by leaving the original image unchanged and varies only the filter size as shown in Fig. 2. More details about feature detection with scale changes can be found in [36], where a framework is presented for generating hypotheses about interesting scale levels in image data by searching for scale-space extrema in the scale normalized Laplacian of Gaussian (LoG).

On the other hand, in the case of an affine transformation the scaling can be different in each direction. The nonuniform scaling has an influence on the localization,

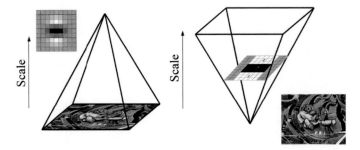

Fig. 2 Constructing the scale space structure

the scale and the shape of the local structure. Therefore, the scale invariant detectors fail in the case of significant affine transformations. Hence, detectors designed to detect the image features under uniform scaling need to be extended to be affine invariant detectors that can detect the affine shape of the local image structures. Thus, these affine invariant detectors can be seen as a generalization of the scale invariant detector.

Generally, affine transformations are constructed using sequences of translations, scales, flips, rotations, and shears. An affine transformation (affinity) is any linear mapping that preserves collinearity and ratios of distances. In this sense, affine indicates a special class of projective transformations that do not move any object from the affine space $\mathbb{R}^3$ to the plane at infinity or conversely. Briefly, the affine transformation of $\mathbb{R}^n$ is a map $f : \mathbb{R}^n \to \mathbb{R}^n$ of the form

$$f(Y) = \mathbb{A}Y + \mathbb{B} \tag{1}$$

for all $Y \in \mathbb{R}^n$, where $\mathbb{A}$ is a linear transformation of $\mathbb{R}^n$. In some special cases, if $\det(\mathbb{A}) > 0$, the transformation is called orientation-preserving, while, if $\det(\mathbb{A}) < 0$, it is orientation-reversing. It is well known that a function is invariant under a certain family of transformations if its value does not change when a transformation from this family is applied to its argument. The second moment matrix provides the theory for estimating affine shape of the local image features. Examples of affine invariant detectors are Harris-affine, Hessian-affine, and maximally stable extremal regions (MSER). It must be borne in mind that the detected features by these detectors are transformed from circles to ellipses.

3 Image Feature Detectors

Feature detectors can be broadly classified into three categories: single-scale detectors, multi-scale detectors, and affine invariant detectors. In a nutshell, single scale means that there is only one representation for the features or the object contours using detector's internal parameters. The single-scale detectors are invariant to image transformations such as rotation, translation, changes in illuminations and addition of noise. However, they are incapable to deal with the scaling problem. Given two images of the same scene related by a scale change, we want to determine whether same interest points can be detected or not. Therefore, it is necessary to build multi-scale detectors capable of extracting distinctive features reliably under scale changes. Details of single-scale and multi-scale detectors as well as affine invariant detectors are discussed in the following sections.

3.1 Single-Scale Detectors

3.1.1 Moravec's Detector

Moravec's detector [37] is specifically interested in finding distinct regions in the image that could be used to register consecutive image frames. It has been used as a corner detection algorithm in which a corner is a point with low self-similarity. The detector tests each pixel in a given image to see if a corner is present. It considers a local image patch centered on the pixel and then determines the similarity between the patch and the nearby overlapping patches. The similarity is measured by taking the sum of square differences (SSD) between the centered patch and the other image patches. Based on the value of SSD, three cases need to be considered as follows:

- If the pixel in a region of uniform intensity, then the nearby patches will look similar or a small change occurs.
- If the pixel is on an edge, then the nearby patches in a parallel direction to the edge will result in a small change and the patches in a direction perpendicular to the edge will result in a large change.
- If the pixel is on a location with large change in all directions, then none of the nearby patches will look similar and the corner can be detected when the change produced by any of the shifts is large.

The smallest SSD between the patch and its neighbors (horizontal, vertical and on the two diagonals) is used as a measure for cornerness. A corner or an interest point is detected when the SSD reaches a local maxima. The following steps can be applied for implementing Moravec's detector:

1. Input: gray scale image, window size, threshold T
2. For each pixel (x, y) in the image compute the intensity variation V from a shift (u, v) as

$$V_{u,v}(x, y) = \sum_{\forall a,b \in window} [I(x + u + a, y + v + b) - I(x + a, y + b)]^2 \quad (2)$$

Fig. 3 Performing the
non-maximum suppression

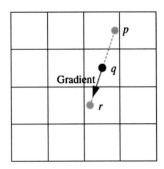

3. Construct the cornerness map by calculating the cornerness measure $C(x, y)$ for
   each pixel (x, y)

$$C(x, y) = \min(V_{u,v}(x, y)) \tag{3}$$

4. Threshold the cornerness map by setting all $C(x, y)$ below the given threshold
   value T to zero.
5. Perform non-maximum suppression to find local maxima. All non-zero points
   remaining in the cornerness map are corners.

For performing non-maximum suppression, an image is scanned along its gradient
direction, which should be perpendicular to an edge. Any pixel that is not a local
maximum is suppressed and set to zero. As illustrated in Fig. 3, p and r are the
two neighbors along the gradient direction of q. If the pixel value of q is not larger
than the pixel values of both p and r, it is suppressed. One advantage of Moravec's
detector is that, it can detect majority of the corners. However, it is not isotropic;
intensity variation is calculated only at a discrete set of shifts (i.e., the eight principle
directions) and any edge is not in one of the eight neighbors' directions is assigned
a relatively large cornerness measure. Thus, it is not invariant to rotation resulting in
the detector is of a poor repeatability rate.

3.1.2 Harris Detector

Harris and Stephens [38] have developed a combined corner and edge detector to
address the limitations of Moravec's detector. By obtaining the variation of the auto-
correlation (i.e., intensity variation) over all different orientations, this results in a
more desirable detector in terms of detection and repeatability rate. The resulting
detector based on the auto-correlation matrix is the most widely used technique.
The 2×2 symmetric auto-correlation matrix used for detecting image features and
describing their local structures can be represented as

$$M(x, y) = \sum_{u,v} w(u, v) * \begin{bmatrix} I_x^2(x, y) & I_x I_y(x, y) \\ I_x I_y(x, y) & I_y^2(x, y) \end{bmatrix} \tag{4}$$

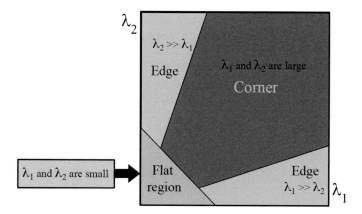

Fig. 4 Classification of image points based on the eigenvalues of the autocorrelation matrix M

where I_x and I_y are local image derivatives in the x and y directions respectively, and $w(u, v)$ denotes a weighting window over the area (u, v). If a circular window such as a Gaussian is used, then the response will be isotropic and the values will be weighted more heavily near the center. For finding interest points, the eigenvalues of the matrix M are computed for each pixel. If both eigenvalues are large, this indicates existence of the corner at that location. An illustrating diagram for classification of the detected points is shown in Fig. 4. Constructing the response map can be done by calculating the cornerness measure $C(x, y)$ for each pixel (x, y) using

$$C(x, y) = \det(M) - K(trace(M))^2 \tag{5}$$

where

$$\det(M) = \lambda_1 * \lambda_2, \quad and \quad trace(M) = \lambda_1 + \lambda_2 \tag{6}$$

The K is an adjusting parameter and λ_1, λ_2 are the eigenvalues of the auto-correlation matrix. The exact computation of the eigenvalues is computationally expensive, since it requires the computation of a square root. Therefore, Harris suggested using this cornerness measure that combines the two eigenvalues in a single measure. The non-maximum suppression should be done to find local maxima and all non-zero points remaining in the cornerness map are the searched corners.

3.1.3 SUSAN Detector

Instead of using image derivatives to compute corners, Smith and Brady [39] introduced a generic low-level image processing technique called SUSAN (Smallest Univalue Segment Assimilating Nucleus). In addition to being a corner detector, it has been used for edge detection and image noise reduction. A corner is detected by placing a circular mask of fixed radius to every pixel in the image. The center pixel is referred to as the nucleus, where pixels in the area under the mask are compared with the nucleus to check if they have similar or different intensity values. Pixels having almost the same brightness as the nucleus are grouped together and the resulting area is termed USAN (Univalue Segment Assimilating Nucleus). A corner is found at locations where the number of pixels in the USAN reaches a local minimum and below a specific threshold value T. For detecting corners, the similar comparison function $C(r, r_0)$ between each pixel within the mask and mask's nucleus is given by

$$C(r, r_0) = \begin{cases} 1, & \text{if } |I(r) - I(r_0)| \leq T, \\ 0, & \text{otherwise,} \end{cases} \tag{7}$$

and the size of USAN region is

$$n(r_0) = \sum_{r \in c(r0)} C(r, r_0) \tag{8}$$

where r_0 and r are nucleus's coordinates and the coordinates of other points within the mask, respectively. The performance of SUSAN corner detector mainly depends on the similar comparison function $C(r, r_0)$, which is not immune to certain factors impacting imaging (e.g., strong luminance fluctuation and noises).

SUSAN detector has some advantages such as: (i) no derivatives are used, thus, no noise reductions or any expensive computations are needed; (ii) High repeatability for detecting features; and (iii) invariant to translation and rotation changes. Unfortunately, it is not invariant to scaling and other transformations, and a fixed global threshold is not suitable for general situation. The corner detector needs an adaptive threshold and the shape of mask should be modified.

3.1.4 FAST Detector

FAST (Features from Accelerated Segment Test) is a corner detector originally developed by Rosten and Drummondn [40, 41]. In this detection scheme, candidate points are detected by applying a segment test to every image pixel by considering a circle of 16 pixels around the corner candidate pixel as a base of computation. If a set of n contiguous pixels in the Bresenham circle with a radius r are all brighter than the intensity of candidate pixel (denoted by I_p) plus a threshold value t, $I_p + t$, or all darker than the intensity of candidate pixel minus the threshold value $I_p - t$, then p is classified as a corner. A high-speed test can be used to exclude a very large number of non-corner points; the test examines only the four pixels 1, 5, 9 and 13. A corner

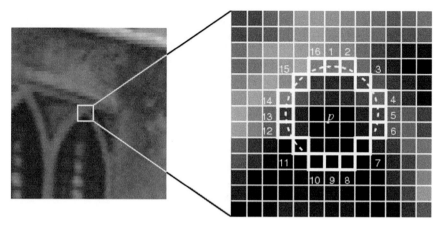

Fig. 5 Feature detection in an image patch using FAST detector [41]

can only exist if three of these test pixels are brighter than $I_p + t$ or darker than $I_p - t$ and the rest of pixels are then examined for final conclusion. Figure 5 illustrates the process, where the highlighted squares are the pixels used in the corner detection. The pixel at p is the center of a candidate corner. The arc is indicated by the dashed line passes through 12 contiguous pixels which are brighter than p by a threshold. The best results are achieved using a circle with $r = 3$ and $n = 9$.

Although the high speed test yields high performance, it suffers from several limitations and weakness as mentioned in [41]. An improvement for addressing these limitations and weakness points is achieved using a machine learning approach. The ordering of questions used to classify a pixel is learned by using the well-known decision tree algorithm (ID3), which speeds this step up significantly. As the first test produces many adjacent responses around the interest point, an additional criteria is applied to perform a non-maximum suppression. This allows for precise feature localization. The used cornerness measure at this step is

$$C(x, y) = max(\sum_{j \in S_{bright}} |I_{p \to j} - I_p| - t, \sum_{j \in S_{dark}} |I_p - I_{p \to j}| - t) \qquad (9)$$

where $I_{p \to j}$ denotes the pixels laying on the Bresenham circle. In this way, the processing time remains short because the second test is performed only on a fraction of image points that passed the first test.

In other words, the process operates in two stages. First, corner detection with a segment test of a given n and a convenient threshold is performed on a set of images (preferably from the target application domain). Each pixel of the 16 locations on the circle is classified as darker, similar, or brighter. Second, employing the ID3 algorithm on the 16 locations to select the one that yields the maximum information gain. The non-maximum suppression is applied on the sum of the absolute difference between the pixels in the contiguous arc and the center pixel. Notice that the corners

detected using the ID3 algorithm may be slightly different from the results obtained
with segment test detector due to the fact that decision tree model depends on the
training data, which could not cover all possible corners. Compared to many existing
detectors, the FAST corner detector is very suitable for real-time video processing
applications because of its high-speed performance. However, it is not invariant to
scale changes and not robust to noise, as well as it depends on a threshold, where
selecting an adequate threshold is not a trivial task.

3.1.5 Hessian Detector

The Hessian blob detector [42, 43] is based on a 2×2 matrix of second-order
derivatives of image intensity $I(x, y)$, called the Hessian matrix. This matrix can be
used to analyze local image structures and it is expressed in the form

$$H(x, y, \sigma) = \begin{bmatrix} I_{xx}(x, y, \sigma) & I_{xy}(x, y, \sigma) \\ I_{xy}(x, y, \sigma) & I_{yy}(x, y, \sigma) \end{bmatrix} \quad (10)$$

where I_{xx}, I_{xy}, and I_{yy} are second-order image derivatives computed using Gaussian
function of standard deviation σ. In order to detect interest features, it searches for a
subset of points where the derivatives responses are high in two orthogonal directions.
That is, the detector searches for points where the determinant of the Hessian matrix
has a local maxima

$$\det(H) = I_{xx}I_{yy} - I_{xy}^2 \quad (11)$$

By choosing points that maximize the determinant of the Hessian, this measure
penalizes longer structures that have small second derivatives (i.e., signal changes)
in a single direction. Applying non-maximum suppression using a window of size
3×3 over the entire image, keeping only pixels whose value is larger than the values
of all eight immediate neighbors inside the window. Then, the detector returns all the
remaining locations whose value is above a pre-defined threshold T. The resulting
detector responses are mainly located on corners and on strongly textured image
areas. While, the Hessian matrix is used for describing the local structure in a neighborhood around a point, its determinant is used to detect image structures that exhibit
signal changes in two directions. Compared to other operators such as Laplacian, the
determinant of the Hessian responds only if the local image pattern contains significant variations along any two orthogonal directions [44]. However, using second
order derivatives in the detector is sensitive to noise. In addition, the local maxima
are often found near contours or straight edges, where the signal changes in only one
direction [45]. Thus, these local maxima are less stable as the localization is affected
by noise or small changes in neighboring pattern.

3.2 Multi-scale Detectors

3.2.1 Laplacian of Gaussian (LoG)

Laplacian-of-Gaussian (LoG), a linear combination of second derivatives, is a common blob detector. Given an input image $I(x, y)$, the scale space representation of the image defined by $L(x, y, \sigma)$ is obtained by convolving the image by a variable scale Gaussian kernel $G(x, y, \sigma)$ where

$$L(x, y, \sigma) = G(x, y, \sigma) * I(x, y) \tag{12}$$

and

$$G(x, y, \sigma) = \frac{1}{2\pi\sigma^2} e^{\frac{-(x^2+y^2)}{2\sigma^2}} \tag{13}$$

For computing the Laplacian operator, the following formula is used

$$\nabla^2 L(x, y, \sigma) = L_{xx}(x, y, \sigma) + L_{yy}(x, y, \sigma) \tag{14}$$

This results in strong positive responses for dark blobs and strong negative responses for bright blobs of size $\sqrt{2}\sigma$. However, the operator response is strongly dependent on the relationship between the size of the blob structures in the image domain and the size of the smoothing Gaussian kernel. The standard deviation of the Gaussian is used to control the scale by changing the amount of blurring. In order to automatically capture blobs of different size in the image domain, a multi-scale approach with automatic scale selection is proposed in [36] through searching for scale space extrema of the scale-normalized Laplacian operator.

$$\nabla^2_{norm} L(x, y, \sigma) = \sigma^2 (L_{xx}(x, y, \sigma) + L_{yy}(x, y, \sigma)) \tag{15}$$

Which can also detect points that are simultaneously local maxima/minima of $\nabla^2_{norm} L(x, y, \sigma)$ with respect to both space and scale. The LoG operator is circularly symmetric; it is therefore naturally invariant to rotation. The LoG is well adapted to blob detection due to this circular symmetry property, but it also provides a good estimation of the characteristic scale for other local structures such as corners, edges, ridges and multi-junctions. In this context, the LoG can be applied for finding the characteristic scale for a given image location or for directly detecting scale-invariant regions by searching for 3D (location + scale) extrema of the LoG function as illustrated in Fig. 6. The scale selection properties of the Laplacian operator are studied in detail in [46].

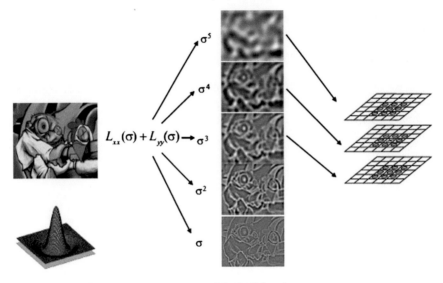

Fig. 6 Searching for 3D scale space extrema of the LoG function

3.2.2 Difference of Gaussian (DoG)

In fact, the computation of LoG operators is time consuming. To accelerate the computation, Lowe [31] proposed an efficient algorithm based on local 3D extrema in the scale-space pyramid built with Difference-of-Gaussian(DoG) filters. This approach is used in the scale-invariant feature transform (SIFT) algorithm. In this context, the DoG gives a close approximation to the Laplacian-of-Gaussian (LoG) and it is used to efficiently detect stable features from scale-space extrema. The DoG function $D(x, y, \sigma)$ can be computed without convolution by subtracting adjacent scale levels of a Gaussian pyramid separated by a factor k.

$$
\begin{aligned}
D(x, y, \sigma) &= (G(x, y, k\sigma) - G(x, y, \sigma)) * I(x, y) \\
&= L(x, y, k\sigma) - L(x, y, \sigma)
\end{aligned}
\tag{16}
$$

Feature types extracted by DoG can be classified in the same way as for the LoG operator. Also, the DoG region detector searches for 3D scale space extrema of the DoG function as shown in Fig. 7. The computation of LoG operators is time consuming. The common drawback of both the LoG and DoG representations is that the local maxima can also be detected in neighboring contours of straight edges, where the signal change is only in one direction, which make them less stable and more sensitive to noise or small changes [45].

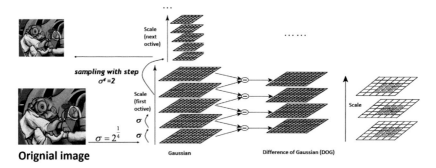

Fig. 7 Searching for 3D scale space extrema in the DoG function [31]

3.2.3 Harris-Laplace

Harris-Laplace is a scale invariant corner detector proposed by Mikolajczyk and Schmid [45]. It relies on a combination of Harris corner detector and a Gaussian scale space representation. Although Harris-corner points have been shown to be invariant to rotation and illumination changes, the points are not invariant to the scale. Therefore, the second-moment matrix utilized in that detector is modified to make it independent of the image resolution. The scale adapted second-moment matrix used in the Harris-Laplace detector is represented as

$$M(x, y, \sigma_I, \sigma_D) = \sigma_D^2 \, g(\sigma_I) \begin{bmatrix} I_x^2(x, y, \sigma_D) & I_x I_y(x, y, \sigma_D) \\ I_x I_y(x, y, \sigma_D) & I_y^2(x, y, \sigma_D) \end{bmatrix} \quad (17)$$

where I_x and I_y are the image derivatives calculated in their respective direction using a Gaussian kernel of scale σ_D. The parameter σ_I determines the current scale at which the Harris corner points are detected in the Gaussian scale-space. In other words, the derivative scale σ_D decides the size of gaussian kernels used to compute derivatives. While, the integration scale σ_I is used to performed a weighted average of derivatives in a neighborhood. The multi-scale Harris cornerness measure is computed using the determinant and the trace of the adapted second moment matrix as

$$C(x, y, \sigma_I, \sigma_D) = det[M(x, y, \sigma_I, \sigma_D)] - \alpha.trace^2[M(x, y, \sigma_I, \sigma_D)] \quad (18)$$

The value of the constant α is between 0.04 and 0.06. At each level of the representation, the interest points are extracted by detecting the local maxima in the 8-neighborhood of a point (x, y). Then, a threshold is used to reject the maxima of small cornerness, as they are less stable under arbitrary viewing conditions

$$C(x, y, \sigma_I, \sigma_D) > Threshold_{Harris} \quad (19)$$

In addition, the Laplacian-of-Gaussian is used to find the maxima over the scale. Where, only the points for which the Laplacian attains maxima or its response is above a threshold are accepted.

$$\sigma_I^2 |L_{xx}(x, y, \sigma_I) + L_{yy}(x, y, \sigma_I)| > Threshold_{Laplacian} \tag{20}$$

The Harris-Laplace approach provides a representative set of points which are characteristic in the image and in the scale dimension. It also dramatically reduces the number of redundant interest points compared to Multi-scale Harris. The points are invariant to scale changes, rotation, illumination, and addition of noise. Moreover, the interest points are highly repeatable. However, the Harris-Laplace detector returns a much smaller number of points compared to the LoG or DoG detectors. Also, it fails in the case of affine transformations.

3.2.4 Hessian-Laplace

Similar to Harris-Laplace, the same idea can also be applied to the Hessian-based detector, leading to a scale invariant detector termed, Hessian-Laplace. At first, we build an image scale-space representation using Laplacian filters or their approximations DoG filters. Then, use the determinant of the Hessian to extract scale invariant blob-like features. Hessian-Laplace detector extracts large number of features that cover the whole image at a slightly lower repeatability compared to its counterpart Harris-Laplace. Furthermore, the extracted locations are more suitable for scale estimation due to the similarity of the filters used in spatial and scale localization, both based on second order Gaussian derivatives. Bay et al. [32] claimed that Hessian-based detectors are more stable than the Harris-based counterparts. Likewise, approximating LoG by DoG for acceleration, the Hessian determinant is approximated using integral images technique [29] resulting in the Fast Hessian detector [32].

3.2.5 Gabor-Wavelet detector

Recently, Yussof and Hitam [47] proposed a multi-scale interest points detector based on the principle of Gabor wavelets. The Gabor wavelets are biologically motivated convolution kernels in the shape of plane waves restricted by a Gaussian envelope function, whose kernels are similar to the response of the two-dimensional receptive field profiles of the mammalian simple cortical cell. The Gabor wavelets take the form of a complex plane wave modulated by a Gaussian envelope function

$$\psi_{u,v}(z) = \frac{||K_{u,v}||^2}{\sigma^2} e^{\left(\frac{||K_{u,v}||^2 ||z||^2}{2\sigma^2}\right)} \left[e^{izK_{u,v}} - e^{\frac{\sigma^2}{2}} \right] \tag{21}$$

where $K_{u,v} = K_v e^{i\phi_u}$, $z = (x, y)$, u and v define the orientation and scale of the Gabor wavelets, $K_v = K_{max}/f^v$ and $\phi_u = \pi u/8$, K_{max} is the maximum frequency, and $f =$

$\sqrt{2}$ is the spacing factor between kernels in the frequency domain. The response of an image I to a wavelet ψ is calculated as the convolution

$$G = I * \psi \tag{22}$$

The coefficients of the convolution, represent the information in a local image region, which should be more effective than isolated pixels. The advantage of Gabor wavelets is that they provides simultaneous optimal resolution in both space and spatial frequency domains. Additionally, Gabor wavelets have the capability of enhancing low level features such as peaks, valleys and ridges. Thus, they are used to extract points from the image at different scales by combining multi orientations of image responses. The interest points are extracted at multiple scales with a combination of uniformly spaced orientation. The authors proved that the extracted points using Gabor-wavelet detector have high accuracy and adaptability to various geometric transformations.

3.3 Affine Invariant Detectors

The feature detectors discussed so far exhibit invariance to translations, rotations and uniform scaling; assuming that the localization and scale are not affected by an affine transformation of the local image structures. Thus, they partially handle the challenging problem of affine invariance, keeping in mind that the scale can be different in each direction rather than uniform scaling. Which in turn makes the scale invariant detectors fail in the case of significant affine transformations. Therefore, building a detector robust to perspective transformations necessitates invariance to affine transformations. An affine invariant detector can be seen as a generalized version of a scale invariant detector. Recently, some features detectors have been extended to extract features invariant to affine transformations. For instance, Schaffalitzky and Zisserman [48] extended the Harris-Laplace detector by affine normalization. Mikolajczyk and Schmid [45] introduced an approach for scale and affine invariant interest point detection. Their algorithm simultaneously adapts location, scale and shape of a point neighborhood to obtain affine invariant points. Where, Harris detector is adapted to affine transformations and the affine shape of a point neighborhood is estimated based on the second moment matrix. This is achieved by following the iterative estimation scheme proposed by Lindberg and Gårding [49] as follows:

1. Identify initial region points using scale-invariant Harris-Laplace detector.
2. For each initial point, normalize the region to be affine invariant using affine shape adaptation.
3. Iteratively estimate the affine region; selection of proper integration scale, differentiation scale and spatially localize interest points.
4. Update the affine region using these scales and spatial localizations.
5. Repeat step 3 if the stopping criterion is not met.

Similar to Harris-affine, the same idea can also be applied to the Hessian-based detector, leading to an affine invariant detector termed as Hessian-affine. For a single image, the Hessian-affine detector typically identifies more reliable regions than the Harris-affine detector. The performance changes depending on the type of scene being analyzed. Further, the Hessian-affine detector responds well to textured scenes in which there are a lot of corner-like parts. However, for some structured scenes, like buildings, the Hessian-affine detector performs very well. A thorough analysis of several state-of-the-art affine detectors have been done by Mikolajczyk and Schmid [50].

There are several other feature detectors that are not discussed in this chapter due to space limitation. Fast Hessian or the Determinant of Hessian (DoH) [32], MSER [51, 52], are some examples. A more detailed discussion of these detectors can be found in [44, 45, 53].

4 Image Feature Descriptor

Once a set of interest points has been detected from an image at a location $p(x, y)$, scale s, and orientation θ, their content or image structure in a neighborhood of p needs to be encoded in a suitable descriptor for discriminative matching and insensitive to local image deformations. The descriptor should be aligned with θ and proportional to the scale s. There are a large number of image feature descriptors in the literature; the most frequently used descriptors are discussed in the following sections.

4.1 Scale Invariant Feature Transform (SIFT)

Lowe [31] presented the scale-invariant feature transform (SIFT) algorithm, where a number of interest points are detected in the image using the Difference-of-Gaussian (DOG) operator. The points are selected as local extrema of the DoG function. At each interest point, a feature vector is extracted. Over a number of scales and over a neighborhood around the point of interest, the local orientation of the image is estimated using the local image properties to provide invariance against rotation. Next, a descriptor is computed for each detected point, based on local image information at the characteristic scale. The SIFT descriptor builds a histogram of gradient orientations of sample points in a region around the keypoint, finds the highest orientation value and any other values that are within 80 % of the highest, and uses these orientations as the dominant orientation of the keypoint.

The description stage of the SIFT algorithm starts by sampling the image gradient magnitudes and orientations in a 16×16 region around each keypoint using its scale to select the level of Gaussian blur for the image. Then, a set of orientation histograms is created where each histogram contains samples from a 4×4 subregion of the original neighborhood region and having eight orientations bins in each. A

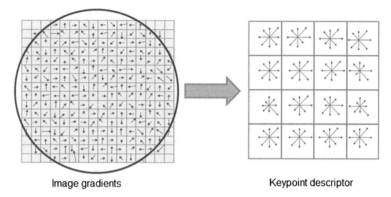

Image gradients   Keypoint descriptor

Fig. 8 A schematic representation of the SIFT descriptor for a 16×16 pixel patch and a 4×4 descriptor array

Gaussian weighting function with σ equal to half the region size is used to assign weight to the magnitude of each sample point and gives higher weights to gradients closer to the center of the region, which are less affected by positional changes. The descriptor is then formed from a vector containing the values of all the orientation histograms entries. Since there are 4×4 histograms each with 8 bins, the feature vector has $4 \times 4 \times 8 = 128$ elements for each keypoint. Finally, the feature vector is normalized to unit length to gain invariance to affine changes in illumination. However, non-linear illumination changes can occur due to camera saturation or similar effects causing a large change in the magnitudes of some gradients. These changes can be reduced by thresholding the values in the feature vector to a maximum value of 0.2, and the vector is again normalized. Figure 8 illustrates the schematic representation of the SIFT algorithm; where the gradient orientations and magnitudes are computed at each pixel and then weighted by a Gaussian falloff (indicated by overlaid circle). A weighted gradient orientation histogram is then computed for each subregion.

The standard SIFT descriptor representation is noteworthy in several respects: the representation is carefully designed to avoid problems due to boundary effects-smooth changes in location, orientation and scale do not cause radical changes in the feature vector; it is fairly compact, expressing the patch of pixels using a 128 element vector; while not explicitly invariant to affine transformations, and the representation is surprisingly resilient to deformations such as those caused by perspective effects. These characteristics are evidenced in excellent matching performance against competing algorithms under different scales, rotations and lighting. On the other hand, the construction of the standard SIFT feature vector is complicated and the choices behind its specific design are not clear resulting in the common problem of SIFT its high dimensionality, which affects the computational time for computing the descriptor (significantly slow). As an extension to SIFT, Ke and Sukthankar [54] proposed PCA-SIFT to reduce the high dimensionality of original SIFT descriptor using the standard Principal Components Analysis (PCA) technique to the normalized gradient image patches extracted around keypoints. It extracts a 41×41 patch at the given scale and computes its image gradients in the vertical and horizontal directions and creates a feature vector from concatenating the gradients in both directions. Therefore, the feature vector is of length $2 \times 39 \times 39 = 3042$ dimensions. The gradient image vector is projected into a pre-computed feature space, resulting a feature vector of length 36 elements. The vector is then normalized to unit magnitude to reduce the effects of illumination changes. Also, Morel and Yu [55] proved that the SIFT is fully invariant with respect to only four parameters namely zoom, rotation and translation out of the six parameters of the affine transform. Therefore, they introduced affine-SIFT (ASIFT), which simulates all image views obtainable by varying the camera axis orientation parameters, namely, the latitude and the longitude angles, left over by the SIFT descriptor.

4.2 Gradient Location-Orientation Histogram (GLOH)

Gradient location-orientation histogram (GLOH) developed by Mikolajczyk and Schmid [50] is also an extension of the SIFT descriptor. GLOH is very similar to the SIFT descriptor, where it only replaces the Cartesian location grid used by the SIFT with a log-polar one, and applies PCA to reduce the size of the descriptor. GLOH uses a log-polar location grid with 3 bins in radial direction (radius is set to 6, 11, and 15) and 8 in angular direction, resulting in 17 location bins as illustrated in Fig. 9. GLOH descriptor builds a set of histograms using the gradient orientations in 16 bins, resulting in a feature vector of $17 \times 16 = 272$ elements for each interest point. The 272-dimensional descriptor is reduced to 128 dimensional one by computing the covariance matrix for PCA and the highest 128 eigenvectors are selected for description. Based on the experimental evaluation conducted in [50], GLOH has

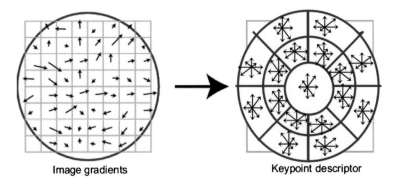

Image gradients   Keypoint descriptor

Fig. 9 A schematic representation of the GLOH algorithm using log-polar bins

been reported to outperform the original SIFT descriptor and gives the best performance, especially under illumination changes. Furthermore, it has been shown to be more distinctive but also more expensive to compute than its counterpart SIFT.

4.3 Speeded-Up Robust Features Descriptor (SURF)

The Speeded-Up Robust Features (SURF) detector-descriptor scheme developed by Bay et al. [32] is designed as an efficient alternative to SIFT. It is much faster, and more robust as opposed to SIFT. For the detection stage of interest points, instead of relying on ideal Gaussian derivatives, the computation is based on simple 2D box filters; where, it uses a scale invariant blob detector based on the determinant of Hessian matrix for both scale selection and locations. Its basic idea is to approximate the second order Gaussian derivatives in an efficient way with the help of integral images using a set of box filters. The 9×9 box filters depicted in Fig. 10 are approximations of a Gaussian with $\sigma = 1.2$ and represent the lowest scale for computing the blob response maps. These approximations are denoted by D_{xx}, D_{yy}, and D_{xy}. Thus, the approximated determinant of Hessian can be expressed as

$$\det(H_{approx}) = D_{xx}D_{yy} - (wD_{xy})^2 \qquad (23)$$

where w is a relative weight for the filter response and it is used to balance the expression for the Hessian's determinant. The approximated determinant of the Hessian represents the blob response in the image. These responses are stored in a blob response map, and local maxima are detected and refined using quadratic interpolation, as with DoG. Finally, do non-maximum suppression in a $3 \times 3 \times 3$ neighborhood to get steady interest points and the scale of values.

The SURF descriptor starts by constructing a square region centered around the detected interest point, and oriented along its main orientation. The size of this window is $20s$, where s is the scale at which the interest point is detected. Then, the interest region is further divided into smaller 4×4 sub-regions and for each sub-region the Harr wavelet responses in the vertical and horizontal directions (denoted

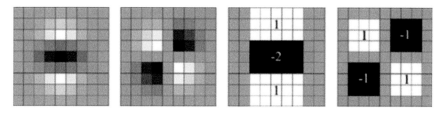

Fig. 10 *Left* to *right* Gaussian second order derivatives in y- (D_{yy}), xy-direction (D_{xy}) and their approximations in the same directions, respectively

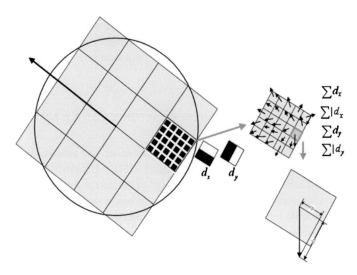

Fig. 11 Dividing the interest region into 4 × 4 sub-regions for computing the SURF descriptor

d_x and d_y, respectively) are computed at a 5 × 5 sampled points as shown in Fig. 11. These responses are weighted with a Gaussian window centered at the interest point to increase the robustness against geometric deformations and localization errors. The wavelet responses d_x and d_y are summed up for each sub-region and entered in a feature vector v, where

$$v = \left(\sum d_x, \sum |d_x|, \sum d_y, \sum |d_y|\right) \qquad (24)$$

Computing this for all the 4 × 4 sub-regions, resulting a feature descriptor of length 4 × 4 × 4 = 64 dimensions. Finally, the feature descriptor is normalized to a unit vector in order to reduce illumination effects.

The main advantage of the SURF descriptor compared to SIFT is the processing speed as it uses 64 dimensional feature vector to describe the local feature, while SIFT uses 128. However, the SIFT descriptor is more suitable to describe images affected by translation, rotation, scaling, and other illumination deformations. Though SURF shows its potential in a wide range of computer vision applications, it also has some shortcomings. When 2D or 3D objects are compared, it does not work if rotation is violent or the view angle is too different. Also, the SURF is not fully affine invariant as explained in [56].

4.4 Local Binary Pattern (LBP)

Local Binary Patterns (LBP) [57, 58] characterizes the spatial structure of a texture and presents the characteristics of being invariant to monotonic transformations of the gray-levels. It encodes the ordering relationship by comparing neighboring pixels with the center pixel, that is, it creates an order based feature for each pixel by comparing each pixel's intensity value with that of its neighboring pixels. Specifically, the neighbors whose feature responses exceed the central's one are labeled as '1' while the others are labeled as '0'. The co-occurrence of the comparison results is subsequently recorded by a string of binary bits. Afterwards, weights coming from a geometric sequence which has a common ratio of 2 are assigned to the bits according to their indices in strings. The binary string with its weighted bits is consequently transformed into a decimal valued index (i.e., the LBP feature response). That is, the descriptor describes the result over the neighborhood as a binary number (binary pattern). On its standard version, a pixel c with intensity $g(c)$ is labeled as

$$S(g_p - g_c) = \begin{cases} 1, & if \ \ g_p \geq g_c \\ 0, & if \ \ g_p < g_c \end{cases} \quad (25)$$

where pixels p belong to a 3×3 neighborhood with gray levels $g_p(p = 0, 1, \ldots, 7)$. Then, the LBP pattern of the pixel neighborhood is computed by summing the corresponding thresholded values $S(g_p - g_c)$ weighted by a binomial factor of 2^k as

$$LBP = \sum_{k=0}^{7} S(g_p - g_c).2^k \quad (26)$$

After computing the labeling for each pixel of the image, a 256-bin histogram of the resulting labels is used as a feature descriptor for the texture. An illustration example for computing LBP of a pixel in a 3×3 neighborhood and an orientation descriptor of a basic region in an image is shown in Fig. 12. Also, the LBP descriptor is calculated in its general form as follows

$$LBP_{RN}(x, y) = \sum_{i=0}^{N-1} S(n_i - n_c).2^i, \quad S(x) = \begin{cases} 1, & x \geq 0, \\ 0, & otherwise \end{cases} \quad (27)$$

where n_c corresponds to the gray level of the center pixel of a local neighborhood and n_i is the gray levels of N equally spaced pixels on a circle of radius R. Since correlation between pixels decreases with the distance, a lot of the texture information can be obtained from local neighborhoods. Thus, the radius R is usually kept small. In practice, the signs of the differences in a neighborhood are interpreted as a N-bit binary number, resulting in 2^N distinct values for the binary pattern as shown in Fig. 13. The binary patterns are called uniform patterns, where they contain at most two bitwise transitions from 0 to 1. For instance, "11000011" and "00001110" are two uniform patterns, while "00100100" and "01001110" are non-uniform patterns.

Several variations of LBP have been proposed, including the center-symmetric local binary patterns (CS-LBP), the local ternary pattern (LTP), the center-symmetric local ternary pattern (CS-LTP) based on the CS-LBP, and orthogonal symmetric local

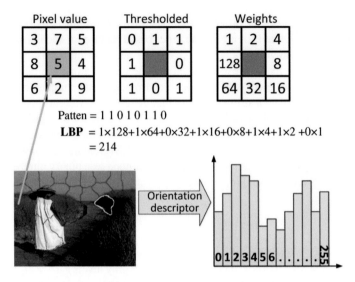

Fig. 12 Computing LBP descriptor for a pixel in a 3×3 neighborhood [59] © 2014 IEEE

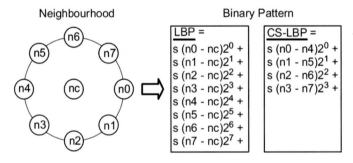

Fig. 13 LBP and CS-LBP features for a neighborhood of 8 pixels [58] © 2009 Elsevier

ternary pattern (OS-LTP) [60]. Unlike the LBP, the CS-LBP descriptor compares gray-level differences of center-symmetric pairs (see Fig. 13). In fact, the LBP has the advantage of tolerance of illumination changes and computational simplicity. Also, the LBP and its variants achieve great success in texture description. Unfortunately, the LBP feature is an index of discrete patterns rather than a numerical feature, thus it is difficult to combine the LBP features with other discriminative ones in a compact descriptor [61]. Moreover, it produces higher dimensional features and is sensitive to Gaussian noise on flat regions.

4.5 Binary Robust Independent Elementary Features (BRIEF)

Binary robust independent elementary features (BRIEF), a low-bitrate descriptor, is introduced for image matching with random forest and random ferns classifiers

[62]. It belongs to the family of binary descriptors such as LBP and BRISK, which only performs simple binary comparison test and uses Hamming distance instead of Euclidean or Mahalanobis distance. Briefly, for building a binary descriptor, it is only necessary to compare the intensity between two pixel positions located around the detected interest points. This allows to obtain a representative description at very low computational cost. Besides, matching the binary descriptors requires only the computation of Hamming distances that can be executed very fast through XOR primitives on modern architectures.

The BRIEF algorithm relies on a relatively small number of intensity difference tests to represent an image patch as a binary string. More specifically, a binary descriptor for a patch of pixels of size $S \times S$ is built by concatenating the results of the following test

$$\tau = \begin{cases} 1, & \text{if } I(P_j) > I(P_i), \\ \\ 0, & \text{otherwise,} \end{cases} \tag{28}$$

where $I(p_i)$ denotes the (smoothed) pixel intensity value at p_i, and the selection of the location of all the p_i uniquely defines a set of binary tests. The sampling points are drawn from a zero-mean isotropic Gaussian distribution with variance equal to $\frac{1}{25}S^2$. For increasing the robustness of the descriptor, the patch of pixels is pre-smoothed with a Gaussian kernel with variance equal to 2 and size equal to 9×9 pixels. The BRIEF descriptor has two setting parameters: the number of binary pixel pairs and the binary threshold.

The experiments conducted by authors showed that only 256 bits, or even 128 bits, often suffice to obtain very good matching results. Thus, BRIEF is considered to be very efficient both to compute and to store in memory. Unfortunately, BRIEF descriptor is not robust against a rotation larger than $35°$ approximately, hence, it does not provide rotation invariance.

4.6 Other Feature Descriptors

A large number of other descriptors have been proposed in the literature and many of them have been proved to be effective in computer vision applications. For instance, color-based local features are four color descriptors based on color information proposed by Weijer and Schmid [63]. The Gabor representation or its variation [64, 65] has been also shown to be optimal in the sense of minimizing the joint two-dimensional uncertainty in space and frequency. Zernike moments [66, 67] and Steerable filters [68] are also considered for feature extraction and description.

Inspired by Weber's Law, a dense descriptor computed for every pixel depending on both the local intensity variation and the magnitude of the center pixel's intensity called Weber Local Descriptor (WLD) is proposed in [28]. The WLD descriptor employs the advantages of SIFT in computing the histogram using the gradient and its orientation, and those of LBP in computational efficiency and smaller support regions. In contrast to the LBP descriptor, WLD first computes the salient micro-patterns (i.e., differential excitation), and then builds statistics on these salient patterns along with the gradient orientation of the current point.

Two methods for extracting distinctive features from interest regions based on measuring the similarity between visual entities from images are presented in [69]. The idea of these methods combines the powers of two well-known approaches, the SIFT descriptor and Local Self-Similarities (LSS). Two texture features called Local Self-Similarities (LSS, C) and Fast Local Self-Similarities (FLSS, C) based on Cartesian location grid, are extracted, which are the modified versions of the Local Self-Similarities feature based on Log-Polar location grid (LSS, LP). The LSS and FLSS features are used as the local features in the SIFT algorithm. The proposed LSS and FLSS descriptors use distribution-based histogram representation in each cell rather than choosing the maximal correlation value in each bucket in the Log-Polar location grid in the natural (LSS, LP) descriptor. Thus, they get more robust geometric transformations invariance and good photometric transformations invariance. A local image descriptor based on Histograms of the Second Order Gradients, namely HSOG is introduced in [70] for capturing the curvature related geometric properties of the neural landscape. Dalal and Triggs [71] presented the Histogram of Oriented Gradient (HOG) descriptor, which combines both the properties of SIFT and GLOH descriptors. The main difference between HOG and SIFT is that the HOG descriptor is computed on a dense grid of uniformly spaced cells with overlapping local contrast normalization.

Following a different direction, Fan et al. [72] proposed a method for interest region description, which pools local features based on their intensity orders in multiple support regions. Pooling by intensity orders is not only invariant to rotation and monotonic intensity changes, but also encodes ordinal information into a descriptor. By pooling two different kinds of local features, one based on gradients and the other on intensities, two descriptors are obtained: the Multisupport Region Order-Based Gradient Histogram (MROGH) and the Multisupport Region Rotation and Intensity Monotonic Invariant Descriptor (MRRID). The former combines information of intensity orders and gradient, while the latter is completely based on intensity orders, which makes it particularly suitable to large illumination changes. Several image features are analyzed

In spite of the fact that, a large number of image feature descriptors have been introduced recently, several of these descriptors are exclusively designed for a specific application scenario such as object recognition, shape retrieval, or LADAR data processing [73]. Furthermore, the authors of these descriptors evaluated their performance on a limited number of benchmarking datasets collected specifically for particular tasks. Consequently, it is very challenging for researchers to choose an appropriate descriptor for their particular application. In this respect, some recent studies compare the performance of several descriptors; interest region descriptors [50], binary descriptors [74], local colour descriptors [75], and the 3D descriptors [76, 77]. In fact, claims that describing image features is a solved problem are overly bold and optimistic. On the other hand, claims that designing a descriptor for general real-world scenarios is next to impossible are simply too pessimistic, given the success of the aforementioned descriptors in several applications. Finally, there is much work to be done in order to realize description algorithms that can be used for general applications. We argue for further research towards using new modalities captured by 3D data and color information. For real time applications, a further path of future research would be the implementation of the algorithms on parallel processing units such as GPU.

5 Features Matching

Features matching or generally image matching, a part of many computer vision applications such as image registration, camera calibration and object recognition, is the task of establishing correspondences between two images of the same scene/object. A common approach to image matching consists of detecting a set of interest points each associated with image descriptors from image data. Once the features and their descriptors have been extracted from two or more images, the next step is to establish some preliminary feature matches between these images as illustrated in Fig. 14.

Without losing the generality, the problem of image matching can be formulated as follows, suppose that p is a point detected by a detector in an image associated with a descriptor

$$\Phi(p) = \{\phi_k(P) \mid k = 1, 2, \ldots, K\} \tag{29}$$

where, for all K, the feature vector provided by the k-th descriptor is

$$\phi_k(p) = (f_{1p}^k, f_{2p}^k, \ldots, f_{n_k p}^k) \tag{30}$$

The aim is to find the best correspondence q in another image from the set of N interest points $Q = \{q_1, q_2, \ldots, q_N\}$ by comparing the feature vector $\phi_k(p)$ with those of the points in the set Q. To this end, a distance measure between the two interest points descriptors $\phi_k(p)$ and $\phi_k(q)$ can be defined as

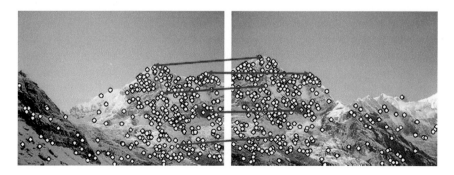

Fig. 14 Matching image regions based on their local feature descriptors [79] © 2011 Springer

$$d_k(p, q) = |\phi_k(p) - \phi_k(q)| \tag{31}$$

Based on the distance d_k, the points of Q are sorted in ascending order independently for each descriptor creating the sets

$$\Psi(p, Q) = \{\psi_k(p, Q) \mid k = 1, 2, \dots, k\} \tag{32}$$

Such that,

$$\psi_k(p, Q) = \left\{(\psi_k^1, \psi_k^2, \dots, \psi_k^N) \in Q \mid d_k(p, \psi_k^i) \leq d_k(p, \psi_k^j), \forall i > j\right\} \tag{33}$$

A match between the pair of interest points (p, q) is accepted only if (i) p is the best match for q in relation to all the other points in the first image and (ii) q is the best match for p in relation to all the other points in the second image. In this context, it is very important to devise an efficient algorithm to perform this matching process as quickly as possible. The nearest-neighbor matching in the feature space of the image descriptors in Euclidean norm can be used for matching vector-based features. However, in practice, the optimal nearest neighbor algorithm and its parameters depend on the data set characteristics. Furthermore, to suppress matching candidates for which the correspondence may be regarded as ambiguous, the ratio between the distances to the nearest and the next nearest image descriptor is required to be less than some threshold. As a special case, for matching high dimensional features, two algorithms have been found to be the most efficient: the randomized k-d forest and the fast library for approximate nearest neighbors (FLANN) [78].

On the other hand, these algorithms are not suitable for binary features (e.g., FREAK or BRISK). Binary features are compared using the Hamming distance calculated via performing a bitwise XOR operation followed by a bit count on the result. This involves only bit manipulation operations that can be performed quickly. The typical solution in the case of matching large datasets is to replace the linear search with an approximate matching algorithm that can offer speedups of several orders of magnitude over the linear search. This is, at the cost that some of the nearest neighbors returned are approximate neighbors, but usually close in distance to the exact neighbors. For performing matching of binary features, other methods can be employed such as [80–82].

Generally, the performance of matching methods based on interest points depends on both the properties of the underlying interest points and the choice of associated image descriptors [83]. Thus, detectors and descriptors appropriate for images contents shall be used in applications. For instance, if an image contains bacteria cells, the blob detector should be used rather than the corner detector. But, if the image is an aerial view of a city, the corner detector is suitable to find man-made structures. Furthermore, selecting a detector and a descriptor that addresses the image degradation is very important. For example, if there is no scale change present, a corner detector that does not handle scale is highly desirable; while, if image contains a higher level of distortion, such as scale and rotation, the more computationally intensive SURF feature detector and descriptor is a adequate choice in that case. For greater accuracy, it is recommended to use several detectors and descriptors at the same time. In the area of feature matching, it must be noticed that the binary descriptors (e.g., FREAK or BRISK) are generally faster and typically used for finding point correspondences between images, but they are less accurate than vector-based descriptors [74]. Statistically robust methods like RANSAC can be used to filter outliers in matched feature sets while estimating the geometric transformation or fundamental matrix, which is useful in feature matching for image registration and object recognition applications.

6 Performance Evaluation

6.1 Benchmarking Data Sets

There are a wide variety of data sets available on the Internet that can be used as a benchmark by researchers. One popular and widely used for performance evaluation of detectors and descriptors is the standard Oxford data set [84]. The dataset consists of image sets with different geometric and photometric transformations (viewpoint change, scale change, image rotation, image blur, illumination change, and JPEG compression) and with different scene types (structured and textured scenes). In the cases of illumination change, the light changes are introduced by varying the camera aperture. While in the case of rotation, scale change, viewpoint change, and blur, two different scene types are used. One scene type contains structured scenes which are homogeneous regions with distinctive edge boundaries (e.g., graffiti, buildings), and the other contains repeated textures of different forms. In this way, the influence of image transformation and scene type can be analyzed separately. Each image set contains 6 images with a gradual geometric or photometric distortion where the first image and the remaining 5 images are compared. Sample images from the Oxford data set are shown in Fig. 15.

6.2 Evaluation Criterion

To judge whether two image features are matched (i.e., belonging to the same feature or not), Mikolajczyk et al. [44] proposed an evaluation procedure based on the repeatability criterion by comparing the ground truth transformation and the detected region overlap. The repeatability can be considered as one of the most important criteria used for evaluating the stability of feature detectors. It measures the ability of a detector to extract the same feature points across images irrespective of imaging conditions. The repeatability criterion measures how well the detector determines corresponding scene regions. In this evaluation procedure, two regions of interest **A** and **B** are deemed to correspond if the overlap error ε is sufficiently small as shown

Fig. 15 Test images *Graf* (viewpoint change, structured scene), *Wall* (viewpoint change, textured scene), *Boat* (scale change + image rotation, structured scene), *Bark* (scale change + image rotation, textured scene), *Bikes* (image blur, structured scene), *Trees* (image blur, textured scene), *Leuven* (illumination change, structured scene), and *Ubc* (JPEG compression, structured scene)

in Fig. 16. This overlap error measures how well the regions correspond under a homography transformation **H**. It is defined by the ratio of the intersection and union of the two regions, that is the error in the image area covered by the two regions,

$$\varepsilon = 1 - \frac{\mathbf{A} \cap (\mathbf{H}^{\mathrm{T}} \ \mathbf{B} \ \mathbf{H})}{\mathbf{A} \cup (\mathbf{H}^{\mathrm{T}} \ \mathbf{B} \ \mathbf{H})} \tag{34}$$

This approach counts the total number of pixels in the union and the intersection of the two regions. Also, a match is correct if the error in the image area covered by two corresponding regions is less than 50 % of the region union, that is, $\varepsilon < 0.5$. The overlap error is computed numerically based on homography **H** and the matrices defining the regions. Thus, to evaluate feature detectors performance, the repeatability score for a given pair of images is computed as the ratio between the number of region to region correspondences and the smaller number of regions in the pair of images.

On the other hand, the performance of a region descriptor is measured by the matching criterion, i.e., how well the descriptor represents a scene region. It is based

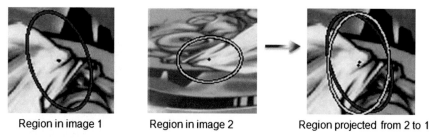

Region in image 1   Region in image 2   Region projected from 2 to 1

Fig. 16 Illustration of overlap error for a region projected onto the corresponding region (*ellipse*)

on the number of correct matches and the number of false matches obtained for the image pair. This is measured by comparing the number of corresponding regions obtained with the ground truth and the number of correctly matched regions. Matches are the nearest neighbors in the descriptor space [50]. In this case, the two regions of interest are matched if the Euclidean distance between their descriptors D_A and D_B is below a threshold τ. The results are presented with *recall* versus *1-precision*. Each descriptor from the reference image is compared with each descriptor from the transformed one and counting the number of correct matches as well as the number of false matches.

$$recall = \frac{No.\ correct\ matches}{Total\ No.\ correspondences}, \tag{35}$$

$$1 - precision = \frac{No.\ false\ matches}{Total\ No.\ all\ matches} \tag{36}$$

where, *No. correspondences* refers to the number of matching regions between image pairs. While, *recall* is the number of correctly matched regions with respect to the number of corresponding regions between two images of the same scene. An ideal descriptor gives a *recall* equal to 1 for any *precision* value. In order to obtain the curve, the value of τ is varied. Practically, *recall* increases for an increasing distance threshold τ because noise introduced by image transformations and region detection increases the distance between similar descriptors. A slowly increasing curve indicates that the descriptor is affected by the image noise. If the obtained curves corresponding to different descriptors are far apart or have different slopes, then the distinctiveness and robustness of these descriptors are different for the investigated image transformation or scene type [50].

7   Conclusions

The objective of this chapter is to provide a straight-forward, brief introduction for new researchers to the image feature detection and extraction research field. It introduces the basic notations and mathematical concepts for detecting and extracting image features, then describes the properties of perfect feature detectors. Various existing algorithms for detecting interest points are discussed briefly. The most frequently used description algorithms such as SIFT, SURF, LBP, WLD,...etc are also discussed and their advantages/disadvantages are highlighted. Furthermore, it explains some approaches to feature matching. Finally, the chapter discusses the techniques used for evaluating the performance of these algorithms.

References

1. Yap, T., Jiang, X., Kot, A.C.: Two-dimensional polar harmonic transforms for invariant image representation. IEEE Trans. Pattern Anal. Mach. Intell. **32**(7), 1259–1270 (2010)
2. Liu, S., Bai, X.: Discriminative features for image classification and retrieval. Pattern Recogn. Lett. **33**(6), 744–751 (2012)
3. Rahmani, R., Goldman, S., Zhang, H., Cholleti, S., Fritts, J.: Localized content-based image retrieval. IEEE Trans. Pattern Anal. Mach. Intell. **30**(11), 1902–1912 (2008)
4. Stöttinger, J., Hanbury, A., Sebe, N., Gevers, T.: Sparse color interest points for image retrieval and object categorization. IEEE Trans. Image Process. **21**(5), 2681–2691 (2012)
5. Wang, J., Li, Y., Zhang, Y., Wang, C., Xie, H., Chen, G., Gao, X.: Bag-of-features based medical image retrieval via multiple assignment and visual words weighting. IEEE Trans. Med. Imaging **30**(11), 1996–2011 (2011)
6. Andreopoulos, A., Tsotsos, J.: 50 years of object recognition: directions forward. Comput. Vis. Image Underst. **117**(8), 827–891 (2013)
7. Dollár, P., Wojek, C., Schiele, B., Perona, P.: Pedestrian detection: an evaluation of the state of the art. IEEE Trans. Pattern Anal. Mach. Intell. **34**(4), 743–761 (2012)
8. Felsberg, M., Larsson, F., Wiklund, J., Wadströmer, N., Ahlberg, J.: Online learning of correspondences between images. IEEE Trans. Pattern Anal. Mach. Intell. **35**(1), 118–129 (2013)
9. Miksik, O., Mikolajczyk, K.: Evaluation of local detectors and descriptors for fast feature matching. In: International Conference on Pattern Recognition (ICPR 2012), pp. 2681–2684. Tsukuba, Japan, 11–15 Nov 2012
10. Kim, B., Yoo, H., Sohn, K.: Exact order based feature descriptor for illumination robust image matching. Pattern Recogn. **46**(12), 3268–3278 (2013)
11. Moreels, P., Perona, P.: Evaluation of features detectors and descriptors based on 3D objects. Int. J. Comput. Vis. **73**(3), 263–284 (2007)
12. Takacs, G., Chandrasekhar, V., Tsai, S., Chen, D., Grzeszczuk, R., Girod, B.: Rotation-invariant fast features for large-scale recognition and real-time tracking. Sign. Process. Image Commun. **28**(4), 334–344 (2013)
13. Tang, S., Andriluka, M., Schiele, B.: Detection and tracking of occluded people. Int. J. Comput. Vis. **110**(1), 58–69 (2014)
14. Rincón, J.M., Makris, D., Uruñuela, C., Nebel, J.C.: Tracking human position and lower body parts using Kalman and particle filters constrained by human biomechanics. IEEE Trans. Syst. Man Cybern. Part B **41**(1), 26–37 (2011)
15. Lazebnik, S., Schmid, C., Ponce, J.: A sparse texture representation using local affine regions. IEEE Trans. Pattern Anal. Mach. Intell. **27**(8), 1265–1278 (2005)

16. Liu, L., Fieguth, P.: Texture classification from random features. IEEE Trans. Pattern Anal. Mach. Intell. **34**(3), 574–586 (2012)
17. Murillo, A., Guerrero, J., Sagues, C.: SURF features for efficient robot localization with omni-directional images. In: International Conference on Robotics and Automation, pp. 3901–3907. Rome, Italy, 10–14 Apr, 2007
18. Valgren, C., Lilienthal, A.J.: SIFT, SURF & seasons: appearance-based long-term localization in outdoor environments. Rob. Auton. Syst. **58**(2), 149–156 (2010)
19. Campos, F.M., Correia, L., Calado, J.M.F.: Robot visual localization through local feature fusion: an evaluation of multiple classifiers combination approaches. J. Intell. Rob. Syst. **77**(2), 377–390 (2015)
20. Farajzadeh, N., Faez, K., Pan, G.: Study on the performance of moments as invariant descriptors for practical face recognition systems. IET Comput. Vis. **4**(4), 272–285 (2010)
21. Mian, A., Bennamoun, M., Owens, R.: Keypoint detection and local feature matching for textured 3D face recognition. Int. J. Comput. Vis. **79**(1), 1–12 (2008)
22. Jain, A.K., Ross, A.A., Nandakumar, K.: Introduction to Biometrics, 1st edn. Springer (2011)
23. Burghardt, T., Damen, D., Mayol-Cuevas, W., Mirmehdi, M.: Correspondence, matching and recognition. Int. J. Comput. Vis. **113**(3), 161–162 (2015)
24. Bouchiha, R., Besbes, K.: Comparison of local descriptors for automatic remote sensing image registration. SIViP **9**(2), 463–469 (2015)
25. Zhao, Q., Feng, W., Wan, L., Zhang, J.: SPHORB: a fast and robust binary feature on the sphere. Int. J. Comput. Vis. **113**(2), 143–159 (2015)
26. Zhang, S., Tian, Q., Huang, Q., Gao, W., Rui, Y.: USB: ultrashort binary descriptor for fast visual matching and retrieval. IEEE Trans. Image Process. **23**(8), 3671–3683 (2014)
27. Tuytelaars, T., Mikolajczyk, K.: Local invariant feature detectors: a survey. Found. Trends Comput. Graph. Vis. **3**(3), 177–280 (2007)
28. Chen, J., Shan, S., He, C., Zhao, G., Pietikäinen, M., Chen, X., Gao, W.: WLD: a robust local image descriptor. IEEE Trans. Pattern Anal. Mach. Intell. **32**(9), 1705–1720 (2010)
29. Viola, P., Jones, M.J.: Robust real-time face detection. Int. J. Comput. Vis. **57**(2), 137–154 (2004)
30. Janan, F., Brady, M.: Shape description and matching using integral invariants on eccentricity transformed images. Int. J. Comput. Vis. **113**(2), 92–112 (2015)
31. Lowe, D.G.: Distinctive image features from scale-invariant keypoints. Int. J. Comput. Vis. **60**(2), 91–110 (2004)
32. Bay, H., Ess, A., Tuytelaars, T., Gool, L.: Speeded-up robust features (SURF). Comput. Vis. Image Underst. **110**(3), 346–359 (2008)
33. Oliva, A., Torralba, A.: Modeling the shape of the scene: a holistic representation of the spatial envelope. Int. J. Comput. Vis. **42**(3), 145–175 (2001)
34. Bianco, S., Mazzini, D., Pau, D., Schettini, R.: Local detectors and compact descriptors for visual search: a quantitative comparison. Digital Signal Process. **44**, 1–13 (2015)
35. Jégou, H., Perronnin, F., Douze, M., Sánchez, J., Pérez, P., Schmid, C.: Aggregating local descriptors into a compact codes. IEEE Trans. Pattern Anal. Mach. Intell. **34**(9), 1704–1716 (2012)
36. Lindeberg, T.: Feature detection with automatic scale selection. Int. J. Comput. Vis. **30**(2), 79–116 (1998)
37. Moravec, H.P.: Towards automatic visual obstacle avoidance. In: 5th International Joint Conference on Artificial Intelligence, pp. 584–594 (1977)
38. Harris, C., Stephens, M.: A combined corner and edge detector. In: The Fourth Alvey Vision Conference, pp. 147–151. Manchester, UK (1988)
39. Smith, S.M., Brady, J.M.: A new approach to low level image processing. Int. J. Comput. Vis. **23**(1), 45–78 (1997)
40. Rosten, E., Drummond, T.: Fusing points and lines for high performance tracking. In: International Conference on Computer Vision (ICCV'05), pp. 1508–1515. Beijing, China, 17–21 Oct 2005

41. Rosten, E., Drummond, T.: Machine learning for high speed corner detection. In: 9th European Conference on Computer Vision (ECCV'06), pp. 430–443. Graz, Austria, 7–13 May 2006
42. Beaudet, P.R.: Rotationally invariant image operators. In: International Joint Conference on Pattern Recognition, pp. 579–583 (1978)
43. Lakemond, R., Sridharan, S., Fookes, C.: Hessian-based affine adaptation of salient local image features. J. Math. Imaging Vis. **44**(2), 150–167 (2012)
44. Mikolajczyk, K., Tuytelaars, T., Schmid, C., Zisserman, A., Matas, J., Schaffalitzky, F., Kadir, T., Gool, L.: A comparison of affine region detectors. Int. J. Comput. Vis. **65**(1/2), 43–72 (2005)
45. Mikolajczyk, K., Schmid, C.: Scale & affine invariant interest point detectors. Int. J. Comput. Vis. **60**(1), 63–86 (2004)
46. Lindeberg, T.: Scale selection properties of generalized scale-space interest point detectors. J. Math. Imaging Vis. **46**(2), 177–210 (2013)
47. Yussof, W., Hitam, M.: Invariant Gabor-based interest points detector under geometric transformation. Digital Signal Process. **25**, 190–197 (2014)
48. Schaffalitzky, F., Zisserman, A.: Multi-view matching for unordered image sets. In: European Conference on Computer Vision (ECCV), pp. 414–431. Copenhagen, Denmark, 28–31 May 2002
49. Lindeberg, T., Gårding, J.: Shape-adapted smoothing in estimation of 3-D shape cues from affine deformations of local 2-D brightness structure. Image Vis. Comput. **15**(6), 415–434 (1997)
50. Mikolajczyk, K., Schmid, C.: A performance evaluation of local descriptors. IEEE Trans. Pattern Anal. Mach. Intell. **27**(10), 1615–1630 (2005)
51. Matas, J., Chum, O., Urban, M., Pajdla, T.: Robust wide baseline stereo from maximally stable extremal regions. In. In British Machine Vision Conference (BMV), pp. 384–393 (2002)
52. Matas, J., Ondrej, C., Urban, M., Pajdla, T.: Robust wide-baseline stereo from maximally stable extremal regions. Image Vis. Comput. **22**(10), 761–767 (2004)
53. Li, J., Allinson, N.: A comprehensive review of current local features for computer vision. Neurocomputing **71**(10–12), 1771–1787 (2008)
54. Ke, Y., Sukthankar, R.: PCA-SIFT: a more distinctive representation for local image descriptors. In: IEEE Conference on Computer Vision and Pattern Recognition (CVPR'04), pp. 506–513. Washington, DC, USA, 27 June–2 July 2004
55. Morel, J., Yu, G.: ASIFT: a new framework for fully affine invariant image comparison. SIAM J. Imaging Sci. **2**(2), 438–469 (2009)
56. Pang, Y., Li, W., Yuan, Y., Pan, J.: Fully affine invariant SURF for image matching. Neurocomputing **85**, 6–10 (2012)
57. Ojala, T., Pietikäinen, M., Mäenpää, M.: Multiresolution gray-scale and rotation invariant texture classification with local binary patterns. IEEE Trans. Pattern Anal. Mach. Intell. **24**(7), 971–987 (2002)
58. Heikkilää, M., Pietikäinen, M., Schmid, C.: Description of interest regions with local binary patterns. Pattern Recogn. **42**(3), 425–436 (2009)
59. Tian, H., Fang, Y., Zhao, Y., Lin, W., Ni, R., Zhu, Z.: Salient region detection by fusing bottom-up and top-down features extracted from a single image. IEEE Trans. Image Process. **23**(10), 4389–4398 (2014)
60. Huang, M., Mu, Z., Zeng, H., Huang, S.: Local image region description using orthogonal symmetric local ternary pattern. Pattern Recogn. Lett. **54**(1), 56–62 (2015)
61. Hong, X., Zhao, G., Pietikäinen, M., Chen, X.: Combining LBP difference and feature correlation for texture description. IEEE Trans. Image Process. **23**(6), 2557–2568 (2014)
62. Calonder, M., Lepetit, V., Özuysal, M., Trzcinski, T., Strecha, C., Fua, P.: BRIEF: computing a local binary descriptor very fast. IEEE Trans. Pattern Anal. Mach. Intell. **34**(7), 1281–1298 (2012)
63. Van De Weijer, J., Schmid, C.: Coloring local feature extraction. In: European Conference on Computer Vision (ECCV), pp. 334–348. Graz, Austria, 7–13 May 2006
64. Subrahmanyam, M., Gonde, A.B., Maheshwari, R.P.: Color and texture features for image indexing and retrieval. In: IEEE International Advance Computing Conference (IACC), pp. 1411–1416. Patiala, India, 6–7 Mar 2009

65. Zhang, Y., Tian, T., Tian, J., Gong, J., Ming, D.: A novel biologically inspired local feature descriptor. Biol. Cybern. **108**(3), 275–290 (2014)
66. Chen, Z., Sun, S.: A Zernike moment phase-based descriptor for local image representation and matching. IEEE Trans. Image Process. **19**(1), 205–219 (2010)
67. Chen, B., Shu, H., Zhang, H., Coatrieux, G., Luo, L., Coatrieux, J.: Combined invariants to similarity transformation and to blur using orthogonal Zernike moments. IEEE Trans. Image Process. **20**(2), 345–360 (2011)
68. Freeman, W., Adelson, E.: The design and use of steerable filters. IEEE Trans. Pattern Anal. Mach. Intell. **13**(9), 891–906 (1991)
69. Liu, J., Zeng, G., Fan, J.: Fast local self-similarity for describing interest regions. Pattern Recogn. Lett. **33**(9), 1224–1235 (2012)
70. Huang, D., Chao, Z., Yunhong, W., Liming, C.: HSOG: a novel local image descriptor based on histograms of the second-order gradients. IEEE Trans. Image Process. **23**(11), 4680–4695 (2014)
71. Dalal, N., Triggs, B.: Histograms of oriented gradients for human detection. In: IEEE Conference on Computer Vision and Pattern Recognition (CVPR'05), pp. 886–893. San Diego, CA, USA, 20–26 June 2005
72. Fan, B., Wu, F., Hu, Z.: Rotationally invariant descriptors using intensity order pooling. IEEE Trans. Pattern Anal. Mach. Intell. **34**(10), 2031–2045 (2012)
73. Al-Temeemy, A.A., Spencer, J.W.: Invariant chromatic descriptor for LADAR data processing. Mach. Vis. Appl. **26**(5), 649–660 (2015)
74. Figat, J., Kornuta, T., Kasprzak, W.: Performance evaluation of binary descriptors of local features. Lect. Notes Comput. Sci. (LNCS) **8671**, 187–194 (2014)
75. Burghouts, G., Geusebroek, J.M.: Performance evaluation of local color invariantss. Comput. Vis. Image Underst. **113**(1), 48–62 (2009)
76. Moreels, P., Perona, P.: Evaluation of features detectors and descriptors based on 3D objects. Int. J. Comput. Vis. **73**(2), 263–284 (2007)
77. Guo, Y., Bennamoun, M., Sohel, F., Lu, M., Wan, J., Kwok, N.M.: A comprehensive performance evaluation of 3D local feature descriptors. Int. J. Comput. Vis. First online, 1–24 (2015)
78. Muja, M., Lowe, D.G.: Scalable nearest neighbor algorithms for high dimensional data. IEEE Trans. Pattern Anal. Mach. Intell. **36**(11), 2227–2240 (2014)
79. Szeliski, R.: Computer Vision: Algorithms and Applications. Springer, USA (2011)
80. Muja, M., David G. Lowe: Fast matching of binary features. In: IEEE Conference on Computer Vision and Pattern Recognition (CVPR'12), pp. 404–410. Toronto, ON, USA, 28–30 May 2012
81. Nister, D., Stewenius, H.: Scalable recognition with a vocabulary tree. In: IEEE Conference on Computer Vision and Pattern Recognition (CVPR'06), pp. 2161–2168. Washington, DC, USA, 17–22 June 2006
82. Yan, C.C., Xie, H., Zhang, B., Ma, Y., Dai, Q., Liu, Y.: Fast approximate matching of binary codes with distinctive bits. Frontiers Comput. Sci. **9**(5), 741–750 (2015)
83. Lindeberg, T.: Image matching using generalized scale-space interest points. J. Math. Imaging Vis. **52**(1), 3–36 (2015)
84. The oxford data set is available at (last visit, Oct. 2015) http://www.robots.ox.ac.uk/~vgg/data/data-aff.html

A Review of Image Interest Point Detectors: From Algorithms to FPGA Hardware Implementations

Cesar Torres-Huitzil

Abstract Fast and accurate image feature detectors are an important challenge in computer vision as they are the basis for high-level image processing analysis and understanding. However, image feature detectors cannot be easily applied in real-time embedded computing scenarios, such as autonomous robots and vehicles, mainly due to the fact that they are time consuming and require considerable computational resources. For embedded and low power devices, speed and memory efficiency is of main concern, and therefore, there have been several recent attempts to improve this performance gap through dedicated hardware implementations of feature detectors. Thanks to the fine grain massive parallelism and flexibility of software-like methodologies, reconfigurable hardware devices, such as Field Programmable Gate Arrays (FPGAs), have become a common choice to speed up computations. In this chapter, a review of hardware implementations of feature detectors using FPGAs targeted to embedded computing scenarios is presented. The necessary background and fundamentals to introduce feature detectors and their mapping to FPGA-based hardware implementations are presented. Then we provide an analysis of some relevant state-of-the-art hardware implementations, which represent current research solutions proposed in this field. The review addresses a broad range of techniques, methods, systems and solutions related to algorithm-to-hardware mapping of image interest point detectors. Our goal is not only to analyze, compare and consolidate past research work but also to appreciate their findings and discuss their applicability. Some possible directions for future research are presented.

Keywords Image interest point · Feature detectors · Hardware implementation · FPGA · Real-time processing · Embedded processing

C. Torres-Huitzil (✉)
LTI, CINVESTAV-Tamaulipas, Ciudad Victoria, Mexico
e-mail: ctorres@tamps.cinvestav.mx

© Springer International Publishing Switzerland 2016
A.I. Awad and M. Hassaballah (eds.), *Image Feature Detectors and Descriptors*, Studies in Computational Intelligence 630, DOI 10.1007/978-3-319-28854-3_3

1 Introduction

Many computer vision tasks rely on the extraction of low-level image features or interest points, which usually represent a small fraction of the total number of pixels in the image. Interest points are salient image pixels that are unique and distinctive, i.e., quantitatively and qualitatively different from other pixels in the image [1]. Such interest points must be robustly detected, meaning that they should retain similar characteristics even after image geometrical transformations and distortions, or illumination changes in the scene. Image features are used in several well-known applications ranging from object recognition [2], texture classification [3], and image mosaicing [4]. Furthermore, image feature detectors and descriptors are the basis for the development of other applications such as image retrieval to assist visual searching in large databases, virtual and augmented reality, and image watermarking or steganography [5]. The overall performance of all those applications relies significantly on both robust and efficient image interest point detection. Local features in every image must be detected in the first processing step of the feature-based perception and recognition pipeline [6], which broadly involves three main steps: (i) the detection of interest points, e.g., corners and blobs, in every frame, (ii) the description of an interest point neighborhood patch through a feature vector, and (iii) the match of descriptor vectors.

A wide variety of feature detectors reported in the literature exist and their output vary considerably as such algorithms make different assumptions on the image content in order to compute their response [7]. To improve reliability in the detection, a recent trend is being made to combine the power of different detectors into a single framework to achieve better results as information of a particular detector can be complemented with the corresponding results of its counterparts [8]. This yields methods that use more significant resources, such as memory and computational elements, usually only available on high performance parallel computing systems. Despite of their inherent differences, most well-known interest point detectors, at low-level processing of a bottom-up approach, are computationally similar as window-based image processing operators are required to be applied locally on every image position or scale [9, 10]. Window-based image processing, in spite of its inherent data parallelism and computation regularity, makes the extraction process computationally intensive as well as high-bandwidth memory demanding, difficult to overcome in real-time embedded applications, such as robot navigation, self-localization, object recognition, online 3D reconstruction, and target tracking. Recall that real-time is a context relative measure, which for image processing and computer vision applications is commonly considered as the processing of at least VGA resolution images at a minimum rate of 30 frames per second.

Motivated for the steady increasing demand for high-performance embedded computing engines, specific custom hardware architectures have been proposed as feature detector accelerators thanks to the inherent parallelism of hardware. Yet, size, weight, and power constraints associated with embedded computing severely limit the implementation choices [11, 12]. In this sense, Field Programmable Gate Array (FPGA) devices appear to fit particularly well computer vision applications thanks to their regular parallel computational structure and the availability of on-chip distributed memory. Furthermore, FPGA technology is always improving in logic density and speed, which constantly increases the complexity of the models that can be implemented on them by software-like techniques, thus facilitating the design space exploration and fast prototyping to build a viable embedded computer vision implementation.

While alternative parallel implementation media such as multicore platforms or Graphics Processing Units (GPUs) have been used to speed up computations by using mainly threads at programming levels [13, 14], major motivating factors for choosing FPGAs are a good power-efficiency tradeoff for embedded applications, and the further possibility to export an FPGA design to an optimized Application Specific Integrated Circuit (ASIC). Far beyond the achievable performance improvements of custom hardware implementations, it is also highly desirable to improve algorithms and propose overall implementation strategies for feature detectors to be more suitable and amenable for embedded platforms. In this context, from a pragmatic point of view, several other aspects should be considered for porting and deployment of image feature detectors on embedded platforms in order to reduce the cost and guarantee a good performance. Factors such processing time, numerical precision, memory bandwidth and size, and power consumption—not easily discernible from sequential algorithmic representations—are particularly important for portable embedded computing platforms. Although current embedded systems are equipped with high performance multicore processors or portable GPUs, feature detectors still represent a computational overhead, as the whole processing power is not fully available at any time for just a single task. Also, recently, the computation needs of image processing and computer vision have dramatically increased since image resolutions are higher and there is a significant demand for processing high frame-rate videos or images and several views derived from multiple cameras in systems with limited computational and energy resources [15]. Thus, efficient hardware implementations of image feature detectors still remains an open challenge as hardware designers should be actively involved in exploring design trade-offs among accuracy, performance, and energy use.

In this chapter, a review of hardware implementations of well-known image interest point detectors using FPGA devices as an implementation media is presented and discussed. The review addresses a broad range of reported techniques, methods, systems, and solutions related to hardware implementations of image interest point detectors to highlight their importance for embedded computing scenarios. The rest of this chapter is organized as follows. In Sect. 2, a background of feature detectors to highlight the two main computational steps involved in most algorithms is presented as well as a brief overview of FPGA technology and its potentials for efficient hardware software codesign. Particularly, this section introduces an abstraction of the computational flow found in well-known feature detectors, namely window-based operators and operator sequencing. In Sect. 3, some relevant works related to image interest point detectors hardware implementation are analyzed, both for simple and scale invariant interest point detectors. Finally, in Sect. 4 some concluding remarks and future directions are presented.

2   Background

2.1   Interest Point Detection

Interest points are simple point features, image pixels that are salient or unique when compared with neighboring pixels. Basically, most interest point detectors include two concatenated processing steps or stages: (i) *detection* that involves a measure of how salient each pixel is, and (ii) *localization* that selects stable points determined by local non-maxima suppression. Quantitative and qualitative criteria of both processing steps are application dependent and specific values of free parameters are used to control the response of a detector. Given the vast diversity of feature detectors reported in the literature and the divergence on its results, a quantitative performance evaluation is an important procedure to assess the quality of image feature detectors under particular conditions [1, 16, 17].

To measure how salient or interesting each pixel x in an image is, an interest point operator [18] is defined using a mapping κ of the form:

$$\kappa(x) : \mathbb{R}^+ \to \mathbb{R} \tag{1}$$

Interest point detectors differ on the nature and complexity of the operator κ that they employ to process neighbors around a pixel x. The κ operator only computes the interest measure for a given pixel based on local information; i.e., neighboring pixels in an image patch. Usually, a detector refers to the complete algorithm that extracts interest points from an input image I to produce an interest image I^*.

Interest point localization is usually performed through non-maxima suppression to eliminate highlighted pixels that are not local maxima. Normally, a thresholding step is applied so that the response given by the initial feature enhancement can be used to distinguish between truly features and non-features. Interest points that are greater than a specified threshold are identified as local maxima of the interest point measure response according to the following equation:

$$\{(x_c, y_c)\} = \{(x_c, y_c) | \kappa(x_c, y_c) > \kappa(x_i, y_i),$$
$$\forall (x_i, y_i) \in W(x_c, y_c), \ \kappa(x_c, y_c) > t\} \tag{2}$$

where $\{(x_c, y_c)\}$ is the set of all interest points found in the image, $\kappa(x, y)$ is the interest point measure response computed at point (x, y), $W(x_c, y_c)$ is a neighborhood centered around the point (x_c, y_c), and t is a specified threshold.

2.2 Overview of Well-Known Interest Point Detectors

Since this chapter is mainly focused on detectors that generate saliency responses for each pixel in the image, this review does not consider some other feature detectors proposed in the literature such as those detectors mainly oriented to highlight regions in the images [7]. The interest point detectors considered in this work are those based on widely used saliency measures such as gradient magnitude, Univalue Segment Assimilating Nucleus (USAN), second moment matrix, among others. On the other hand, scale-space interest point detectors are based in measures such as the Hessian matrix, Laplacian of Gaussian (LoG), and Difference of Gaussian (DoG). Such detectors are better known as blob detectors, which do not necessarily make highly selective features, i.e., single salience pixels. In practice, a scale-space is represented as an image pyramid in which an image is successively filtered by a family of smoothing kernels at increasing scale factors. Extrema detection or non-maxima suppression to identify interest points is performed by searching for local extrema using each scale in an image pyramid. Such image pyramid based scale-space representation requires a huge amount of memory and incurs heavy window-based operations to produce an interest image.

The gradient magnitude is a measure used in several edge detection techniques such as the Canny detector [19]. The USAN is used for edge and corner detection in the Smallest USAN (SUSAN) detector [20], which places a circular mask over the pixel to be tested (the nucleus). Every pixel is compared to the nucleus and the response of SUSAN is given in terms of similar pixels to the nucleus. Harris detector [21] is a well-known corner detector based on the eigenvalues of the second moment matrix, which is often used for feature detection and for describing local image structures. The FAST (Features from Accelerated Segment Test) feature detector, proposed by Rosten and Drummond [22], is very fast to compute. In this detector, a feature is detected at a given pixel if the intensities of at least 9 contiguous pixels of a circular arc of 16 pixels are all below or above the intensity of the pixel by a given threshold. However, these detectors are not invariant to scale and therefore not very stable across scale changes; scale-space measures are used instead. The Hessian matrix is the foundation of the detection step for scale-space SURF feature detector [23]. Difference of Gaussian is the basic detection step for SIFT features [24]. However, as exposed previously, computing window-based operators such as convolutions at several image scales is computationally expensive, and in such detectors non-maxima suppression is performed over the scale space.

The above mentioned detectors, summarized in Table 1, are a representative set of detectors widely used to identify points of interest or blobs for object recognition and other applications that have been addressed from a hardware implementation perspective in different embedded scenarios.

Table 1 Summary of selected well-known image interest point detectors that have been addressed for embedded hardware implementations

Detector	Saliency measure principle	Feature type	Scale invariance
Canny [19]	Gradient magnitude	Corner	No
SUSAN [20]	Univalue segment assimilating nucleus	Corner	No
Harris [21]	Second moment matrix	Corner	No
FAST [22]	Similarity of contiguous pixels	Corner	No
SURF [23]	Hessian matrix	Blob	Yes
SIFT [24]	Difference of Gaussian	Blob	Yes

2.3 Underlying Principles of Interest Point Detectors

Low-level interest point detectors attempt to isolate image pixels that contain visually important data, such as edges, corners, or blobs. Some detectors are used as stand-alone systems for low-level image analysis and others feed the results into other systems performing further computational steps for higher-level image understanding tasks [15]. Many feature-based image embedded applications have real-time constraints and they would benefit from being able to detect features in strict time bounds in resource constrained computing platforms.

In spite of their differences, most feature detector algorithms perform similar data-parallel window-based computations on all pixels of the input image to create a resultant or intermediate image. Moreover, such window-based operators, probably with different kernel sizes and larger memory requirements, might be repeatedly applied in more complex and time consuming algorithms to the input images through different processing steps, or at different scales, until a final output interest image is generated. In order to assist in the understanding of the advances on hardware implementation of image interest point detectors, this section presents two underlying computational principles of feature detector algorithms that have been and should be exploited for an efficient embedded hardware implementation [9]: (i) *window-based image operators* to compute both the initial response of the detector and non-maxima suppression, and (ii) *operator sequencing* as a buffering and sequencing mechanism between intermediate window-based processing stages.

The first principle, *window-based image operators*, means that we have parallelism to be exploited by replicating computational elements at hardware level. Managing the inherent data level parallelism and spatial locality of window-based operators, under an appropriate computational model, are essential factors for extracting high performance from modern and future custom hardware architectures. The second one, *operator sequencing*, means that communication optimizations are primary and essential concerns that must be addressed to exploit efficiently the potential parallelism of hardware structures and to reduce memory bandwidth requirements. Communication latency and overhead can dominate the critical path of the computation, and interconnect throughput can be the main performance bottleneck in embedded implementations. Furthermore, the parallelism can be exploited effectively by minimizing latency and data traffic by a careful selection of the location of operators and data in space, i.e., the datapath construction [25].

2.3.1 Window-Based Image Processing Operators

A window-based operation is performed when a window of $k \times k$ pixels or neighborhood is extracted from the input image and it is transformed according to a window mask or kernel, denoted by K, based on mathematical functions to produce a result in the output image [10]. Usually a single output data is produced by each window operation and it is stored in the same position as the central pixel of the window. A window-based operator is conceptually shown in Fig. 1 over an input image I to produce an interest image Y'. An output window W is computed by operating the extracted pixels against the kernel values. This window is then reduced to a single output pixel at location (r, c) in the output image Y'. Figure 2 shows some typical

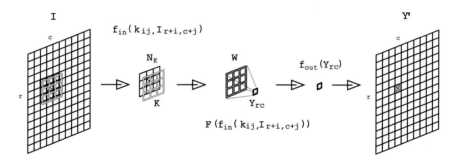

Fig. 1 A conceptual view of a window-based operator over an input image I with a kernel K, adapted from [9, 10]. An output window W is computed by operating the extracted pixels against the kernel values, which are overlapped on every extracted neighborhood centered at location (r, c) in the image. This window is then reduced to a single output pixel at location (r, c) in the output image Y'

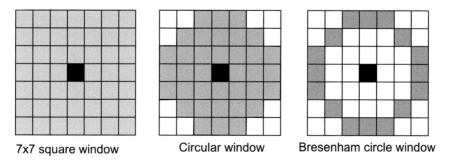

7x7 square window   Circular window   Bresenham circle window

Fig. 2 Some typical windows used for interest point detection. *Gray pixels* are used in the computation of the window-based operator around the *dark pixel*

windows used for interest point detection. Gray pixels are used in the computation of the window-based operator around the dark pixel.

Window-based image operations can be formalized as follows. Let I be the $M \times N$ input image, Y the output image, and K a $k \times k$ kernel. A window operation can be defined according to the following equation:

$$Y_{rc} = F(f_{in}(k_{ij}, I_{r+i,c+j})), \forall (i, j) \in K, \forall (r, c) \in I \qquad (3)$$

where k_{ij} represents a value or coefficient from the kernel K, $I_{r+i,c+j}$ a pixel from a $k \times k$ neighborhood N_K around the (r,c) pixel in the input image, f_{in} defines a scalar function, and F defines the local reduction function. Normally the output value Y_{rc} is combined with other scalars, or compared to a threshold value by means of a scalar function f_{out} to produce a final response Y'_{rc}:

$$Y'_{rc} = f_{out}(Y_{rc}) \qquad (4)$$

Common scalar functions, usually two-input operands, include relational and arithmetic-logic operations. Typical local reduction functions used in window-based operators are accumulation, maximum/minimum, and absolute value, which operate on multiple input operands related to the size of the kernel. The scalar and local reduction functions form the image algebra upon which window-based image operators rely. The scalar and local reduction functions for some of the interest point detectors, and for the interest point localization algorithm considered in this work, are summarized in Table 2. These sets of functions are either necessary nor sufficient but they do incorporate some of the most common basic operations that are commonly used by well-known interest point detectors.

The fundamental parts of window-based operations are shown graphically in Fig. 1. Note that three concatenated computational elements can be identified in the processing flow of a window operation [9]. These elements are organized in a three-piece computational component, called henceforth the threefold operator, which can be defined in terms of the scalar functions f_{in} and f_{out}, and the local reduction function F. According to Fig. 1, from left to right, the first computational element operates two windows of scalar values, by means of a set of scalar functions f_{in}, to produce concurrently a set of scalar values. The operands to this computational element are an image window or neighborhood of pixels extracted from the input

Table 2 Common scalar and local reduction functions for Harris and SUSAN algorithms, and non-maxima supression

Algorithm	f_{in}	F	f_{out}
Harris	$\times$	$+$	$+, -, \times$
SUSAN	$\leq, >$	$+$	$<, \geq, -$
Localization	$\times$	max	$>$, and, $=$

image at every pixel, and a kernel, denoted as N_K and K, respectively. The resulting output set of scalar values is denoted by W. This set of intermediate values is the input operand to the following computational element of the threefold operator, which applies the local reduction function F onto the window W to produce the scalar value Y_{rc}. This output value is then operated by a scalar function f_{out} to produce the final output value $Y'(r,c)$.

The sequencing order of the elements in the threefold operator shows the natural and regular data flow in the window-based operator. Data can be regarded at two levels of granularity: scalar values, and windows of scalar values that result from grouping $k \times k$ scalar values. The threefold operator is a building block or a primitive for describing more elaborated forms of image processing. Communication channels and buffering schemes among window-based operators must be supported so that they can be sequenced. In the following section, the mechanism that allows communication for a sustained data flow among threefold operators is presented.

2.3.2 Operator Sequencing

Window-based image operators such as convolution and non-maxima suppression are key components in image processing that by themselves receive considerable attention by the community so as to propose efficient hardware implementations [10, 26]. However, window-based operators are rarely used isolated as they usually work in cascade to produce an output result in more complex applications.

For instance, the computations in Harris and SUSAN detectors can be described as a sequence of threefold operators [9]. Figure 3 shows the main computational steps to compute the Harris measure and how data flows between window-based operators. Here we take the representative Harris corner detector as an example to illustrate the idea of window-based operator sequencing. Harris corner detection is based on the auto-correlation of gradients on shifting windows as it can be seen from Fig. 3. The first order partial derivatives of the image (Step 1) are calculated and then smoothed with a Gaussian kernel G_σ (Step 2). For each pixel in the image at position (x,y) the Harris matrix is then computed:

$$\mathcal{H}(x,y) = \begin{bmatrix} a & c \\ c & b \end{bmatrix} \tag{5}$$

where a, b and c are the scalar values that result from convolving a Gaussian kernel against the first order partial derivatives of an image I over x and y directions, that is, $I_x = \frac{\partial I}{\partial x}$ and $I_y = \frac{\partial I}{\partial y}$. The values of the elements in the Harris matrix are calculated by the following expressions: $a = G_\sigma * I_x^2$, $b = G_\sigma * I_y^2$, $c = G_\sigma * (I_x \times I_y)$. The equation for computing the Harris measure (Step 3) at (x,y), is given by:

$$\kappa(x,y) = (a \times b - c^2) - k \times (a + b)^2 \tag{6}$$

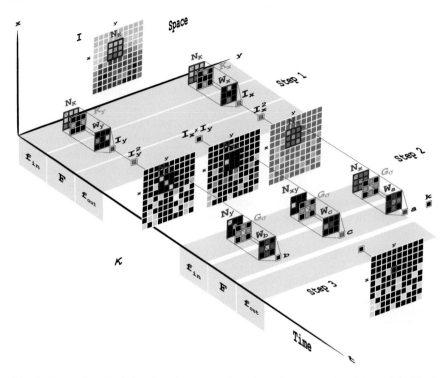

Fig. 3 Sequencing of window-based operators throughout the computational flow of the Harris detector measure, adapted from [9]

and k is a constant with a typical value of 0.04. The Harris measure is then used to decide if a corner ($\kappa \gg 0$), edge ($\kappa \ll 0$) or flat region ($\kappa \approx 0$) is found.

According to Fig. 3, to compute concurrently the Harris algorithm response, connection and temporal storage among threefold operators must be supported so as to avoid the use of external memory and to allow data to be rhythmically propagated from one stage to the other. This connection can be provided by a stream storage component that continually holds and groups scalar values to later pump them as full-accessible windows to exploit data parallelism. The purpose of the stream storage component is two-fold: it provides a mechanism for extracting neighborhoods, or windows of pixels, from an input image, and it makes possible to sequence window operators allowing a cascaded connection for stream processing. Yet, a coarse-grain inter-operator pipeline can be exploited.

The conceptual representation of the storage component and its desirable functionality to extract 5×5 neighborhoods from the input image is shown in Fig. 4. All the pixels covered by the window are stored in registers or flip-flops and they are all individually accessible, meanwhile remaining pixels of image rows are stored in First-in-First-out (FIFO) structures. As an incoming pixel is shifted in, the oldest pixel currently in the FIFO is shifted out; this mechanism allows the window to be

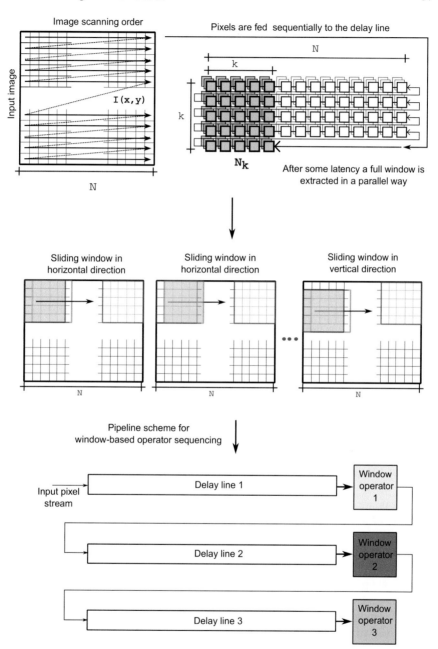

Fig. 4 Graphic representation of the delay line used to extract image windows in a row-based image scanning order (*top*), the sliding window effect in horizontal and vertical directions (*center*), and a pipelined scheme for three window-based operator sequencing using delay lines (*bottom*)

moved or slided to the next position by reading a new pixel and moving all others one step in the FIFO [27] as conceptually shown in Fig. 4. Considering a kernel of size of k × k values and an input image of width N, the required storage space L in the component is L = (k − 1) × N + k. Moreover, window-based operators can be organized in a coarse-grain pipeline for fast computation as shown Fig. 4. Eventually, after some latency, the resultant pixels of the last windows-based operator can be produced successively clock by clock.

Sliding window operations and image interest point detectors are prone to boundary problems. These occur when the window reaches outside of the boundaries of the image. There two general methods to address this issue: the most straightforward is simply omitting these values from the calculation and another used method is to insert extra pixels around the image boundary, which is called padding [28].

2.4   Overview of FPGA Technology

Configurable hardware devices such as FPGAs are cheap and flexible semiconductor devices that offer a good compromise between the hardware efficiency of custom digital ASICs and the flexibility of a rather simple software-like handling for describing computations [29], allowing fast prototyping or reducing the time-to-market of a digital system. State-of-the-art FPGAs offer high-performance, high-speed and high capacity programmable logic that enhance design flexibility and their computational capabilities to be applied in various and diverse application fields such as image and signal processing [27, 29], computer arithmetics [30], mobile robotics [31], industrial electrical control systems [32], space and aircraft embedded control systems [33], neural and cellular processing [34, 35]. According to [36], since their introduction, FPGAs have grown in capacity by more than a factor of 10^4, in performance by a factor of 10^2, and cost and energy per operation have both decreased by more than a factor of 10^3.

FPGAs are digital devices whose architecture is based on a matrix or regular structure of fine grain computational blocks known as Configurable Logic Blocks (CLBs). Figure 5 shows a conceptual view of a generic FPGA architecture and its associated design flow. Each CLB in the regular array is able to implement combinational logic functions (usually four to six input functions) as Look-up-Tables (LUTs) and provides some multiplexers and a few elementary memory components (flip-flops or latches) to implement sequential logic [37]. The CLBs can be efficiently connected to neighboring CLBs as well as distant ones thanks to a rich configurable and segmented routing structure. Also, the configurable communication structure can connect CLBs to border Input/Output Blocks (IOBs) that drive the chip input/output pads. Memory cells control the logic blocks as well as the connections so that the components can fulfill the required specifications by automatic tools that map the application, specified at a high-level of abstraction, onto the chip, following the general design flow shown in Fig. 5. FPGAs provide a fully customizable platform where any kind of custom operation, either complex or simple, can be implemented, but the

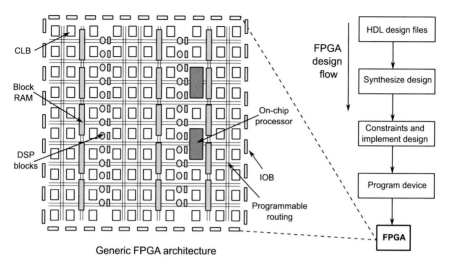

CLB

Block
RAM

DSP
blocks

FPGA
design
flow

On-chip
processor

IOB

Programmable
routing

Generic FPGA architecture

HDL design files

Synthesize design

Constraints and
implement design

Program device

FPGA

Fig. 5 Generic architecture of a current FPGA device and the main steps of the design flow to map a high level specification into an FPGA

design of custom complex systems could be very challenging task that is still carried out manually or in a semiautomatic way.

Recently, FPGA architectures have evolved to a higher level of abstraction and some dedicated and specialized blocks such as embedded multiport RAM, Digital Signal Processing (DSP) accelerators, embedded hard processor cores, such as the *PowerPC* or *ARM*, and soft processor cores such as *Nios* or *Microblaze* are available on the same chip, transforming FPGAs into truly Systems-on-Chip (SoC). This architectural evolution has its origin in the recent advances in VLSI technology and boosted by the development of appropriate design tools and methods that allow current FPGA-based implementations to be mapped from high-level specifications onto new improved FPGAs [32]. Modern FPGA toolsets include high-level synthesis compilation from C, CUDA and OpenCL to logic or to embedded microprocessors. As a consequence, an embedded processor, intellectual property (IP), and an application IP can now be developed and downloaded into the FPGA to construct a system-on-a-programmable-chip (SoPC) environment, allowing users to design a SoPC module by mixing hardware and software in one FPGA chip under a hardware/software codesign approach. The modules requiring fast processing but simple and regular computations are suitable to be implemented by dedicated hardware datapaths in the FPGA running at hundred of MHz, and the complex algorithm parts with irregular computations can be realized by software on the embedded processors in the FPGA. The results of the software/hardware codesign increase the programmability and the flexibility of the designed digital system, enhance the system performance by parallel processing, and reduce the development time. The fast increase of complexity in SoPC design has motivated researchers to seek design abstractions with better productivity than Register Transfer Level (RTL). Electronic System-Level (ESL) design automation has been widely identified as the next productivity boost for the semiconductor industry, where High-Level Synthesis (HLS) plays a central role, enabling the automatic synthesis of high-level untimed specifications, to low-level cycle-accurate RTL specifications for efficient implementation in FPGAs [38].

In spite of these architectural advances, current FPGA devices, however, are optimized for regular computations and fixed point arithmetic, and they are not well suited to floating point arithmetic. Moreover, although FPGAs have reasonably large amounts of on-chip memory, for many applications is not enough, and it is often required to have good interfaces to off-chip memory [39]. This is particularly important for algorithms characterized by dependencies, such window-based operator algorithms, as it is not possible to ensure that all the output values of an intermediate computational step are directly available in a subsequent step. Thus, it might happen that some of them have to be stored for later use introducing a memory overhead.

3   Interest Point Detectors on FPGAs

Recently, significant efforts have been made directed toward the increase of the performance of image processing and computer vision applications, specifically in the acceleration of image interest point and feature detectors algorithms. The motivation for this effort oriented to FPGA-based solutions relies on the high computation requirements needed for achieving real-time or near-real-time processing capabilities. Several works have shown that FPGAs are a real opportunity for efficient embedded hardware implementations of image feature detectors but specific problems need to be solved by means of architectural innovations. Modern FPGA devices incorporate embedded resources that facilitate architecture design and optimization. For instance, the embedded DSP blocks, multipliers and adders, enable important speedups in window-based image operators, and small internal RAMs, both distributed and block RAM, can be used as circular buffers to cache image rows for local processing and data reuse [14].

FPGA implementations of image interest point detectors require some algorithmic modifications to map efficiently the algorithms onto the hardware logic. Due to the high cost of on-chip logic required for floating point units, a constrained fixed-point arithmetic is preferable for computations on FPGAs. A systematic methodology is required to evaluate different precision choices in the development phase. Most of the reported works exploits the idea of designing in a way that reflects properly the goal of parallel execution on FPGAs by fully exploiting the advantages of the available parallel computational resources. Although FPGAs allow a great degree of parallelism in the implementation of such algorithms, important data dependencies exist, and it is necessary to reduce the total amount of hardware, keeping the final cost at a reasonable point. Scheduling of processing units, techniques for hardware sharing, pipelining, and accuracy versus resource utilization trade-offs should be evaluated as well [14]. On the other hand, image interest point detectors performance depends to a some extent on high clock frequencies and high bandwidth external memory. FPGA devices are limited in these two aspects, and specialized memory management units have to be developed to optimize the scheduling to the external memory accesses in the FPGA.

This section is focused on reviewing some relevant works of the state-of-the-art that deal with the efficient implementation of image interest point algorithms onto FPGA devices. The review follows the traditional categorization and refer the edges, corners, regions as the important visual features. This section has been divided into two main subsections devoted to review the existing FPGA-based implementations for simple interest point detectors and some scale-invariant feature detectors algorithms, which due to the nature of computational complexity and the huge demanding of memory consumption require different implementation strategies. On the other hand, a fair comparison of performance and resources utilization between the different hardware implementations is not straightforward, because a different FPGA technologies and devices are used. Hence, it is not intended to compare the implementations in terms of basic resource such as LUT, register, DSP block, and BRAM available in modern FPGA devices, but to highlight the implementation strategies, simplifications and the achieved overall performance results.

3.1 Simple Interest Point Detectors

Several hardware architectures have been proposed for implementing simple interest point operators, mostly based on delay-lines to extract image windows to be processed on arrays of neighboring processing elements (PEs). This well-known approach usually supports small windows, since the memory requirement for the delay-lines, to store the pixels to be reused, is proportional to the size for the maximum supported window, and the image width. Most of these implementations employ a pipeline technique in which a raster-scan image is sequentially fed into a PE array and the window-based operations are carried out in parallel in each PE. Table 3 shows a summary of FPGA-based hardware implementations of simple image interest point detectors, highlighting image resolution, the operating frequency, the achieved frame rate and the target device. The FPGA resource utilization in terms of LUT, register, DSP block, and BRAM are shown Table 4; it is important to point out that the internal component for Xilinx and Altera FPGA architectures are not completely equivalent and some works do not report the details of the resources used in their implementations.

Hernandez-Lopez et al. [9] propose a flexible hardware implementation approach for computing interest point extraction from gray-level images based on two different detectors, Harris and SUSAN, suitable for robotic applications. The FPGA-based architecture implements two different feature detectors by abstracting the fundamental components of window-based image processing while still supporting a high frame per second rate and low resource utilization in a single datapath. It provides a unified representation for feature detection and localization on an FPGA without altering in any way the nature of the algorithms, keeping a reliable hardware response that is

Table 3 Summary of selected FPGA-based hardware implementation of simple image interest point detectors

First author	Detector	Image size	Frequency (MHz)	Frames per second	Device
Hernandez-Lopez [9]	SUSAN	640 × 480	50	161	XC6VLX240t
	Harris	640 × 480	50	161	XC6VLX240t
Torres-Huitzil [40]	SUSAN	512 × 512	60	120	XCV50
Possa [41]	Canny	512 × 512	242	909	Arria V 5AGXFB3
	Harris	512 × 512	232	869	Arria V 5AGXFB3
Xu [42]	Canny	512 × 512	100	1386.9	XC5VSX240T
Lim [43]	Harris	640 × 480	–	–	Altera Cyclone IV
	FAST	640 × 480	–	–	Altera Cyclone IV
Kraft [44]	FAST	512 × 512	130	500	Spartan-3 XC3S200-4
Hsiao [45]	Harris	640 × 480	–	46	Altera Cyclone II 2C35

Table 4 Comparison of hardware resource utilization of selected FPGA-based hardware implementation of simple image interest point detectors

First author	Detector	LUTs	Registers	DSP	BRAM
Hernandez-Lopez [9]	SUSAN/Harris	24189 (1 %)	4347 (1 %)	41 (5 %)	0
Torres-Huitzil [40]	SUSAN	685 (89 %)	540 (71 %)	–	0
Possa [41]	Canny	3406 (2 %)	6608 (1.2 %)	28 (3 %)	553 Kb(3 %)
	Harris	8624 (6 %)	17137 (3.1 %)	76 (7 %)	863 Kb (5 %)
Xu [42]	Canny	82496 (65 %)	40640 (32 %)	224 (25 %)	16184 Kb (87 %)
Lim [43]	Harris	–	–	–	–
	FAST	–	–	–	–
Kraft [44]	FAST	2368 (62 %)	1547 (40 %)	0	216 Kb (100 %)
Hsiao [45]	Harris	35000 (23 %)	–	35 (83 %)	430 Kb (35 %)

not compromised by simplifications beyond the use of fixed-point arithmetic. The design is based on parallel and configurable processing elements for window operators and a buffering strategy to support a coarse-grain pipeline scheme for operator sequencing. When target- ed to a Virtex-6 FPGA, a throughput of 49.45 Mpixel/s (processing rate of 161 frames per second of VGA image resolution) is achieved at a clock frequency of 50 MHz.

Torres-Huitzil et al. [40] present an FPGA-based hardware architecture for high speed edge and corner detection based on the SUSAN algorithm using a 7×7 mask to compute the USAN area. The architecture design was centered on the minimization on the number of accesses to the image memory and avoiding the use of delay-lines. The architecture employs a novel-addressing scheme, column-based scan order of the image pixels, that significantly reduces the memory access overhead and makes explicit the data parallelism at a low temporal storage cost. Interestingly, internal storage requirements to extract image windows are only dependent on the mask size but not on the image size. The design is based on parallel modules with internal pipeline operation in order to improve its performance. The computational core of the architecture is organized around a configurable 7×7 systolic array of elemental processing elements, which can provide throughputs over tenths of Giga Operations per Second (GOPs). The proposed architecture was implemented on an XCV50 FPGA clocked at 60 MHz to process a 512×512 images at rate of 120 frames per second.

Possa et al. [41] present flexible parameterizable architectures for the Canny edge and the Harris corner detectors. The architectures, with reduced latency and memory requirements, contain neighborhood extractors and threshold operators that can be parameterized at runtime to process a streamed image or sequence of images with variable resolutions. Algorithm simplifications are proposed to reduce mathematical complexity, memory requirements, and latency without losing reliability. One of the main computational blocks used to implement the detectors is the neighborhood extractor (NE), which provides a sliding window with a fixed dimension to the subsequent processing block. The NE design supports images with variable resolutions and automatically handles the image borders, keeping a reduced memory requirement and minimizing the latency. The basic structure of the NE is a set of cascaded line buffers connected to register arrays from where it is possible to read the current and two or more previously stored pixels. In the Harris architecture the partial derivatives and Gaussian filtering are computed using neighborhoods of 3×3 and 5×5, respectively. The Harris' output values were truncated to preserve an 8 bit-width datapath but an additional Gaussian filtering step using neighborhoods of size 5×5 is needed to compensate for the saturated values and to enhance localization. Then, non-maxima suppression is applied on neighborhoods of size 9×9. The architecture is implemented on an FPGA clocked at a frequency of 242 MHz to process a 512×512 image in 1.1 ms.

Xu et al. [42] propose a strategy to implement the Canny algorithm at the block level without any loss in edge detection performance compared with the original frame-level Canny algorithm. The original Canny algorithm uses frame-level statistics to predict the high and low thresholds and as a consequence its latency is proportional to the frame size. In order to reduce the latency and meet real-time requirements, authors presented a distributed Canny edge detection algorithm which has the ability to compute edges of multiple blocks at the same time. To support this strategy, an adaptive threshold selection method is proposed that predicts the high and low thresholds of the entire image while only processing the pixels of an individual local block, yielding three main benefits: (1) a significant reduction in the latency; (2) better edge detection performance; (3) the possibility of pipelining the Canny edge detector with other block-based image codecs. In addition to this implementation strategy, a low complexity non-uniform quantized histogram calculation method is proposed to compute the block hysteresis thresholds. The proposed algorithm is scalable and has very high detection performance. The proposed algorithm is implemented using a 32 computing engine architecture and is synthesized on the Xilinx Virtex-5 FPGA. The architecture takes 0.721 ms to detect edges of 512×512 images, when clocked at 100 MHz.

Lim et al. [43] propose FPGA implementations of two corner detectors, Harris and FAST algorithms. The design solution for the Harris implementation is based on the sliding window concept and internal buffers build from delay lines. The operations involved in the Harris measure are computed using a combination of addition, subtraction, multiplication (using hardware multipliers) and bitwise shift operations. The Gaussian coefficients were chosen as powers of 2, so that convolution can be done with bitwise shift operations but at the cost of some precision. The FAST implementation requires 6 image rows to be buffered in a FIFO structure to extract 7×7 window. The 16 pixels in a Bresenham circle together with the center pixel are passed to the corner classification module. Two binary vectors are computed using subtraction and comparators to classify pixels as one of the following: brighter, darker or similar in intensity. In the first binary vector, brighter elements are assigned 1, while similar pixels are assigned 0. The second binary vector assigns 1 to darker pixels, and 0 to similar pixels. By using two binary vectors, the corner classification is simplified to a search for 9 consecutive 1s (in either vector), and this is efficiently done using LUTs available in FPGAs. LUTs perform a function equivalent to *and* operations on every possible segment of 9 consecutive pixels. There are only 16 patterns that correspond to a corner. An *or* operation is applied to their outputs to determine if such a pattern is found. The proposed designs were implemented in a low-end Altera Cyclone IV FPGA, however performance is not reported in terms of frames per second and the operating frequency is not clearly specified. The stream processing architecture allows the corner classification stage to be done within a single clock cycle. Previously, Kraft et al. [44] present a complete FPGA architecture implementing corner detection based on the FAST algorithm. BlockRAM memories along with read/write address generation logic were used as FIFO delay buffers. The FIFO depth is equal to the horizontal resolution of the image. The design is divided into modules: the thresholder, the contiguity module, the corner score module and the non-maximum suppression module. The proposed solution is capable of processing the incoming image data with the speed of nearly 500 frames per second for a 512×512, 8-bit gray-scale image using a Xilinx's Spartan 3 family, namely the XC3S200-4, at a clock frequency of up to 130 MHz.

Finally, Hsiao et al. [45] analyze the data flow of multilayered image processing, sequenced window-based operators, to avoid waiting for the result from every previous steps to access the memory which occurs in many applicable algorithms such as the Harris corner detector. By combining the parallel and pipelined strategies to eliminate unnecessary delays in the algorithm dataflow, authors propose a visual pipeline architecture and the use of FPGA to implement efficiently their hardware scheme. Basically, authors chain together window-based operators in order to wait until the result from previous process has been generated. This sequencing type of image processing algorithms is named as multilayered image processing. By analyzing the Harris algorithm, authors obtain that the multilayered image processing architecture has four layers in which all processes can be timely triggered and paralleled without waiting for the end of previous processes. The multiscale Harris corner detector was validated on a platform with a FPGA chip of Altera Cyclone II 2C35, achieving a processing rate of 46 frames per second of 640 × 480 images.

3.2 Scale-Invariant Detectors

Scale-invariant interest point detectors detect features at different scales by using window-based operators in multi-scale filter banks. Although conceptually simple, the computation of window-based operators with large kernels is computational demanding for current multicore architectures. For instance, this implies that more than 12.41 Giga Operations per Second (GOPs) are required to support a real-time processing rate of 1280 × 720 30 frames per second HD video with a single 15 × 15 kernel. Clearly, the computational load and the complexity of memory access grow exponentially as the kernel dimensions increase.

In general terms, the multiscale nature of the algorithms affects architecture mapping to the FPGA, it requires more on-chip memory resources than usually available, and since the image pyramid construction is inherently sequential, it must be completed before the computation of interest point measures. However, a large amount of fine grain parallelism can be exploited within each scale since each pixel can be processed independently. On one hand, embedded memories represent a considerable cost of designs, typically limiting FPGA deployment in embedded environments. One the other hand, the scalability is a major concern in implementations since the architectures need major modifications when the kernel size increases.

Scale-invariant detectors receive great interest from the research community and their computational challenges that must be faced, explains why this is still a topical subject. For that reason a considerable number of recent papers on hardware implementations can be found in the literature. Some hardware implementation works propose implementation techniques and novel architectures for optimizing area/performance trade-off or for offering higher design/operation flexibility, mainly by (i) taking advantage of the separability of filters or properties of the coefficient kernels that allows to make some simplifications, and (ii) a reformulation of the most computationally expensive phase of the algorithms so as to reduce the complexity of computations by analyzing the operations and data involved [46].

A common used implementation technique employed as an alternative to speed up computations in scale-invariant detectors is to decompose large kernels into linear or simpler ones and then implement efficiently the simpler ones. For instance, to reduce the computational complexity of the 2D Gaussian filtering, the separability property of the Gaussian kernel is exploited. Under such approach, the 2D image convolution with a Gaussian filter can be carried out by first convolving the image with a horizontal projection of the 2D filter in the horizontal direction and then with a vertical projection of the 2D filter in the vertical direction or vice versa as shown in Fig. 6. As a consequence, a two-pass processing over image data is required to perform 2D filtering.

As another example of an efficient implementation technique, which allow fast computation of any box-type convolution filter like the one use to obtain an approximation of the Hessian matrix in SURF, is the use of integral images [47]. An integral image is a representation proposed by Viola and Jones that allows to compute the sum of all values within any rectangle in constant time. Figure 7 shows a conceptual view of the integral image, and mathematically can be expressed as follows:

$$II(x, y) = \sum_{i=0}^{y} \sum_{j=0}^{x} I(i, j) \tag{7}$$

The integral image can be implemented effectively with sequential accumulations, $R_I(x, y) = R_I(x - 1, y) + I(x, y)$ and $II(x, y) = II(x, y - 1) + R_I(x, y)$, where R_I is the sum of row, and the initial values $R_I(-1, y) = 0$ and $II(x, -1) = 0$. After constructing an integral image, the inner sum of any rectangle can be evaluated simply with one addition and two subtractions, as shown in Fig. 7. The sum with the highlighted box can be evaluated as $II(D) + II(A) - II(B) - II(C)$. Although the usage of integral image can accelerate computation on interesting point detection and description, the computation of integral image itself may introduce a memory overhead that is proportional to the input image size.

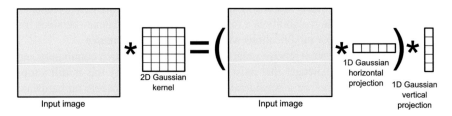

Fig. 6 Exploiting the separability property of the Gaussian kernel so that the 2D convolution can be carried out by first convolving the image with horizontal projection and then with the vertical projection of the kernel

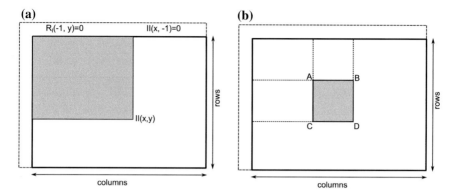

Fig. 7 **a** Illustration of the integral image value at (x,y) computed as the sum of all pixels above and to the left of such point. **b** The sum with the highlighted box can be evaluated in constant time as $II(D) + II(A) - II(B) - II(C)$

Similarly to simple image interest point detectors, Table 5 shows a summary of FPGA-based hardware implementations of scale-invariant image interest point detectors, highlighting image resolution, the operating frequency, the achieved frame rate and the target device. As these hardware implementations are more resource demanding in terms of memory, for reference, the resource utilization in terms of LUT, register, DSP block, and BRAM are shown Table 6, However, the comparison of resources utilization is not straightforward between the different implementations, because different FPGA technologies are used and not all the solutions implement the descriptor computation on custom hardware.

Jiang et al. [48] present a high-speed full FPGA-based hardware implementation of the SIFT detector based on a parallel and pipelined architecture able of real-time extraction of image features. Task-level parallelism and a coarse-grain pipeline structure are exploited between the main hardware blocks, and data-level parallelism and pipelining are exploited inside each block architecture. Two identical random access

Table 5 Summary of selected FPGA-based hardware implementation of scale-invariant image interest point detectors

First author	Detector	Image size	Frequency (MHz)	Frames per second	Device
Jiang [48]	SIFT	512 × 512	50–100	152.67	Virtex-5 LX330
Wang [49]	SIFT	1280 × 720	159.160	60–120	XC5VLX110T
Chang [50]	SIFT	320 × 240	145.122	900	XC2VP30
Bonato [51]	SIFT	320 × 240	50–100	30	EP2S60F672
Zhong [52]	SIFT	320 × 256	106.57	100	XC4VSX35
Krajnik [53]	SURF	1024 × 768	75	10	V5FXT70-G

Table 6 Comparison of hardware resource utilization of selected FPGA-based hardware implementation of scale-invariant image interest point detectors

First author	Device	LUTs	Registers	DSP	BRAM
Jiang [48]	Virtex-5 LX330	26398 (12.73 %)	10310 (4.97 %)	89 (46.35 %)	7.8Mb (75.23 %)
Wang [49]	XC5VLX110T	17055 (25 %)	11530 (17 %)	52 (81 %)	4.605Mb (91 %)
Chang [50]	XC2VP30	6699 (24 %)	5676 (20 %)	–	1.958Mb (79 %)
Bonato [51]	EP2S60F672	43366 (90 %)	19100 (37 %)	64 (22 %)	1.138Mb (52 %)
Zhong [52]	XC4VSX35	18195 (59 %)	11821 (38 %)	56 (29 %)	2.742Mb (81 %)
Krajnik [53]	V5FXT70-G	15271 (34 %)	16548 (36.9 %)	40 (31.25 %)	1.54Mb (29 %)

memories are adopted with pingpong operation to execute the key point detection module and the descriptor generation module in task-level parallelism. While speeding up the key point detection module of SIFT, the descriptor generation module has become the bottleneck of the system's performance. Authors propose an optimized descriptor generation algorithm based on a novel window-dividing method with square subregions arranged in 16 directions, and the descriptors are generated by reordering the histogram instead of window rotation. Therefore, the main orientation detection block and descriptor generation block can be computed in parallel. The proposed system was implemented on an FPGA and the overall time to extract SIFT features for a 512×512 image is 6.55 ms, and the number of feature points can reach up to 2900.

Wang et al. [49] propose an FPGA-based embedded system architecture for SIFT feature detection, as well as binary robust independent elementary features (BRIEF) feature description and matching. The proposed system is able to establish accurate correspondences between consecutive frames for 720-p (1280×720) video through an optimized FPGA architecture for the SIFT feature detection. The architecture aims at reducing the utilization of FPGA resources. The SIFT key-point detection component consists of the DoG scale space construction module and the stable key-point detection module. The DoG module is driven by the image stream directly from the camera interface and it performs 2-D Gaussian filtering and image subtraction taking advantage of the fact that Gaussian kernels are separable and symmetrical. In this method, the 2-D convolution is performed by first convolving with an 1-D Gaussian kernel in the horizontal direction and then convolving with another 1-D Gaussian kernel in the vertical direction. The performance of the proposed system was evaluated on the Xilinx XUPV5-LX110T development board. The proposed system achieves feature detection and matching at 60 frame/s for 720-p video at a clock frequency of around 160 MHz. It is important to mention, however, that architecture components might run at different clock frequencies, and as for other cases it is not possible to report a single operating clock frequency.

Chang et al. [50] proposed a hardware architecture for the SIFT detector. In this work, part of the algorithm was reformulated taking into account the potential for exploitation of data parallelism. To decrease the amount of multiplication-accumulation operations and thanks to the separability property of Gaussian kernel authors used the separable convolution. They introduced octaves processing interleaving, which allowed to perform all convolution operations for a given scale in a single processing unit. The main contribution of this architecture and the algorithm that it implements is that as the number of octaves to be processed is increased, the amount of occupied device area remains almost constant. This phenomenon is due to the fact that all octaves for the same scale, no matter how many, will be processed in the same convolution block. The proposed architecture was modeled and simulated using Xilinx System Generator 10.1 and Simulink, and it was synthesized in a Xilinx Virtex II Pro (XC2VP30-5FF1152) at a maximum frequency of 145.122 MHz. With the achieved throughput it is possible to process high-definition video (1080×1280 pixels) at a 50 frames per second (fps) rate.

Bonato et al. [51] present an FPGA-based architecture for SIFT feature detection. Their implementation uses a hardware/software co-design strategy; except the generation of descriptors, which is executed on a NIOS-II software processor, the remaining stages of SIFT are implemented in hardware. This architecture consists of three hardware blocks, one for the generation of DoG scale-space, one for the calculation of the orientation and magnitude, and one for the location of key-points. This implementation operated at 30 frame/s on 320×240 images. The feature description part of SIFT was on the NIOS takes 11.7 ms per detected feature, which makes it infeasible to perform as a full real-time SIFT implementation. As a single image may have hundreds of features, it is still far from satisfactory for the real-time performance. The validation platform was centered around a Stratix II FPGA and the operating frequencies for key point detection and descriptor computation components were 50 and 100 MHz, respectively.

Zhong et al. [52] presents a low-cost embedded system based on an architecture that integrates FPGA and DSP for SIFT on 320×256 images. It optimizes the FPGA architecture for the feature detection step of SIFT to reduce the resource utilization, and optimizes the implementation of the feature description step using a high-performance DSP. This hardware/software system detects SIFT features for 320×256 images within 10 ms and takes merely about 80 μs per feature to form and extract the SIFT feature descriptors. The feature detection part of their design can achieve real-time performance. However, the feature description part of their system was implemented in DSP, and it is not possible to guarantee real-time performance when the number of features in an image reaches 400 or more. The architecture was prototyped on a XC4VSX35 device at a frequency around 100 MHz.

Krajnik et al. [53] present a complete hardware and software solution of an FPGA-based computer vision embedded module capable of carrying out SURF image features extraction algorithm. Aside from the custom implementation of the main computations of the detector, the module embeds a Linux distribution that allows to run programs specifically tailored for particular applications built upon SURF. The module is based on a Virtex-5 FXT FPGA which features powerful configurable logic and an embedded PowerPC processor. Authors describe the hardware module as well as the custom FPGA image processing cores that implement the algorithm's most computationally expensive process, the interest point detection. Since the Fast-Hessian detector is computed in hardware, the determinant calculation is done in integer arithmetic with a limited precision for a specific number of octaves and scale intervals and limited image size. The achieved frame rate for $1,024 \times 768$ pixel images is about 10 frames per second at a 75 MHz clock frequency. The architecture power consumption is approximately 6 W.

4 Concluding Remarks and Future Directions

The final goal of embedded computer vision, focusing on efficiency, is often real time processing at video frame rates or dealing with large amounts of image data. In spite of the computation power of computing platforms increases rapidly over time, image feature detection is not the final step, but just an intermediate one in a processing chain of the computer vision pipeline, followed by matching, tracking, object recognition, etc. Efficiency is therefore one of the major issues that should be considered when designing or selecting a feature detector for a given application. Motivated by the demand for high-speed performance, difficult to overcome on sequential processors, alternative hardware architectures have been used as feature detector accelerators so as to speed up computations. Thanks to the fine grain massive parallelism, flexibility of software-like methodologies and a good power-efficiency tradeoff, FPGA devices have become a common choice for embedded computer vision applications.

In this chapter, a review of FPGA-based hardware implementations of image interest point detectors has been presented. An overview of some of the most widely used interest point detectors and their FPGA-based hardware implementation were presented so as to provide a starting point to the readers interested in techniques, methods and solutions related to algorithm-to-hardware mapping of image interest point detectors. It is noteworthy that in spite of their algorithmic differences, there are natural and tight connections between the computational principles of such feature detectors. The review highlights the notion of window-based operators and operator sequencing underlying in the algorithmic nature of interest point detectors as computational primitives that should be supported and exploited to achieve efficient hardware implementations. By taking into account these principles and combining fixed-point arithmetics with parallel and pipelined implementation strategies to eliminate unnecessary delays and overlap operations, efficient hardware architectures that use FPGAs can be designed. Most of the reviewed research works on image interest point detectors implementations sacrifice the accuracy by avoiding floating-point arithmetics and/or altering the original detector algorithm so as to ease the hardware implementation of hardware-greedy arithmetic operators, for instance division and square root, at the cost of some precision. Internal arithmetic operations are usually done with adequate precision, obtained experimentally, using fixed-point number representation and the two's complement format.

Reviewed works show that FPGA technology is appropriate for embedded hardware implementations that provide real-time performance thanks to the fine-grain parallel processing performed in the device and the on-chip memory facilities for internal storage that promote internal data reuse. On the other hand, the smaller operating clock frequency, compared to high-end embedded multicore processors, and low power requirements naturally lead FPGA devices to be a very suitable stand-alone platform for embedded applications. Interestingly, power consumption is rarely reported in detail in the reviewed works and much work is still needed to properly address this issue. On the other hand, comparing speed and resource utilization among FPGA implementations for the same algorithm reported in the literature should be done with caution, as different devices would have different speed grades that may enable the same design to be faster, and a fair comparison should involve a kind of normalization in terms of the used technology. Among the reviewed detectors, SUSAN and FAST detectors are competitive with the standard, more computationally expensive feature detectors and according to the presented results they are hardware compliant as they require fewer FPGA resources without important modifications or simplifications compared to the others. SIFT is one of the most memory demanding detectors that benefits of separability of filters or properties of the kernel's coefficients that allows to make some simplifications so as to make feasible its FPGA implementation. Furthermore, the review shows that for a full embedded implementation of scale-invariant feature detectors, a hardware/software co-design is preferable. The computational-intensive detection principle is usually mapped into an specialized parallel hardware architecture, meanwhile the more irregular computations involved in the descriptor are implemented on optimized software running on embedded processors, taking advantage of the system-on-a-programmable-chip (SoPC) platform offered by current FPGA devices.

Despite the observed encouraging results of using FPGA technology to implement image interest point detectors, further work is still needed to improve the hardware accelerators portability across FPGA-based platforms in more realistic proof-of-concept applications with shorter design cycles. Overall, further improvements in hardware implementations depend on the ability to automatically extract the parallelism from high level specifications and as a consequence it might become a limitation to explore faster the design space. Recent trends suggest that the integration of both software and hardware functionalities in a single chip using an embedded processor, Intellectual Property (IP) cores, and some customized peripherals, providing a so-called system-on-programmable-chip (SoPC) solution. Might boost FPGA-based embedded computing applications. On the other hand, the FPGA's performance for feature detectors can be increased by utilizing Double Data Rate (DDR)-based external memory banks or newer memory technology, as currently most prototyping platforms contain only one such memory bank. Moreover such memory banks should math FPGA embedded memory size and organization (memory block size(s), memory banking, and spacing between memory banks) that fits the needs of the application. This is particularly important for achieving high performance for the multiscale interest point detectors due to a large extent to their iterative nature and the related memory management challenges.

It is clear that porting algorithms that have been tailored to CPU-like architectures to an FPGA is a difficult task, and the modifications and simplifications undertaken in this endeavor might even affect the robustness of the original algorithm. In general, this approach might be acceptable for some specific applications, but it is neither suitable as a general-purpose standalone module, nor acceptable for many other vision applications. Thus, real-time embedded hardware designs that may be used as a stand-alone multi-detector module that can be easily adapted to diverse computer vision applications are highly desirable and they should be further explored in the future so as to define truly generic IP modules that can be customized and ported to meet different user environment and system requirements. Far beyond the achievable performance improvements of custom hardware implementations, it is also highly desirable to improve algorithms and propose overall implementation strategies for feature detectors to be more suitable and amenable for embedded platforms since factors such processing time, numerical precision, memory bandwidth and size, and power consumption are not easily discernible from sequential algorithmic representations.

Acknowledgments Author wants to acknowledge the comments and suggestions from the editors and reviewers to improve the quality of this chapter. Also, the author kindly acknowledges the partial support received from Conacyt, Mexico, through the research grants No. 99912 and 237417.

References

1. Schmid, C., Mohr, R., Bauckhage, C.: Evaluation of interest point detectors. Int. J. Comput. Vision **37**(2), 151–172 (2000)
2. Dorko, G., Schmid, C.: Selection of scale-invariant parts for object class recognition. In: Ninth IEEE International Conference on Computer Vision. Proceedings, vol. 1, pp. 634–639 (2003)
3. Lazebnik, S., Schmid, C., Ponce, J.: Affine-invariant local descriptors and neighborhood statistics for texture recognition. In: Ninth IEEE International Conference on Computer Vision. Proceedings, vol. 1, pp. 649–655 (2003)
4. Brown, M., Lowe, D.G.: Automatic panoramic image stitching using invariant features. Int. J. Comput. Vis. **74**(1), 59–73 (2007)
5. Wang, X., pan Niu, P., ying Yang, H., li Chen, L.: Affine invariant image watermarking using intensity probability density-based harris laplace detector. J. Vis. Commun. Image Represent. **23**(6), 892–907 (2012)
6. Ebrahimi, M., Mayol-Cuevas, W.: Adaptive sampling for feature detection, tracking, and recognition on mobile platforms. IEEE Trans. Circuits Syst. Video Technol. **21**(10), 1467–1475 (2011)
7. Li, Y., Wang, S., Tian, Q., Ding, X.: A survey of recent advances in visual feature detection. Neurocomputing **149**, Part B(0), 736–751. http://www.sciencedirect.com/science/article/pii/S0925231214010121 (2015)
8. Guerra-Filho, G.: An iterative combination scheme for multimodal visual feature detection. Neurocomputing **120**, 346–354 (2013)
9. Hernandez-Lopez, A., Torres-Huitzil, C., Garcia-Hernandez, J.: Fpga-based flexible hardware architecture for image interest point detection. Int. J. Adv. Robot. Syst. **12**(93), 1–15 (2015)

10. Torres-Huitzil, C., Arias-Estrada, M.: FPGA-based configurable systolic architecture for window-based image processing. EURASIP J. Adv. Sig. Proc. **2005**(7), 1024–1034 (2005)
11. Lim, Y., Kleeman, L., Drummond, T.: Algorithmic methodologies for FPGA-based vision. Mach. Vis. Appl. **24**(6), 1 (2013)
12. Tippetts, B., Lee, D.J., Archibald, J.: An on-board vision sensor system for small unmanned vehicle applications. Mach. Vis. Appl. **23**(3), 403–415 (2012)
13. Park, I.K., Singhal, N., Lee, M.H., Cho, S., Kim, C.: Design and performance evaluation of image processing algorithms on GPUs. IEEE Trans. Parallel Distrib. Syst. **22**(1), 91–104 (2011)
14. Pauwels, K., Tomasi, M., Diaz Alonso, J., Ros, E., Van Hulle, M.: A comparison of FPGA and GPU for real-time phase-based optical flow, stereo, and local image features. IEEE Trans. Comput. **61**(7), 999–1012 (2012)
15. Andreopoulos, Y., Patras, I.: Incremental refinement of image salient-point detection. IEEE Trans. Image Process. **17**(9), 1685–1699 (2008)
16. Awrangjeb, M., Lu, G., Fraser, C.: Performance comparisons of contour-based corner detectors. IEEE Trans. Image Process. **21**(9), 4167–4179 (2012)
17. Bostanci, E., Kanwal, N., Clark, A.: Spatial statistics of image features for performance comparison. IEEE Trans. Image Process. **23**(1), 153–162 (2014)
18. Olague, G., Trujillo, L.: Interest point detection through multiobjective genetic programming. Appl. Soft Comput. **12**(8), 2566–2582 (2012)
19. Canny, J.: A computational approach to edge detection. IEEE Trans. Pattern Anal. Mach. Intell. **8**(6), 679–698 (1986)
20. Smith, S.M., Brady, J.M.: SUSAN—a new approach to low level image processing. Int. J. Comput. Vis. **23**, 45–78 (1997)
21. Harris, C., Stephens, M.: A combined corner and edge detector. In: In Proceedings of the 4th Alvey Vision Conference, pp. 147–151 (1988)
22. Rosten, E., Drummond, T.: Machine learning for high-speed corner detection. In: Proceedings of the 9th European Conference on Computer Vision—Volume Part I, pp. 430–443. ECCV'06, Springer, Berlin (2006)
23. Bay, H., Ess, A., Tuytelaars, T., Gool, L.V.: Speeded-up robust features (SURF). Comput. Vis. Image Underst. **110**(3), 346–359 (2008), similarity Matching in Computer Vision and Multimedia
24. Lowe, D.: Object recognition from local scale-invariant features. In: The Proceedings of the Seventh IEEE International Conference on Computer Vision, vol. 2, pp. 1150–1157 (1999)
25. Delorimier, M., Kapre, N., Mehta, N., Dehon, A.: Spatial hardware implementation for sparse graph algorithms in graphstep. ACM Trans. Auton. Adapt. Syst. **6**(3), 17:1–17:20. http://doi.acm.org/10.1145/2019583.2019584 (2011)
26. Toledo-Moreo, F.J., Martinez-Alvarez, J.J., Garrigos-Guerrero, J., Ferrandez-Vicente, J.M.: FPGA-based architecture for the real-time computation of 2-D convolution with large kernel size. J. Syst. Architect. **58**(8), 277–285 (2012)
27. Torres-Huitzil, C.: Fast hardware architecture for grey-level image morphology with flat structuring elements. Image Process., IET **8**(2), 112–121 (2014)
28. Hedberg, H., Kristensen, F., Wall, V.: Low-complexity binary morphology architectures with flat rectangular structuring elements. IEEE Trans. Circuits Syst. I: Regul. Pap. **55**(8), 2216–2225 (2008)
29. Girau, B., Torres-Huitzil, C.: Massively distributed digital implementation of an integrate-and-fire LEGION network for visual scene segmentation. Neurocomputing **70**(79), 1186–1197 (2007)
30. Rangel-Valdez, N., Barron-Zambrano, J.H., Torres-Huitzil, C., Torres-Jimenez, J.: An efficient fpga architecture for integer nth root computation. Int. J. Electron. **102**(10), 1675–1694 (2015)
31. Barron-Zambrano, J.H., Torres-Huitzil, C.: FPGA implementation of a configurable neuromorphic CPG-based locomotion controller. Neural Netw. **45**, 50–61 (2013)
32. Monmasson, E., Cirstea, M.: FPGA design methodology for industrial control systems—a review. IEEE Trans. Industr. Electron. **54**(4), 1824–1842 (2007)

33. Hartley, E., Jerez, J., Suardi, A., Maciejowski, J., Kerrigan, E., Constantinides, G.: Predictive control using an FPGA with application to aircraft control. IEEE Trans. Control Syst. Technol. **22**(3), 1006–1017 (2014)
34. Chappet De Vangel, B., Torres-Huitzil, C., Girau, B.: Randomly spiking dynamic neural fields. J. Emerg. Technol. Comput. Syst. **11**(4), 37:1–37:26 (2015)
35. Torres-Huitzil, C., Delgadillo-Escobar, M., Nuno-Maganda, M.: Comparison between 2D cellular automata based pseudorandom number generators. IEICE Electron. Express **9**(17), 1391–1396 (2012)
36. Trimberger, S.: Three ages of FPGAs: a retrospective on the first thirty years of FPGA technology. Proc. IEEE **103**(3), 318–331 (2015)
37. Kuon, I., Tessier, R., Rose, J.: FPGA architecture: survey and challenges. Found. Trends Electron. Des. Autom. **2**(2), 135–253. http://dx.doi.org/10.1561/1000000005 (2008)
38. Cong, J., Liu, B., Neuendorffer, S., Noguera, J., Vissers, K., Zhang, Z.: High-level synthesis for fpgas: from prototyping to deployment. IEEE Trans. Comput. Aided Des. Integr. Circuits Syst. **30**(4), 473–491 (2011)
39. Kumar Jaiswal, M., Chandrachoodan, N.: FPGA-based high-performance and scalable block LU decomposition architecture. IEEE Trans. Comput. **61**(1), 60–72 (2012)
40. Torres-Huitzil, C., Arias-Estrada, M.: An FPGA architecture for high speed edge and corner detection. In: Fifth IEEE International Workshop on Computer Architectures for Machine Perception. Proceedings, pp. 112–116 (2000)
41. Possa, P., Mahmoudi, S., Harb, N., Valderrama, C., Manneback, P.: A multi-resolution FPGA-based architecture for real-time edge and corner detection. IEEE Trans. Comput. **63**(10), 2376–2388 (2014)
42. Xu, Q., Varadarajan, S., Chakrabarti, C., Karam, L.: A distributed canny edge detector: algorithm and FPGA implementation. IEEE Trans. Image Process. **23**(7), 2944–2960 (2014)
43. Lim, Y., Kleeman, L., Drummond, T.: Algorithmic methodologies for FPGA-based vision. Machine Vis. Appl. **24**(6), 1197–1211. http://dx.doi.org/10.1007/s00138-012-0474-9 (2013)
44. Kraft, M., Schmidt, A., Kasinski, A.J.: High-speed image feature detection using FPGA implementation of fast algorithm. In: Ranchordas, A., Arajo, H. (eds.) VISAPP (1), pp. 174–179.INSTICC—Institute for Systems and Technologies of Information, Control and Communication (2008)
45. Hsiao, P.Y., Lu, C.L., Fu, L.C.: Multilayered image processing for multiscale harris corner detection in digital realization. IEEE Trans. Industr. Electron. **57**(5), 1799–1805 (2010)
46. Torres-Huitzil, C.: Resource efficient hardware architecture for fast computation of running max/min filters. Sci. World J. **2013**, 1–10 (2013)
47. Viola, P., Jones, M.: Rapid object detection using a boosted cascade of simple features. In: Proceedings of the 2001 IEEE Computer Society Conference on Computer Vision and Pattern Recognition. CVPR 2001, vol. 1, pp. I-511-I-518 (2001)
48. Jiang, J., Li, X., Zhang, G.: SIFT hardware implementation for real-time image feature extraction. IEEE Trans. Circuits Syst. Video Technol. **24**(7), 1209–1220 (2014)
49. Wang, J., Zhong, S., Yan, L., Cao, Z.: An embedded system-on-chip architecture for real-time visual detection and matching. IEEE Trans. Circuits Syst. Video Technol. **24**(3), 525–538 (2014)
50. Chang, L., Hernndez-Palancar, J., Sucar, L., Arias-Estrada, M.: FPGA-based detection of SIFT interest keypoints. Mach. Vis. Appl. **24**(2), 371–392 (2013)
51. Bonato, V., Marques, E., Constantinides, G.: A parallel hardware architecture for scale and rotation invariant feature detection. IEEE Trans. Circuits Syst. Video Technol. **18**(12), 1703–1712 (2008)
52. Zhong, S., Wang, J., Yan, L., Kang, L., Cao, Z.: A real-time embedded architecture for SIFT. J. Syst. Architect. **59**(1), 16–29 (2013)
53. Krajník, T., Šváb, J., Pedre, S., Čížek, P., Přeučil, L.: FPGA-based module for SURF extraction. Mach. Vis. Appl. **25**(3), 787–800 (2014)

Image Features Extraction, Selection and Fusion for Computer Vision

Anca Apatean, Alexandrina Rogozan and Abdelaziz Bensrhair

Abstract This chapter addresses many problems: different types of sensors, systems and methods from the literature are briefly revised, in order to give a recipe for designing intelligent vehicle systems based on computer vision. Many computer vision or related problems are addressed, like segmentation, features extraction and selection, fusion and classification. Existing solutions are investigated and three different data-bases are presented to perform typical experiments. Features extraction is aimed for finding pertinent features to encode information about possible obstacles from the road. Feature selection schemes are further used to compact the feature vector in order to decrease the computational time. Finally, several approaches to fuse visible and infrared images are used to increase the accuracy of the monomodal systems.

Keywords Obstacle detection · Features extraction and selection · Fusion · Classification · Visible and infrared images · Intelligent vehicles

1 Introduction

Computer Vision (CV) applications aim at finding correspondence between two images of the same scene or the same object, 3D reconstruction, image registration, camera calibration, object recognition, and image retrieval, just to mention few among them. In recent years, the continuous progress in image processing and CV algorithms has attracted more and more attention in research areas approaching object detection, recognition and tracking. Autonomous intelligent vehicles,

A. Apatean (✉)
Communication Department, Technical University of Cluj-Napoca, 400027 Cluj, Romania
e-mail: anca.apatean@com.utcluj.ro

A. Rogozan · A. Bensrhair
PSI Department, INSA Rouen, Rouen, France
e-mail: alexandrina.rogozan@insa-rouen.fr

A. Bensrhair
e-mail: abdelaziz.bensrhair@insa-rouen.fr

© Springer International Publishing Switzerland 2016
A.I. Awad and M. Hassaballah (eds.), *Image Feature Detectors and Descriptors*,
Studies in Computational Intelligence 630, DOI 10.1007/978-3-319-28854-3_4

intelligent robots, personal assistants are just few of the applications where such operations converge. Advanced techniques in features extraction, selection and fusion have been successfully applied and extended therein.

Nowadays, machine learning have gained popularity in various applications, with a significant contribution to equipping robots and vehicles with seeing capability, i.e. vision by CV, but also with communication skills [1, 2]. Real-time robust systems have been built thanks to the advancement of relevant CV techniques, some used in combination with similar processing in speech or language domain, for voice commands detection and interpretation among others.

Robots represent the ultimate challenge for real-time systems engineers because they combine image, sound and text processing, artificial intelligence, and electro-mechanical mechanisms, all collaborating. Nao [1] and Asimo [2] are really great robots, able to detect persons and their faces and recognize thus humans identity; they ask questions and make decisions based on answers; they move, play a ball, jump or protect themselves at falling. Asimo can even act like a host, by serving you the preferred drink. Advances in robot technology will further change how common people interact with robots. Even today's robots are generally perceived to perform repetitive tasks (except the ones like Nao or Asimo), human-robot interaction will soon become a necessity. Tomorrow visionary assume that sensors, communications and other operational technologies will work together with information technologies, to create intelligent industrial products.

Even more, a large number of small, communicating real-time computers found in most smartphones, appliances, wearables, etc. promise to transform the way human beings interact with their environment. Moreover, people are talking more and more about the large adoption (at an industrial level) of the imminent technology called Internet of Things (IoT) and a lot of sensation has been made around this subject, assuming it will revolutionize the world just as Industry revolution and then Internet revolution did [3–6]. This trend will encourage the development of low-cost solutions involving more or less knowledge of machine learning, data or text mining algorithms. These low-cost platforms for real-time intelligent applications generally integrate multiple sensors and are able to adapt their functioning automatically to the user/system behavior/functioning. Such systems have to continuously monitor not only the surroundings, but also the user/system state and behavior. This is generally accomplished with different types of sensors, vision systems, microphones and so on, all needed to improve or generalize the IoT functionality. One such possible device could help the driver, e.g. to detect that she/he closed the eyes, so fall asleep or it has been too long without looking at the windshield or detect a users emotion and predict her/his future affect state (healthy issues), among others.

The imminent IoT promise a world in which intelligent machines not only connect, but cooperate with each other. Thus, a plenty of applications to improve health and well-being of children and elderly will be available and this will be obtained with a significant contribution of CV techniques.

In the Intelligent Vehicle (IV) field, the majority of research projects concentrated at the exterior of the vehicle as more surveys prove it [7–16]; thus, the cameras were mostly oriented to the road. Still, recently, there are also research teams having the main interest to the interior of the car, proposing solutions to monitor the driver [17–19]. Smart environments and monitoring systems will be available by the large adoption of the IoT technologies, but what about if a generic activity monitoring system or simpler, a robot, will be able to completely observe and understand the driver affect state? How far is then an intelligent autonomous car in the industrial sector, to accomplish constraints such as real time functioning, low cost implementation, robustness in all possible driving situations, or even more?

This chapter provides a background overview, addressing many CV problems: after a short introduction in Sect. 1 in the IV field, different types of sensors, systems and methods from the literature are surveyed in Sect. 2. Next, the proposed recipe for designing intelligent systems for IV is presented in Sect. 3. The CV related aspects, like features extraction and selection, fusion and classification in the frame of applications from IV field are approached in Sect. 4. After reviewing more possible solutions from the literature for these problems, our proposed solutions is provide as a guide to obtain possible systems. To highlight each module from the proposed solution, three databases are briefly presented and some typical experiments are described.

2 State of the Art in Intelligent Vehicles

2.1 Autonomous Intelligent Vehicles

The interest for the IV field has been increased during the last years to assure the safety of both the driver and the other traffic participants. Leading car developers such as Daimler Chrysler, Volkswagen, BMW, Honda, Renault, Valeo, among others, have recorded significant contributions regarding this field. The developed consistent research has shown that IVs could senses the environment by using active sensors, like radars or laser scanners, combined with passive sensors, like VISible (VIS) or InfraRed (IR) spectrum cameras. Systems combining multiple types of active and passive sensors have been shown efficient for both obstacle detection and automatic navigation during the DARPA Grand Challenge 2007, also known as the "Urban Challenge" [20]. In the frame of VIAC, i.e. VisLab Intercontinental Autonomous Challenge [21], four electric and driverless vehicles navigated a 13,000 km test from Parma, Italy, to Shanghai, China. The VIAC vehicles were equipped with 4 lidars and 7 digital cameras, providing frontal, lateral and rear sensing. The prototype BRAiVE [22] is another VisLab contribution; being equipped with 10 cameras, 5 laser scanners, 1 radar, 1 GPS and inertial measurement unit and one emergency stop system, it is able to detect obstacles, follow an in-front vehicle, warn for possible collisions, recognize traffic signs, detect parking slot among others. Another relevant example is the Google driverless car which considers the environmental data taken with roof-mounted lidar, uses machine-vision techniques to identify road geometry and obstacles, and controls the cars throttle, brakes and steering mechanism, all being performed in real-time [23].

Honda has implemented the first intelligent night vision system [24] using two far infrared cameras, installed in the front bumper of the vehicle. It provides visual and audio cautions (when it detects pedestrians in or approaching the vehicles path) for the driver in order to help during the night driving. Another recent innovation to help driver see better at night and in the most diverse weather conditions, is the BMW Night Vision system [25]. With its long range detection capability (up to 300 m for a human being), their system assist the drivers, by providing more time to react and avoid accidents. Similar systems equip today also some vehicles from Mercedes and Audi.

The autonomous systems previously mentioned exist but they are prototypes, generally used for research purposes and their implementation costs make them unapproachable for a series vehicle. Such series vehicles, by a large adoption, could really decrease to zero the number of accidents from the traffic road areas. Still, night vision systems implemented on-board of more vehicles today proves their support, attracting a larger segment of drivers every day. Even with all this commercial success, existing on-board systems are far from an autonomous series intelligent vehicle.

2.2   The More, The Better

Is more, better? Not even in real life scenarios this is always valid, so the same is when equipping a vehicle on the road. Each new sensor add its influence to the system cost, possible interference problems, computational workload, difficulties in interpreting and storing the raw data, etc. By now, (semi)autonomous vehicles have been developed, and they could also demonstrate intelligence, by seeing, understanding or even interpreting the scene and taking decisions on how to react on this (i.e. detect the road, the obstacles, even recognize the obstacle type, on daytime or night-time). But these systems, generally either use some mixture of expensive sensors, or the processing of data has been conducted on a powerful expensive workstation server-like (ultra-high speed processing, multiple cores, large storage, etc.) or both. These system setups could not be applicable to industrial sector due to the cost, space and weight constraints.

To be a solution for equipping a series vehicle, the system should have sensors with low cost, no interference issues, function in real time and prove robustness in diverse atmospheric and illumination situations. Thus, to answer the question from the beginning of the subsection, not every time more means better. In the frame of IV systems, more and diverse sensors could add their benefits to improve the system functionality, but the implementation cost generally constrict this to only a few, a single type or even a single one.

2.3 Sensors and Systems in Intelligent Vehicles

Many systems developed in the IV field and employing one or multiple sensors are reviewed in literature [9, 10]. In [26], common sensors used in the IV field are examined: some sensors may have many advantages, but also some strong limitations, which will make them to be not-so-properly for the implementation of a general *Obstacle Detection and Recognition (ODR)* system. In such an ODR system, the hypothesis generation phase is first occurring, when *Obstacle Detection (OD)* or ROI (Region Of Interest) estimation is accomplished. *Obstacle Recognition* (OR) or hypothesis verification component follows, which would, first, help the detection by discarding the false alarms, and second, it would assist the system in making the appropriate decision in different situations according to the obstacle type. An unsolved issue of the existing ODR systems is their limited ability to ensure a precise and robust OR in real conditions.

What type of sensor is best? Generally, the sensors used in the IV field can be classified according to different criteria: first, as concerns the perception about the environment, i.e. the type of the measured information, they are *proprioceptive* and *exteroceptive*; second, they can be classified about the *spectrum* position of the radiation they use to function and third, they can be classified as *active* or *passive* concerning the presence or absence of a radiation required in their functioning. Further details and information can be found in [26].

One sensor which could provide enough information to detect obstacles (even those occluded) in any illumination or weather situation, to recognize them and to identify their position in the scene does not exist (at least today, as the authors surveyed). In the IV domain there is no such a perfect sensor to handle all these concerned tasks, but there are systems employing one or many different sensors in order to perform obstacles detection, recognition or tracking or some combination of them. Thus, there is not a single best sensor which could equip a vehicle to provide autonomous driving functionality, but generally a combination of not-so-expensive, complementary sensors is aimed.

Active or passive sensors? Many systems use *radars*, due to their strongest advantage: insensitivity to atmospheric changes, like rain, snow or fog. Being an optical sensor, *laser scanner* is affected by critical weather conditions (fog, snow), its functioning and detection range being limited. Active sensors, like radars and laser scanners, are mainly used for OD due to their capability to provide distance to possible obstacles, and generally not adopted for OR. On the other side, passive sensors, like cameras, provide richer frontal and lateral information of the road scene allowing thus an efficient OR. Moreover, while IR cameras may be used for OD by objects temperature, the OD task with VIS cameras could be implemented by stereo-vision (allowing for depth information), and/or by an optical flow process (providing motion information). Although the image processing could be computationally expensive, fast algorithms and electronics are proposed all the time, allowing for real time implementation of such CV-based systems. In this way, the implementation cost of a system comprising only passive sensors could be a proper solution for equipping a series vehicle.

To conclude, the sensors proposed for the implementation of an ODR system, as motivated in [27], are only passive ones: their counterparts, the active sensors, are extremely invasive, susceptible of interference problems in cluttered traffic and present high acquisition price.

A possible CV-based system could use a VIS stereo-vision sub-ensemble, augmented with a monocular IR camera, in order to benefit from the complementary characteristics of both types of passive sensors, as proposed in [27]. In the frame of INSA laboratory, a VIS stereo-vision system was developed for OD of vehicles [28] and pedestrians [29] and we aimed to continue this work by enhancing the system with an OR component based on VIS—IR weighted fusion. The passive sensors basic functioning is generally improved by fusion of multiple types of sensors or different sensors of the same type; this can be accomplished at many possible levels: sensors, raw-data, features, SVM-kernels, scores, classifiers, decisions, as presented in what follows.

In the CV domain, images are mainly processed for detection, recognition or tracking. Moreover, the algorithmic components of an ODR system implemented on a generic platform can be decomposed into acquisition, preprocessing, detection or segmentation, classification and tracking, although specific system implementations might not have one or more of these components. These steps could even be generalized to many application domains [30].

The hardware implementation of these operations (from the CV domain) generally aim an FPGA (Xilinx Virtex) platform, which cost could be relatively low for a series vehicle. For example, in [31] the optimized Speeded-Up Robust Feature (SURF) algorithm, which is specific to CV applications, was implemented on such a VLSI architecture (supporting more tasks to be realized in parallel).

3   A Possible ODR System

Similarly to the way humans use their senses to relate to the world around them, a computer system has to interpret the environmental data, and this is generally accomplished by machine perception task. This also imply CV, with methods for acquiring, processing, analyzing, and understanding images and, in general, processing high-dimensional data from the real world. This will further produce numerical or symbolic information, e.g., in the form of decisions. Computer vision has many applications already in use today in the IV field: road detection and following, scene understanding, pedestrian or vehicle detection, obstacles recognition, tracking, night vision, geographical modeling, etc.

To address such applications, generally Machine Learning (ML) is used to build computer systems that learn from experience or data. These systems require a learning process that specify how they should respond (as a result of experiences or examples they have been exposed to) to new examples, unknown.

Almost all categorization systems developed by now in the IV field employ an OD step followed by an OR module. Very often there is also a third module in which the recognized obstacles are tracked in their trajectory until they are no longer viewed in the scene. The OR module is supposed to support the OD module (which mostly is based on stereo-vision) by its ability to deal with the great variability of obstacles (mostly pedestrians and vehicles, but also cyclists or other type of obstacles), appearances and occlusions in any illumination and weather conditions. Besides, the variety of obstacles appearances (different type, shape, size, viewing angle, texture and color, they could be occluded or not, etc.) combined with the outdoor environment, and the moving vehicle constraints make the ODR a challenging task. Also a problematic task in the IV domain is the development of an *intelligent* automatic pilot (for an autonomous vehicle) which could therefore entirely control the vehicle like a human (or even in an improved way). Such a system has not only to recognize the road trajectory and to detect any possible obstacle which may appear near or on the road, but also to identify/recognize the obstacle type, to estimate its behavior and propose certain actions. In addition, it could also monitor the driver state and in critical situations, take over the vehicle control. This is generally accomplished by ML mechanisms, and two directions for such systems addressing the ODR task can be distinguished:

(a) most of them aim the detection of a *particular type of obstacle* (pedestrian or vehicle most often) and perform thus a binary classification (it is or it is not the obstacle they meant), while

(b) very few systems consider as *obstacle* anything *obstructing the host vehicle path* and the detected obstacle enter in the recognition module where its type is predicted by multiclass classification.

3.1 Main Processing

In a real system, when classification is implied, object detection or ROI estimation/identification is generally followed or related by an object recognition process; this latter can also be divided into two tasks: *object verification* and *object identification*. The former seeks to verify a new object against a stated object model (it implies a one-to-one comparison) and it is often used to verify if the object is the one a hypothesis claims it is—for example the ROI provided by an OD module for a specific type of object (case a). In the latter, i.e. object identification, the task is more general: it has to discover the object type, category or class. The sample of the current object is compared with the existing models of the objects in the database (one-to-many comparison). Using decision logic (generally with a threshold value), a model matching the current object is chosen (case b). Assuming that the database contains a sort of model for all the objects being identified, defines a closed set of objects, but practically it is hard to have precise models for all objects that may appear in a real scenario. In the IV domain, generally there are two possible situations:

- to use an *open set* of objects, this corresponding to unsupervised methods, where the identification models are trained using data from only the object it represents; generally, objects present low intraclass variability, e.g. the case of recognizing pedestrians—each possible pose represent a class, or
- to use an approximation of *closed set*, i.e. some representative models are used to construct the database and this is the case of supervised methods; generally, objects present high intraclass variability, e.g. recognizing all types of possible pedestrians as a single class.

The most frequent case is to recognize objects by their specific appearance (e.g. recognizing pedestrians from the frontal/lateral view) or behavior (e.g. recognizing pedestrians by the human walk) and these are supervised methods because they imply training and the availability of data.

The criteria used in the ODR task depends on the definition of what the obstacle is. In some systems, the detection of obstacles is limited to the localization of specific shapes corresponding to obstacles (like vehicles, pedestrians, cyclists), which is based on a search for specific patterns, such as shape, symmetry, edges, pedestrians head or vehicles lights. This search for patterns common or not to multiple obstacle classes generally lead to the determination of a Bounding Box (BB). These approaches are generally based on some knowledge about the obstacle type, and they will be referred in what follows as *knowledge-based*. A more general definition of an obstacle, which leads to more complex algorithmic solutions, identifies as an obstacle any object that obstructs the path the host vehicle is driving on. In this case, instead of recognizing specific patterns, the OD task is reduced to identifying the area in which the vehicle can safely move and anything rising out significantly from the road surface would be considered as obstacle. Due to the general applicability of this definition, the problem is using more complex techniques, like those based on the processing of two or more images, which are: the analysis of optical flow field or the processing of nonmonocular (i.e. stereo) images. The *optical flow-based technique* requires the analysis of a sequence of two or more images: a two-dimensional vector is computed, encoding the horizontal and vertical components of the velocity of each pixel. The obtained motion information can be used to compute ego-motion and moving obstacles can be detected and/or tracked in the scene by analyzing the difference between the expected and real velocity fields and by removing background changes. On the other hand, the processing of *stereo images* requires identifying correspondences between pixels in a pair of left and right images. Stereo-based approach is generally more robust than the optical flow-based one, especially when both host-vehicle and obstacles have small or null speeds.

To conclude, knowledge-based methods employ a priori knowledge about the obstacle to hypothesize its locations in the image. Motion-based methods use optical flow techniques, while stereo-based approaches generally use an v-disparity method or Inverse Perspective Mapping (IPM) to estimate the locations of obstacles in images.

3.2 CV-Based Systems in Intelligent Vehicles

The most developed systems are specific for one kind of object detection, either pedestrian or vehicle. These dedicated systems are looking for obstacles in the scenes using either *an active sensor*, like radar or laser scanner which will provide the distance to the respective object, or *a passive one* like cameras. Besides, employing active technologies, which are efficient and robust, but too expensive, is not the frequent case when a series vehicle is aimed. In this chapter, the use of only passive sensors (i.e. cameras) which are quite cheap, but their functioning still possible to be improved (as they struggle in the presence of occlusions and in difficult lighting conditions) are considered.

When the OD task is limited to the localization of specific patterns corresponding to obstacles, e.g. in the knowledge-based approaches, the processing can be based on the analysis of *a single still image*, in which relevant features are searched for. The other systems, in which a more general definition of obstacle is exploited and all types of obstacles are searched at a time, the OD assignment is reduced to identifying the area in which the vehicle should safely move. Generally, in the frame of these type of systems, the road detection is performed by a monocular camera, but the localization of possible obstacles on the vehicle path is realized employing *two cameras* instead of a single one (for stereo approaches) or by using *a video sequence* of images (for motion-based approaches).

3.3 Background Overview

3.3.1 The Obstacle Detection Task

The majority of developed systems have used in the OD step one of the following three methods: (1) knowledge-based, (2) motion based, or (3) stereo-based.

Motion-based methods detect objects based on their relative motion, extracted from the optical flow information. Unfortunately they are quite slow, since they need to analyze a sequence of frames. These methods have been shown more appropriated for fixed cameras, employed for parking surveillance, than mounted on a vehicle [32, 33].

Being designed according to the model of human perception of objects in space, *stereo-vision methods* [34] are able to detect all types of objects, even the occluded, static or moving ones, based on their distance with respect to the system. These methods are mainly based on v-disparity and IPM algorithms. Generally, stereo-vision is used for OD task, but there are systems employing it also to provide the size/scale of the potentially objects for OR.

There are few directions for searching possible obstacles in the scenes, in the case of *knowledge-based methods*: (1) to use sliding windows encoding an obstacle specific shape of different sizes over the entire image or on some areas from the image (determined from perspective constraints), or (2) to look for areas presenting symmetries or textures specific to possible obstacles, (3) to try to detect specific parts of the proposed obstacles, e.g. pedestrians legs or head, the wheels, headlights or the shadow produced by a vehicle. The simplest knowledge-based technique is the one of sliding window, where windows at different scales and locations are shifted over the image. This method is used nowadays with Haar, Haar-like and/or HOG [35] features. Even the process is slow, there are speed-up variants with cascade classifiers [36]. Template matching with the obstacles contour is another possible approach, the templates generally being the head, legs or other parts of human body shape; headlamps or tires of vehicle. The information is generally provided as an image or as a set of features extracted from that obstacle samples. These methods are application-specific, and could employ some constraints, like aspect-ratio, geometry or even scene appearance area, which limit their generalization to multiple obstacles detection [37]. The methods employing symmetry or edges detection allow efficient OD only if the objects do not present strong occlusions (they should match an heuristic obstacle model). A major drawback is that they use an inflexible model, fixed by an important number of parameters that could affect the system robustness [38]. Methods performing the segmentation of an obstacle in sub-regions are quite promising, since they not only simplify the representation, but they can also be used to detect partially occluded objects [39, 40]. Another promising OD method is based on local features matching and recognition using SIFT (Scale Invariant Feature Transform) to detect and describe local features in images [41], and SURF (Speed Up Robust Features) based on sums of Haar wavelet applied on integral images. In [42] pedestrian hypotheses are generated using a Hierarchical Codebook which is a compact representation of local appearance of pedestrian heads in FIR (Far IR) images. Then, BBs are constructed and overlapping ones are merged together by clustering.

3.3.2 The Obstacle Recognition Task

For the OR task, different features extractors together with classification algorithms have been tested during the last years, in order to solve the obstacle categorization problems. In [43] motion and appearance information (sum of pixels in various rectangles subtracted from each other similar with Haar wavelets), were classified with an AdaBoost cascade approach. Simple detectors (with a smaller number of features) were placed earlier in the cascade, whereas complex detectors were activated later. In [35] HOG features are processed by an SVM to detect pedestrians. In [44] multiple cues (i.e. hierarchical shape matching and texture information) were combined within a neural network. Another combination of features, i.e. HOG and Haar wavelets proves efficient within an AdaBoost classifier [45], as it has been shown as the top performer on Caltech database for pedestrian detection. Still, the existing CV-based methods are generally difficult to compare because they are rarely tested on a common data set and with common experimental setup. Just recently some authors compared the results previously obtained by them or by other research teams on the same database [46–48].

Very few systems treat the problem of VIS—IR fusion. Among them, [49] have used a four-camera system (stereo color visible and stereo infrared) for pedestrian detection, based on v-disparity for the OD task and on an SVM classification of HOG features for the OR. A local appearance model based on SURF features, combined with an SVM classifier, within a multimodal VIS—IR fusion framework is used in [42] to recognize both pedestrians and vehicles.

3.4   The Proposed ODR Module

The main component of an ODR system is the OD module, but because it has not yet reach a robust and acceptable accuracy when working alone in an autonomous system, almost all existing systems from the IV literature provide a second component, the OR module. The main purpose of the recognition module is to identify the type or class of the detected obstacle, and to eliminate the false alarms, i.e. to reject them. The OD with visible spectrum (VIS) cameras could be improved by *stereo-vision*, allowing for depth information, and/or by *an optical flow process* providing motion information. A system based on stereo-vision, augmented with a monocular IR camera, seems in [27] the best solution for an ODR task. The system uses both VIS and IR to benefit from the complementary characteristics of both types of passive sensors and thus to assure a proper functioning in more and diverse situations. The architecture of the system presented in [27], and given in Fig. 1 is based on three passive cameras: a stereo-vision pair for OD (using the stereo matching method developed at INSA laboratory), and an IR camera for OR to remove the false alarms. The BBs generated by the OD component, as potential obstacles, have first to be projected on the IR image and then they are given, together with the BB in VIS spectrum, as inputs to the OR component.

The system has to discriminate between usual road obstacles like pedestrians, vehicles and cyclists, but also to distinguish these types of obstacles from other objects belonging to the background, as traffic signs, barriers, or regions of the image scene without any particular significance. Moreover, the results from the OD

Fig. 1 The proposed obstacle detection and recognition system

Fig. 2 A possible
segmentation in visible
domain

step are not trivial: it could provide several BBs for the same obstacle, the BBs could
be centered or not on the obstacle, or they even could overlap each other, providing
thus much more situations which have to be considered.

3.4.1 The Proposed Obstacle Detection Module

On the VIS domain, stereo configuration is required because the segmentation is hard
or even impossible using monocular VIS spectrum vision in the context of a cluttered
background and real time functioning. The detection part on VIS was already treated
in the frame of INSA laboratory, e.g. like in [28, 29] and it is still a work in progress.
Like other systems from the IV literature, the OD module has registered some false
alarms: possible examples of detected BBs, including unwanted false alarms, are
presented in Fig. 2.

On the IR domain, due to the pixels emphasis in intensity corresponding to hot
areas, even a simple threshold-based segmentation will provide good results, as it
can be noticed in Fig. 3, where the results are after applying a simple binarization
operation. The method has been developed in a first attempt and it did not received
too much attention due to the small efforts in its development. Still, being based
on a simple intensity threshold-based segmentation in IR domain, it is a simple and
rapid way of segmenting obstacles in normal situations of day or night. It has to
be mentioned that, like other detection systems from the IV literature, the proposed
OD module will select the objects being very closed to each other as a single one,
i.e. as belonging to the same BB. Thus, it could be considered that those multiple
obstacles belong to the same class, or they are parts of the same obstacle. One
possible inconvenient is that in the evaluation stage, if the coordinates of the manually
annotated BB does not fit those provided by the OD module, the OR module could
be penalized; but this is more about the experimental setup.

Fig. 3 A simple intensity threshold-based segmentation in infrared domain

The OD module from VIS could be separable from the OD module from IR, and thus independent and parallel detection tasks could be accomplished. Diverse illumination and weather situations will have to be studied and the modules calibrated on some specific situations; different tests for more possible scenarios are needed due to the adaptation to illumination and weather, e.g. hot day-normal day in summer, day-night, summer-winter, etc.

The OD module functioning would be an intelligent one, due to the fact that the system should recognize different possible setup scenarios and act according to that specific illumination or weather situation. For example, during night, when less obstacles are expected to be met on the road, the system could rely more on the IR sub-ensemble. This is also required due to the lack of information in VIS on night. On the other side, the IR based OD module could also have an important credit during daytime, not only the VIS one. Still, on a hot summer day, the IR based OD module could present some drawbacks due to the fact that even the pavement could be detected as possible obstacle. By a proper and intelligent combination of both the VIS and IR detection modules, the fused OD results will contain all the obstacles from the scene, but also possible false alarms, as it is the case of other systems from the literature. Next, the OR module will have to enter in action and eliminate these false alarms as soon as possible. The remaining ROIs will be thus real obstacles from the road and their type/class will be also known; thus, possible actions of obstacles may be anticipated and the IV system would be prepared to intervene.

3.4.2   The Proposed Obstacle Recognition Module

The efficiency of the OD module could be improved by the use of complementary VIS and IR information and by computing a compact, but pertinent bimodal signature of obstacles. This could be accomplished via some extracted features from the BBs corresponding to the obstacle. Thus, the authors concentrated on ML algorithms to implement the OR stage: from images corresponding to possible models for each class to be detected, pertinent families of features have been extracted and Feature Vectors (FV) have been constructed.

In the remaining part of this chapter, the OR component is more emphasized, with the following being presented: the image databases on which the proposed schemes have been experimented, the measures by which the performances of these schemes have been evaluated, but also how the FV that will characterize/define each instance within the system was composed. Our main purpose was not to develop a system on the whole, but only the OR module which was intended to be based on fusion in order to exploit the complementary information of VIS and IR cameras. Therefore, in our work we intended to verify if it worth to perform the fusion: Will the VIS—IR fusion bring in benefits from the OR point of view, besides the advantages it implies in the OD step?

3.4.3 Offline and Online Setup

An obstacle recognition system consists in two main parts, as Fig. 4 shows. In the *training step*, a database with different BBs enclosing possible obstacles (manually annotated) from the road is used. In the OR stage, there is also information provided by the sensors, like in the OD case, but here it takes the form of a training and/or testing database. The *testing part* comprises the same pre-processing module like the training one, but here the test image provided by the OD module is aimed, because the system runs on-line. A Features Extraction (FE) and Features Selection (FS)

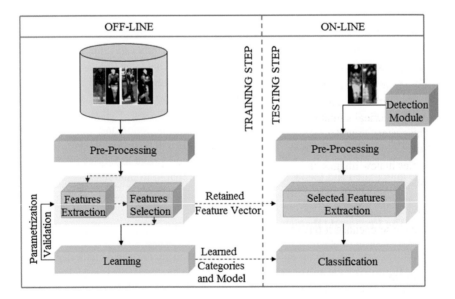

Fig. 4 Training and testing steps in the frame of an OR system

module follows which together with the last module, i.e. the learning one, has to accomplish the system parametrization and validation. This latter operation consists in choosing the most pertinent features to compute an optimized FV which will best characterize the data from the training database, but also in establishing the classifier which will best learn the instances from the training set. In the testing step, the FV used to characterize the test data will comprise the same features determined as being relevant in the training step.

To conclude, our ODR system belongs to that category of systems based on pattern recognition and consists of three main modules: (1) sensors (VIS and IR cameras) that gather the observations to be classified, also including the pre-processing module, (2) a FE mechanism (often attended by a FS operation) that digitize the observations, and (3) a classification or description scheme that does the actual job of classifying or describing observations, relying on the previously obtained information.

Applications based on pattern recognition aims to classify data (also called patterns) based either on a priori knowledge or on some statistical information extracted from the patterns. How well computers succeed in recognizing patterns depend on a multitude of factors: how comprehensive is the training set (Does it cover all possible situations in which objects can appear?); How efficient is the classifier to be used (Does it succeed in learning well all the objects from the training set and then experiments performed on the test set provided high accuracies? What about the classification time? Is its value satisfactory from the viewpoint of a real time system?). In the frame of our system, we tried to develop an OR module to give affirmative responses to all these questions.

4   Features Extraction, Selection and Fusion

In the frame of CV applications, the image visual content represents the only available source of information. To describe the content of an image, usually some numerical measures with different possibilities to represent the information could be used. The images numerical signature is via some extracted features (also called attributes).

The system must learn different classes of objects (like pedestrian, cyclist, vehicle) from several available examples, and then it should be able to recognize similar examples in new images, unknown. This imply the extraction of some relevant information (features) from images, so that the characterization of an image to be made with as less features as possible to shorten the computing time required by the system to learn and then to recognize the objects. However, this operation of FE is a very sensitive one because it has to select a number of features great enough to assure a proper/good characterization of images, but also low enough to obtain a low processing time. Next, the obtained FVs could be compacted, i.e. optimized, by using a FS mechanism.

For our system, different families of texture-based features together with statistical moments and size-related features have been extracted from each BB hypothesis provided by the stereo-vision OD module. We have chosen to extract features as rich and diverse as possible in order to take advantage of their complementarity, but if some features will prove redundant, they would be eliminated by the FS mechanism.

In order to further improve the performances of the system, but also to adapt the system to various weather or illumination situations, different fusion schemes, combining VIS and IR information, were proposed. Before presenting how these steps have been accomplished in the developed experiments, the databases will be described.

4.1 Image Databases for Our ODR System

4.1.1 The Tetravision Image Database

A visible-infrared image database (i.e. the Tetravision) was used in the most recent experiments to recognize the type of obstacle previously determined as ROI by a stereo-vision OD module. In the frame of the IV field there are very few systems based only on passive sensors and even less performing VIS and IR information fusion. Among them, the Tetravision system proposed at VisLab [50, 51] at the University of Parma, was designed for pedestrian detection from four stereo correlated VIS—IR images. A Tetra-vision configuration, comprising stereo CCD cameras and stereo un-cooled FIR cameras (working in the $7–14\,\mu$m spectrum) was used.

The annotated database was small, having only 1164 objects, but it is a very difficult one, because of the high intra-class variability; for each class of objects, there are three types of poses: Entire (E), Occluded (O) and Group (G) and two viewing angles: Frontal (F) and Lateral (L) (Fig. 5). We performed experiments for the classification problems with 4 (P,V,C,B) and 8 (PE, PO, PG, VE, VO, VG, C, B) classes of objects. The database was randomly divided into a training set (80 %) and a testing set (20 %), the class instances being well balanced between the training and testing sets; unfortunately, there is not a balanced distribution of objects in classes: the cyclist class represents only 4.8 % of objects in our database. The Field of View (FOV) of the stereo-vision cameras was with almost 0.10 rad greater on each axis than that of the FIR camera, so the obstacles does not have the same size in VIS and IR images. Further details and information can be found in [27].

4.1.2 The Robin Image Database

Several companies and research centres, like Bertin technologies, CNES, Cybernetix, DGA, EADS, INRIA, ONERA, MBDA, SAGEM, THALES have engaged in 2006 in the Robin project. This competition was for the evaluation of object detection, object recognition and image categorisation algorithms. There were six datasets, with two main competitions for each dataset: some detection tasks (object on a patch) and some classification tasks (assign a category to a patch containing a centred object).

Fig. 5 Different images from the tetravision database

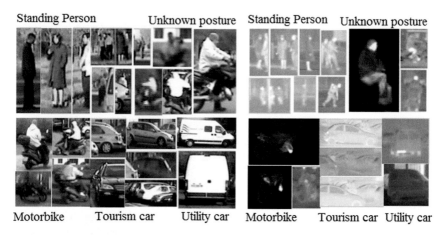

Fig. 6 Objects belonging to the 5 classes (standing person, unknown posture, motorbike, tourism car, utility car) from VIS and IR domains of Robin database

We subscribed for the dataset produced by Bertin-Cybernetix, where the proposed dataset was made of colour and infrared images of vehicles and pedestrians (Fig. 6).

The VIS and IR images (with a resolution of 128×128 pixels) were not correlated each other, which means that a scene captured in the VIS domain does not necessary have a correspondent in the IR domain. Our task was the discrimination of humans and vehicles, so the goal was to assign the correct label to a patch which may contain an element of a class or some backgrounds. At the contest, the class of detected object for each BB has to be decided. The Bertin-Cybernetix dataset contains a lot of images, and every image represented a possible road scene, with one or more objects therein (groups). The experiments performed on the Robin database comprised two possible scenarios: classifying with 2 classes (P, V) or with 5 classes (Standing Person, Unknown posture, Motorbike, Tourism car and Utility car).

Fig. 7 Different cars and
pedestrians from the images
database Caltech

4.1.3   The Caltech Image Database

The images from the first experiments we developed were most of them provided
by the Caltech Database. A lot of indoor and outdoor images containing different
objects could be found there, but we considered only images with cars or pedestrians
in different arbitrary poses, like Fig. 7 shows. The images were manually selected,
cropped and then resized at a dimension of 256×256 pixels. In order to increase the
processing speed of the entire algorithm, we considered images in gray level format.

The first experiments we have realized were developed on the Caltech database,
thus only on VIS images. Next, as we obtained the Robin dataset, with also IR
images, even not-correlated with the VIS ones, we developed the first experiments
on IR domain. Finally, when the Tetravision database has been obtained, the fusion
schemes we proposed could be tested.

4.2   Features Extraction

First, the features used to represent the image content in a digital or numeric format are obtained as FVs. The extracted features could be then compacted to reduce
the size of the image space representation (in the entire image database) by a FS
procedure. Numerical attributes generally describe the colorimetric and/or the geometric properties of the images or of some regions within the images. The choice
of these attributes influences the classification results and the recognition process.
Transforming the visual information (which humans observe easily in images) in
some numerical values, features or attributes of low level (primitives) is not an easy
thing, due to the fact that there are no studies indicating what particular type of
attribute is good (i.e. it succeed in capturing the most relevant information) in any

object recognition problem. Also, there is little or no research to indicate that types of feature families are more appropriate on different specific modalities (i.e. in the VIS and the IR domains). For computing the features, different families of features, which then could be combined to ensure a wider representation of the image content, have been aimed. Generally, different types of attributes capture different information from the images (this is valid even for attributes belonging to the same family of features). Thus, to represent the image content, some intuitive, generic and low level features were used, such as color, texture and shape.

Color is a commonly used feature in the CV domain, especially to recognize objects from nature, due to the multitude of colors that can represent different objects; therefore, it can help in the segmentation or classification process. In the context of IV applications, where IR images are represented by different gray levels, and VIS images also suffer a reduction of information due to the existence of situations like fog, night, etc. images on a single channel (in gray levels) have been considered.

Shape attributes are very useful for representing objects when some a priori information is known about the shape of the object. For example, there are a multitude of applications that use shape features specific to the pedestrian class (it is known that a pedestrian should have a roughly circular area representing the head; also a pedestrian must fall into certain patterns concerning the ratio height/width). By extracting some features that characterize objects in a general manner, i.e. globally, we believe better results can be obtained than those based on shape (symmetry, snakes, template matching), in which all shape of the object must be included in the BB in order to be recognized in the OR stage.

Since FE is desired to be fast for real-time constraints, the performances of the entire system depend heavily on the chosen features. We choose to extract obstacle shape independent but fast to compute features, so we have concentrated on different texture-based features. We did not select symmetries or edges because they are slower and it might not work very well for obstacles with arbitrary poses or presenting occlusions.

4.3 Features Extraction Experiments

In the experiments we performed, different families of texture-based features were investigated for VIS and respectively IR images for monomodal (Caltech), or bimodal systems with the possibilities that images were correlated eachother (Tetravision) or not (Robin).

The experiments were conducted to find out if some features could be better suited for the VIS domain and others better suited for the IR domain, because the final goal was to prepare for the VIS—IR fusion. Also, their combination was considered, i.e. bimodal FVs, to improve the recognition performance, considering their complementarity. First, the purpose was to find which features are more appropriate for VIS and respectively for IR modality. Next, which of them are less time consuming, and finally how to combine them in a proper manner to achieve best results?

To extract the features characterizing an obstacle, texture-based features were aimed, because they are shape independent and fast to compute; some statistical moments and size-related features were also added.

4.3.1 Extracted Feature Families

Width and height of the initial BB enclosing the object were chosen to be part of the FV because some of the applied transformations deformed the image by a resize operation. In order to preserve the initial size of the BB, we retained width and height (2 features for size, denoted *geom*). A vehicle will have a height approximately equal to the width, or lower, while for a pedestrian these characteristics would be exactly the opposite. However, considering that in the image-databases we used there are also cyclists and backgrounds, or different kind of vehicles, and objects could be occluded (so not the entire shape of the object will be comprised in the BB), or grouped (so there will be more objects belonging to the same class in a single BB), unfortunately these 2 features will not have the discrimination power that one may think.

The mean, median, mode, variance, standard deviation, skewness and kurtosis are the *statistic moments* (*7stat*) concerning the gray level information we have also used.

Moments of Hu were also aimed, due to the fact that global properties of the respective image could be exploited. A significant work considering moments for pattern recognition was performed by Hu [52] by deriving a set of seven invariant moments, using non-linear combinations of geometric moments. These invariants remain the same under image translation, rotation and scaling.

The *wavelet families* were more extensive experimented in our work. In a first attempt, wavelet families were aimed to construct the FV and they have been tested with different mother wavelet and different scales on the first database we achieved, i.e. the Caltech database. Wavelet transform was a relatively new analysis technique and replaces the Fourier transform sinusoidal waves by a family generated by translations and dilations of a window called mother wavelet. A two-dimensional Discreet Wavelet Transform (DWT) leads to a decomposition of approximation coefficients at level j in four components: the approximations at level j+1 and the details in three orientations (horizontal, vertical, and diagonal). Different wavelet families, like Daubechies, Coiflet and biorthogonal wavelets, but also fractional B-spline functions were used to compute different FVs. Different types of fractional B-splines wavelets also have been investigated: causal, anti-causal, symmetric and generalized. By varying a parameter of the mother wavelet, a direct control over a number of key wavelet properties can be obtained: the parametric form of the basis functions, their smoothness, their space-frequency localization, but also the size of the basis functions. The DWT of a signal was calculated by passing it through a series of filters (high and low pass filters) and then downsampled. At each level, the signal was decomposed into low and high frequencies, and this decomposition halved the resolution since only half the number of samples were retained to characterize the obstacle. To apply the wavelet decomposition, for some mother wavelets a resize operation was need. Generally, a 3, 4 and 5 level decomposition was performed: for one image of 128×128 pixels, if a Haar wavelet transform was used, then at the first level of decomposition resulted 64×64 pixels, at the second level 32×32 pixels, and so on.

Next, besides features like Haar wavelet, the *Gabor wavelet* have also been considered, because both types of wavelets offer complementary information about the pattern to be classified and have proved good performance in other systems [53]. The mean and the standard deviation of the magnitude of the Gabor coefficients were calculated for 4 scales and 4 orientations, obtaining thus 32 *gbr* features.

The *Discrete Cosine Transform* (DCT) tends to concentrate information, being intensively used for image compression applications. The first nine DCT coefficients are suggested to be used as texture features, but inspired by [54] the base component was ignored. Therefore, we obtained a number of 8 *dct* features.

For the grayscale images, the co-occurrence matrix characterizes the texture of the image and the generated coefficients are often called Haralick features. Only 4 of 7 are generally proposed to be used: the homogeneity, entropy, contrast and correlation. The *Gray Level Co-occurrence Matrix* (GLCM) is used to explore the spatial structure of the texture and it captures the probability that some pixels appear in pairs with the same level of gray but with different orientations. We performed the computation in 4 different directions: $0°$, $45°$, $90°$ and $135°$ as it is proposed in [54]. In this manner, we obtained a number of 16 *cooc* features.

The *Run Length Encoding* (RLE) method works by reducing the physical size of a repeating string of characters, i.e. sequences in which the same data value occurred in many consecutive data elements are stored as a single data value and counted. For a given image, the proposed method defines a run-length matrix as number of runs (i.e. the number of pixel segments having the same intensity) starting from each location of the original image in a predefined direction. Short run emphasis, long run emphasis, gray-level distribution, run-length distribution and run percentage are the five features proposed by Galloway. Two supplementary measures (low gray-level run emphasis and high gray-level run emphasis) have also been considered. Thus, a set of 7 features obtained in one direction have been chosen, but performed at $0°$ and $90°$ as proposed in [54] yield a number of 14 *rle* features.

Some signal processing techniques are based on texture filtering and analyze the frequency contents in the spatial domain. Laws have suggested a set of 5 convolution masks for FE based on texture. From these 5 masks, a set of 25 two-dimensional masks have been further obtained and based on these 2D masks, 14 laws features are reached. These features are then reported to the elements from the first diagonal, in the following manner: the first 10 features are normalized with the first element from the diagonal, and the rest of 4 features are normalized with the remaining 4 diagonal elements. To these 14 features, the mean and the standard deviation have been applied as it is suggested in [54], resulting thus a number of 28 *laws* features.

4.3.2 Constructing Different Feature Vectors

In the developed experiments, in this stage our main purpose was to obtain a small
FV and a good classification rate (above 90 %).

Comparing the results we obtained, the level 4 of wavelet decomposition gives the
best solution for the obstacle recognition problem considering the size of the FV and
the achieved accuracy rates (recognition rates) on the Caltech database [52]. Also, the
results were better in the case of using 2 classes for the classification (the variability
between classes was smaller than in the case with 5 classes) when experimenting with
the Robin database [55]. When comparing different types of wavelets, the accuracy
given by Daubechies and Biorthogonal wavelets were better than their fractional
counterparts, while causal and generalized fractional wavelets were better than the
anti-causal and symmetric ones [56, 57].

To construct the FV, the wavelet features were combined with moments (i.e. within
a features fusion). The mother wavelet used to compute the wavelet decomposition
were causal and generalized B-spline functions (22 causal and 22 generalized), with
different scaling and translation parameters in [52]. Thanks to the fractional B-spline
functions the FVs dimension were reduced from 18×18 features (corresponding to
the 4th level of wavelet decomposition) to $10 \times 10 + 7 + 7$ features (corresponding
to the 5th level of wavelet decomposition combined with the 7 statistical moments
and the 7 moments of Hu) for the same classification percentage or even better. The
proposed FVs were tested using a Bayes classifier, a Bayes Net and a Radial Basis
Function (RBF) Network with a normalized Gaussian RBF. By adding moments,
one level of wavelet decomposition has been reduced, i.e. from a FV comprising 324
features, the new FV was having only 114 features, so a third the size of the initial
FV for the same or even better accuracy. In a further experiment, presented in [58]
Daubechies, Coiflet and biorthogonal wavelets were added, resulting in a number of
29 supplementary functions from which to compose the FVs. Like in the previous
case, when testing on Caltech database, the moments demonstrated their power to
increase the accuracy.

The FV from [55] corresponds to the height and width of the original BB of
the object, the 64 Haar wavelet coefficients, the 7 statistical moments, the DCT
coefficients, the GLCM coefficients and the Gabor coefficients. The recognition
rates obtained when experimenting on Robin database were the best in the case of
combining all features together. If compare the accuracy given by the KNN and SVM
classifiers with a 10f-CV test mode, the SVM was much better than the KNN in all
cases. Also the accuracy rates for 5 classes of objects were lower than for 2 classes
of objects, due to the fact that the variability between classes was higher.

When considering the Tetravision database, we concentrated on the Haar wavelet,
obtaining thus 64 features from each modality, VIS and IR. Because the images
representing objects have very small size (especially in VIS), the wavelet decomposition was chosen to be performed at level one (for VIS images) or level two
(for IR images), and finally we obtained a number of 8×8 wavelet coefficients for
both types of images. To perform the fusion, for the image representation we choose
the width and height of the original BB, the 7 statistical moments, the wavelet and

the gabor transforms, the *dct*, *cooc*, *rle* and *laws* coefficients. Therefore, a number of 171 features have been extracted from each modality VIS and IR: VIS171 and IR171 feature vectors were finally retained, as presented in [27]. Due to the fact that the information is extracted individually from the VIS and IR images, the provided FVs are called monomodals.

4.3.3   Feature Vectors Evaluation

Different algorithms of FE provide different characteristics (as we already mentioned, grouped in feature-families) which can be combined in different FVs, representing the inputs into the classifier. The accuracy of the classifier depends on how well these features succeed in representing the information and it is not necessary proportional with their number (or FV dimension). Is it possible that the same FE algorithm applied on the VIS and on the IR domains to deliver distant results, i.e. to exist some features better suited for the VIS domain and others better suited for the IR domain. Also, their combination can bring in some improvements from the viewpoint of the recognition performance, depending on how complementary they are when representing the information. There are FE algorithms consuming less time than others at the extraction of these features from images. There are also families of features that can be separable (when calculating the coefficients of a family, they can be calculated individually, and do not need to be calculated all if we do not need all of them) and this will influence the extraction time of those coefficients.

To assess the performance representation of the numerical attributes, in this section we present the results of an experiment in which we tested, using a simple classifier KNN the representation ability of the visual content of each family of attributes. It does not need a model-selection stage, as the SVM does, because it is not having multiple parameters to be optimized before the usage. Still, because SVM is more parameterizable and therefore better adapted to any classification problem, it is expected that the recognition rates to be higher by the use of the SVM. First, the concern was not to optimize the classifier on each family or combination of feature-families, but to evaluate their individual importance.

Few questions were foreseen when preparing for fusion: 1. Are several features better adapted for VIS and other better adapted for IR? Or, if a family is behaving well on VIS, it will be also good on IR? 2. The number of features of one family influences the classification rate? A family with many features will provide a greater recognition rate compared to another family having less features? 3. Are the chosen features pertinent for the learning process? Or they will suffer of overfitting (will provide good results on the training set, but they would not predict very well the test data)?

In the following, the importance of these coefficients but also the individual performance of each family of features has been evaluated. To maximize the performance of individual descriptors, new vectors have been formed as combinations of feature families. Thus, we have combined the texture descriptors in a single FV of texture (*Text*), comprising *haar*, *dct*, *cooc*, *gbr*, *rle*, *laws* and including 162 characteristics for each VIS and IR modalities. Adding the 7 *stat*istical moments, a new vector called (*StatText*) is obtained. If in addition, we add the 2 *geom*etrical features, the maximum size vectors (denoted *AllFeatures*) of 171 features has been obtained. In order to answer these questions, we used all the vectors comprising families or combination of feature-families and we performed 2 experiments for the classification problems with 4 and 8 classes of objects. In a first experiment we considered only the training dataset (932 objects), and the obtained accuracies for the classification problem with 4 classes of objects was with approximately 10 % higher than those obtained for the classification problem with 8 classes of objects (due to the reduced number of instances per each object class).

In a first attempt, we used a KNN classifier, because the model selection was avoided on purpose (the features complexity was aimed in this first set of experiments). The obtained classification results on IR domain were slightly better than those from the VIS one, but this could be due to the fact that in the dataset the images from IR domain have a higher resolution compared with their VIS counterparts. We have also noticed that if a family of features behaves well on IR images, it was providing also good results on VIS ones. From the obtained results, presented in Table 1, can be observed that the families *haar*, *gbr*, *laws* and *dct* are better than *stat*, *cooc*, *geom* and *rle*. However, the first group of families (except *dct*) had the largest number of features, therefore the increased accuracies could be of that reason. In order to highlight this aspect, we proposed a further careful investigation of individual and correlated features contribution, by using several FS techniques. The obtained results indicate that a finer selection process had to be performed, at the features level (not at their families).

Table 1 Performance representation of monomodal FVs obtained using 10f-CV on the training set for the classification problem with 4 classes of objects

Input vector			Accuracy using 10f-CV with KNN classifier		Inputs by decreasing bAcc for VIS		
Attributes					Input vector	Accuracy with KNN classifier	
Name		Number	VIS	IR		VIS	IR
geom		2	47.50	47.50	haar	77.00	79.60
stat		7	58.98	66.03	gbr	72.60	81.55
Texture	haar	64	77.00	79.60	laws	67.38	69.75
	gbr	32	72.60	81.55	dct	65.65	75.00
	dct	8	65.65	75.00	stat	58.98	66.03
	cooc	16	54.23	66.10	cooc	54.23	66.10
	rle	14	42.95	55.00	geom	47.50	47.50
	laws	28	67.38	69.75	rle	42.95	55.00
Text		162	83.75	87.13			
StatText		169	83.45	87.90			
AllFeatures		171	83.67	88.00			

On IR, although the Haar wavelet coefficients are most numerous, they are exceeded in their performance by Gabor features which are only half as concerning their number. From the viewpoint of the performance, after *gbr* and *haar* features, on the 3rd and 4th positions are *dct* and *laws* features, accounting 5 %, respectively 16 % of the total vector, followed by *cooc* and *stat*, and finally *rle* and *geom*. The *geom* features do not have to be ignored, because with only 1 % of the features, they succeed to obtain an accuracy of about 50–60 %, as presented in Table 1. The conclusion is that the number of features of one family does not necessary assure a proportional higher classification rate: there are families with fewer features providing a greater recognition rate than another family having more features (e.g. gbr vs. haar).

The danger of overfitting is to find features that explain well the training data, but have no real relevance or no predictive power (for the test set). Generally, one can notice that the accuracies obtained in the learn-test (LT) stage overperformed (or are very closed to) the values obtained using the 10f-CV procedure, so our data is not presenting overfitting. Therefore, we can say we have chosen some general features, which are capable to retain the pertinent information from both VIS and IR individual domains.

Next, when running the experiments with the SVM classifier, we have focused on the feature vector *AllFeatures*, incorporating all the 171 features, because only after the FS process we will drop some features if they did not help in the classification process, i.e. if they have been found as being not relevant.

4.4 Features Selection

In the experiments, we have chosen to extract features as rich and diverse as possible in order to take advantage of their complementarity, but we did not ignore the possibility of some redundant information, which therefore has to be eliminated in order to decrease: the learning complexity and the classification time, but also the extraction time. In fact, even the number of features is not very high for a given BB, when considering all the BB hypotheses within an image, it could reach a significant value and increase significantly the extraction time. The FS step has thus to find a compact, relevant and consistent set of features for the classification task.

Different FS methods have been tested in order to, first, evaluate the pertinence of features individually allowing for a ranked list of features and, second, to evaluate sub-sets of features in order to take into account the correlation between features. Chi Squared, Information Gain, ReliefF, Significance and Symmetrical Uncertainty have been first considered, as single-attribute evaluators (Rankers), while for the second round Correlation and Consistency-based subset-attribute evaluators were used. In the latter set of FS methods, the Best First, Linear Forward and Genetic Search were used as subsets generation methods, while for the Ranker ones a thresholding step was needed.

Usually, the FS methods are applied only once on the whole training set, but for a better robustness, we propose to apply the FS methods also according to a cross-validation scheme. Further details can be found in [27].

4.5 Features Selection Experiments

The FS could be applied independently on VIS and respectively IR vectors or on the correlated bimodal VISIR vector.

From all tested FS methods, we have retained only those which accomplished a bi-level optimization criteria, allowing higher classification accuracy on a smaller number of features: a Ranker one (FsR_{75}) and a Search one (FsS_{CV}), as shown in Table 2. The FsR_{75} uses Information Gain to evaluate each feature independently before providing a ranked list of features. After combining in a single list of relevance, a percentage of 75 % features have been further selected. The retained features are the ones which appeared at least in one of the 100 selection process (by using a 100f-CV scheme). Even if a family of features provide good results when all features were considered in a FV, it is possible that no feature of that family will be retained in the FS process, like is the case of *dct* family on the VIS modality with the FsS_{CV} method. Once the selected FVs obtained, the VISIR fusion process followed.

4.6 Fusion

We compared from the viewpoint of accuracy and robustness, three different probabilistic fusion schemes: at the feature level, at the SVM kernels level and respectively at the matching-scores level. A first fusion scheme is proposing the fusion to be performed at the features level, therefore at a low-level. This fusion would be obtained in the frame of the module which realize the FE and FS operations, and for this reason, it could be performed in two possible ways: between the two modules or after both of them. Another proposed fusion scheme is a high level one, being realized at the outputs of the VIS and IR classifiers, combining thus matching-scores. Two possible ways could be reached here too: a not-adaptive fusion and an adaptive fusion. The last proposed fusion scheme realizes the combination of the VIS and IR information at an intermediate level, i.e. at the SVM kernels. Further details and information about the theory part we considered for the fusion schemes can be found in [27].

For the early *features fusion*, the results obtained with the bimodal *AllFeatures* and selected FVs are shown in Table 3, where also the characteristics of the optimized kernel are given. Single Kernel (SK) is the simple, i.e. classical kernel within the SVM classifier, possible types being polynomial, RBF, sigmoidal, etc. The parameters of the SK are also given: (SKtype and its hyper-parameters SKpar, and C), where C is the complexity parameter. The results of a static adaptation of features fusion (sAFF) scheme having the optimal modality-feature relative weight $\alpha*$ are also given. All

Table 2 Comparison of feature representations

FS method	No. of selected features	Domain	Selection-percentage on each family of features							
			2geom (%)	7stat (%)	64haar (%)	32gbr (%)	8dct (%)	16cooc (%)	14rle (%)	28laws (%)
FsR_{75}	256: $VIS_{141} + IR_{115}$	VIS	50	85.7	71.9	**100**	75	87.5	**100**	75.0
		IR		28.6	45.9	**100**	**100**	75	85.7	67.9
FsS_{CV}	126: $VIS_{65} + IR_{61}$	VIS	50	28.6	35.9	43.8	0	25	64.3	42.9
		IR		14.3	40.6	50	62.5	18.8	42.9	14.3
bAcc for KNN (n = 1) on all features for the respective family			47.5	59.0	77.0	72.6	65.7	54.2	43.0	67.4
			47.5	66.0	79.6	**81.6**	75.0	66.1	55.0	69.8

Table 3 Classification performance for the proposed fusion schemes

System input vectors		Monomodal systems		Bimodal systems					
		NO Fusion		Features fusion		Kernels fusion		Scores fusion	
Pb.	Modality	$\overline{bAcc\%}$	Optimal SK	$\overline{bAcc\%}$ sAFF	Optimal SK	$\overline{bAcc\%}$ sAFK	Optimal BK	$\overline{bAcc\%}$ sAFSc	$\overline{bAcc\%}$ dAFSc
AllFeat	VIS_{171}	94.7	SK1	97.3	SK6	96.7	BK1	**98.7**	98.4
	IR_{171}	94.9	SK2		$\alpha^* = 0.4$		$\alpha^* = 0.5$	$\overline{\alpha^*} = 0.5$	–
	$VISIR_{340}$	–	–					$\sigma = 0.13$	–
FsR_{75}	$sVIS_{141}$	95.6	SK3	97.0	SK7	96.1	BK2	98.4	98.3
	sIR_{115}	95.5	SK2		$\alpha^* = 0.4$		$\alpha^* = 0.6$	$\overline{\alpha^*} = 0.4$	–
	$sVISIR_{256}$	–	–					$\sigma = 0.12$	–
FsS_{CV}	$sVIS_{65}$	**96.2**	SK4	**97.7**	SK7	**96.9**	BK3	**98.7**	**98.7**
	sIR_{62}	95.2	SK5		$\alpha^* = 0.5$		$\alpha^* = 0.5$	$\overline{\alpha^*} = 0.5$	–
	$sVISIR_{126}$	–	–					$\sigma = 0.13$	–

the results obtained with the early-fusion scheme outperformed those obtained with the monomodal system. The best classification performance (bAcc of 97.7 %) was obtained with the most compact bimodal sVISIR126 representation determined with our FsS_{CV} method.

The *kernel fusion* has been also experimented, and it implies learning a BK (bimodal kernel), as a linear combination of monomodal SKs, involving thus some parameters to be optimized: BKtype, BKparamVIS, BKparamIR, C, α. Since there are much more parameters to be optimized, the kernel fusion is more flexible than the early fusion, and it is thus more promising, because it should fit better to VISIR heterogeneous data. As it could be noticed from Table 3, the best performance (96.9 %) was obtained again with the very compact bimodal FsS_{CV} representation with a static adaptation fusion of kernels (sAFK) scheme. The results obtained with the intermediate-fusion scheme are better than those obtained with the monomodal system, but they are unfortunately lower than those obtained with the early-fusion system, we believe due to the fact that in our image set there is not enough data to train such a complex fusion model.

The *matching-scores fusion* involves learning, in an independent manner, two monomodal SKs and their hyper-parameters, before estimating the relative modality-score weight α. For the dynamic dAFSc adaptation case, the optimal value α could intervene with different values. The static adaptation weight was learned on the validation set, while the dynamic one was estimated for each test object, on the fly, during the classification process, in Table 3 being given the average optimal value. The matching-score fusion scheme outperforms monomodal VIS and IR systems, but also feature and kernel-fusion bimodal systems, with the best performance (98.7 %) being obtained for both static and dynamic adaptation approaches, with the sVISIR126 representation.

In the case of monomodal systems, the following single kernels were selected: SK1 = (RBF,2^{-15},2^7), SK3 = (RBF,2^{-19},2^{11}) and SK4 = (POL,1,0.5) for VIS domain, and SK2 = (RBF,2^{-19},2^9) and SK5 = (RBF,2^{-17},2^9) for IR domain. The single kernels SK6 = (RBF,2^{-19},2^7) and SK7 = (RBF,2^{-17},2^5) corresponds to bimodal systems, but also the bimodal kernels BK1 = (RBF-RBF,2^{-19},2^{-17},2^7), BK2 = (RBF-RBF,2^{-19},2^{-17},2^8) and BK3 = (RBF-RBF,2^{-17},2^{-19},2^7).

The obtained results emphasis the fact that our FsS_{CV} feature selection method allows us to determine a compact, but pertinent signature of an object, which improves not only the precision and the robustness, but also the computation time of the classification process. Even that both matching-score and kernel-fusion require the estimates of two kernels and their hyper-parameters, the problem is easier for the late fusion because the optimization for one SK and its hyper-parameters is independent for the other SK, which is not the case of the kernel-based fusion scheme, being based on a BK instead of two SKs. We believe the latter is more promising, because it use two correlated kernels and theirs hyper-parameters are necessary, thus requiring a greater and better balanced image database. In the Tetravision database we used for the experiments, there are no environmental illumination or weather changes which could require a dynamic adaptation scheme, explaining thus the small standard deviation for the dynamic weight dAFSc and the value of 0.5 for the optimal static adaptation weights of feature fusion (sAFF), of kernel fusion (sAFK) and respectively of scores fusion (sAFSc).

4.6.1 Details About the Experiments

Weka (stands for the Waikato Environment for Knowledge Analysis) includes a collection of ML algorithms for data mining tasks, and it is developed in Java (belongs to University of Waikato in New Zealand). We use Weka in order to apply a learning method, i.e. a classifier (a supervised one) to our dataset and analyze its output. In Weka, the performance of all classifiers is measured by a common evaluation module, by a confusion matrix. Before applying any classification algorithm to the data, it must be converted to ARFF form with type information about each attribute. The summary of the results from the training data ends with the confusion matrix, showing how many instances of a class have been assigned to each possible classes. If all instances are classified correctly, only the diagonal elements of the matrix are non-zero. There are several different levels at which Weka can be used: it provides implementations of state-of-the-art learning algorithms that can be applied to a dataset or a dataset can be preprocessed, fed into a learning scheme, and analyze the resulting classifier and its performance.

Weka is very easy to use, without writing a line of code. Still, in many situations writing some lines of code could help, especially when multiple combinations of datasets and algorithms are desired to be compared. In this situation, we recommend in addition the use of Matlab or Python as a platform to write code sequences and to run Weka algorithms from there. The Weka class library can be thus accessed from an own Matlab/Python/Java program, and new ML algorithms could also be implemented.

In the experiments we performed, we used MATLAB Version for LIBSVM [59] and our own developed MATLAB Interface for WEKA [60].

In a 10-fold cross-validation process (10f-CV), the original sample of data is partitioned into 10 sub-samples. From these 10 sub-samples, a single sub-sample is retained (the validation or testing set), and the remaining 9 sub-samples are used to train the classifier (the training set). The cross-validation process is then repeated 10 times and a combination of the 10 results (generally the average) is performed in order to obtain the accuracy parameter.

4.7 Conclusion

In this paper, the features extraction, features selection and fusion techniques were applied on different VIS and IR databases of images, separately or combined, in order to encode the information about possible obstacles from the road. Feature selection methods were used to select only the more discriminant features from the entire feature vector. The probabilistic fusion schemes were used to further improve the results at the features level, at the SVM kernels level and respectively at the matching-scores level. All these proposed schemes are less susceptible to the un-calibrations of on-board cameras compared to the ones performed at the lowest level of information (i.e. at data or pixel level).

Tomorrow vision is that sensors, communications and other operational technologies will work together with information technologies, to contribute to a new revolution of systems belonging to IoT. A common data model and sensing and control architecture that supports the flow of insights is desired, because devices will be more intelligent. However, a new breed of robots is envisioned, and they are supposed to even understand human emotion, estimate a driver or patient ill state and act accordingly. Thus, in this chapter a possible recipe has been given for CV-based applications, where machine learning algorithms proved their efficiency. The proposed system could learn from available databases some features about the obstacles and new obstacles could be further recognized based on it. To accomplish such a complex task, methods for features extraction, selection and fusion have been briefly presented, but also state of the art solutions from the intelligent vehicles domain were given.

The proposed chapter provide an insight on the available background needed to construct intelligent computer vision applications for the imminent era of Internet of Things as concerns scene understanding but also human-computer interaction from the vision side.

References

1. Who is NAO? https://www.aldebaran.com/en/humanoid-robot/nao-robot. Accessed 30 Aug 2015
2. Asimo of Honda, http://asimo.honda.com/. Accessed 30 Aug 2015
3. Atzori, L., Iera, A., Morabito, G.: The internet of things: a survey. J. Comput. Netw., Elsevier **54**(15), 2787–2805 (2010)
4. daCosta, F.: Rethinking the Internet of Things. Apress, A Scalable Approach to Connecting Everything (2014). ISBN 13: 978-1-4302-5740-0, 192 Pages
5. Evans, P.C., Annunziata, M.: Industrial internet: pushing the boundaries of minds and machines, General Electric. http://www.ge.com/docs/chapters/IndustrialInternet.pdf (2012). Accessed 30 Aug 2015
6. General Electric and Accenture, Industrial Internet insights report (2014). Accessed 30 Aug 2015
7. Bertozzi, M., Broggi, A., Cellario, M., Fascioli, A., Lombardi, P., Porta, M.: Artificial Vision in Road Vehicles, vol. 90, pp. 1258–1271. IEEE Proc. (2002)
8. Li, L., Song, J., Wang, F.Y., Niehsen, W., Zheng, N.N.: New developments and research trends for intelligent vehicles. IEEE Intell. Syst. **4**, 10–14 (2005)
9. Bu, F., Chan, C.Y.: Pedestrian detection in transit bus application-sensing technologies and safety solutions. IEEE Intell. Veh. Symp. 100–105 (2005)
10. Chan, C.Y., Bu, F.: Literature review of pedestrian detection technologies and sensor survey. Tech. rept. California PATH Institute of Transportation Studies, Berkeley, CA. Midterm report (2005)
11. Cutler, R., Davis, L.S.: Robust Real-Time Periodic Motion Detection, Analysis, and Applications, vol. 22, pp. 781–796, PAMI (2000)

12. Enzweiler, M., Gavrila, D.M.: Monocular Pedestrian Detection: Survey and Experiments, vol. 31, p. 2179–2195. IEEE TPAMI (2009)
13. Gandhi, T., Trivedi, M.M.: Pedestrian collision avoidance systems: a survey of computer vision based recent studies. In: IEEE Conference on Intelligent Transportation Systems, pp. 976–981 (2006)
14. Scheunert, U., Cramer, H., Fardi, B., Wanielik, G.: Multi sensor based tracking of pedestrians: a survey of suitable movement models. In: IEEE International Symposium on Intelligent Vehicles, pp. 774–778 (2004)
15. Sun, Z., Bebis, G., Miller, R.: On-road Vehicle Detection: A Review, vol. 28, pp. 694–711. IEEE TPAMI (2006)
16. McCall, J.C., Trivedi, M.M.: Video-based lane estimation and tracking for driver assistance: survey, system, and evaluation. IEEE Trans. Intell. Transp. Syst. **7**(1), 20–37 (2006)
17. Bishop, R.: A survey of intelligent vehicle applications worldwide. In: Intelligent Vehicles Symposium (2000)
18. Baldwin, K.C., Duncan, D.D., West, S.K.: Johns Hopkins APL technical digest. 25th ed. Johns Hopkins University Applied Physics Laboratory. The driver monitor system: a means of assessing driver performance, pp. 1–10 (2004)
19. Dong, Y., Zhencheng, H., Uchimura, K., Murayama, N.: Driver inattention monitoring system for intelligent vehicles: a review. IEEE Trans. Intell. Transp. Syst. **12**(2), 596–614 (2011)
20. DARPA Grand Challenge, Urban Challenge. http://archive.darpa.mil/grandchallenge/index.asp (2007)
21. Broggi, A., Medici, P., Zani, P., Coati, A., Panciroli, M.: Autonomous vehicles control in the VisLab intercontinental autonomous challenge. Annu. Rev. Control **36**, 161–171 (2012)
22. A VisLab prototype, www.braive.vislab.it/
23. Guizzo, E.: How google self-driving car works, IEEE spectrum online. http://spectrum.ieee.org/automaton/robotics/artificial-intelligence/how-google-self-driving-car-works (2011)
24. Honda. http://world.honda.com/HDTV/IntelligentNightVision/200408/
25. FLIR Application Story, BMW incorporates thermal imaging cameras in its cars. www.flir.com/uploadedFiles/ApplicationStory_BMW.pdf (2009)
26. Discant, A., Rogozan, A., Rusu, C., Bensrhair, A.: Sensors for obstacledetection—a survey. In: 30th International Spring Seminar on Electronics Technology. Cluj-Napoca, Romania (2007)
27. Apatean, A., Rogozan, A., Bensrhair, A.: Visible-infrared fusion schemes for road obstacle classification. J. Transp. Res. Part C: Emerg. Technol. **35**, 180–192 (2013)
28. Toulminet, G., et al.: Vehicle detection by means of stereo vision-based obstacles features extraction and monocular pattern analysis. IEEE Trans. Image Process. **15**(8), 2364–2375 (2006)
29. Miron, A., Besbes, B., Rogozan, A., Ainouz, S., Bensrhair, A.: Intensity self similarity features for pedestrian detection in far-infrared images. In: IEEE Intelligent Vehicles Symposium (IV), pp. 1120–1125 (2012)
30. Emerich, S., Lupu, E., Rusu, C.: A new set of features for a bimodal system based on on-line signature and speech. Digit. Signal Process. Arch. **23**(3), 928–940 (2013)
31. Zhang, W.L., Liu, L.B., Yin, S.Y., Zhou, R.Y., Cai, S.S., Wei, S.J.: An efficient VLSI architecture of speeded-up robust feature extraction for high resolution and high frame rate video. Sci. China Inf. Sci. **56**(7), 1–14 (2013)
32. Enzweiler, M., Kanter, P., Gavrila, D.M.: Monocular pedestrian recognition using motion parallax. In: IEEE Intelligent Vehicles Symposium, pp. 792–797 (2008)
33. Poppe, R.: Vision-based human motion analysis: an overview. Comput. Vis. Image Underst. **108**, 4–18 (2007)
34. Keller, C., Enzweiler, M., Gavrila, D.M.: A new benchmark for stereo-based pedestrian detection. In: IEEE Intelligent Vehicles Symposium (2011)
35. Viola, P., Jones, M.: Robust real-time face detection. Int. J. Comput. Vis. **57**(2), 137–154 (2004)
36. Dalal, N., Triggs, B., Schmid, C.: Human detection using oriented histograms of flow and appearance. In: European Conference Computer Vision, pp. 428–441 (2006)

37. Gavrila, D.M.: A Bayesian, exemplar-based approach to hierarchical shape matching. IEEE Trans. Pattern Anal. Mach. Intell. **29**(8), 1408–1421 (2007)
38. Broggi, A., Cerri, P., Antonello, P.C.: Multi-resolution vehicle detection using artificial vision. In: IEEE Intelligent Vehicles Symposium, pp. 310–314 (2004)
39. Mikolajczyk, K., Schmid, C., Zisserman, A.: Human detection based on a probabilistic assembly of robust part detectors. In: European Conference on Computer Vision, pp. 69–81 (2004)
40. Wu, B., Nevatia, R.: Detection and tracking of multiple, partially occluded humans by Bayesian combination of edgelet based part detectors. Int. J. Comput. Vis. **75**(2), 247–266 (2007)
41. Lowe, D.G.: Distinctive image features from scale-invariant keypoints. Int. J. Comput. Vis. **60**(2), 1–110 (2004)
42. Besbes, B., Rogozan, A., Bensrhair, A.: Pedestrian recognition based on hierarchical codebook of SURF features in visible and infrared images. In: IEEE Intelligent Vehicles Symposium, pp. 156–161 (2010)
43. Viola, P., Jones, M., Snow, D.: Detecting pedestrians using patterns of motion and appearance. Int. J. Comput. Vis. **63**(2), 153–161 (2005)
44. Gavrila, D.M., Munder, S.: Multi-cue pedestrian detection and tracking from a moving vehicle. Int. J. Comput. Vis. **73**(1), 41–59 (2007)
45. Wojek, C., Schiele, B.: A performance evaluation of single and multi-feature people detection. In: Proceedings of the 30th DAGM Symposium on Pattern Recognition, pp. 82–91. Springer,Berlin (2008)
46. Geronimo, D., Lopez, A.M., Sappa, A.D., Graf, T.: Survey of pedestrian detection for advanced driver assistance systems. IEEE Trans. Pattern Anal. Mach. Intell. **32**(7), 1239–1258 (2010)
47. Dollar, P., Wojek, C., Schiele, B., Perona, P.: Pedestrian detection: an evaluation of the state of the art. IEEE TPAMI **34**(4), 743–761 (2012)
48. Miron, A., Rogozan, A., Ainouz, S., Bensrhair, A., Broggi, A.: An evaluation of the pedestrian classification in a multi-domain multi-modality setup. Sensors **15**(6), 13851–13873 (2015)
49. Krotosky, S.J., Trivedi, M.M.: On color-, infrared-, and multimodal stereo approaches to pedestrian detection. IEEE Trans. Intell. Transp. Syst. **8**(4), 619–629 (2007)
50. Bertozzi, M., et al.: Lowlevel pedestrian detection by means of visible and far infra-red tetravision. In: IEEE Intelligent Vehicles Symposium, pp. 231–236 (2006)
51. Bertozzi, M., et al.: Multi stereo-based pedestrian detection by daylight and far-infrared camera. In: Hammoud, R., Augmented Vision Perception in Infrared: Algorithms and Applied Systems. Springer Inc., pp. 371–401 (Ch16) (2009)
52. Apatean, A., Emerich, S., Lupu, E., Rogozan, A., Bensrhair, A.: Ruttier obstacle classification by use of fractional B-spline wavelets and moments. In: IEEE Region 8 Conference Computer as a Tool (2007)
53. Sun, Z., Bebis, G., Miller, R.: Monocular precrash vehicle detection: features and classifiers. IEEE Trans. Image Process. **15**, 2019–2034 (2006)
54. Florea, F.: Annotation automatique dmages mdicales en utilisant leur contenu visuel et les rgions textuelles associates. Application dans le contexte dn catalogue de sant en ligne. Ph.D. thesis, Institut National Des Sciences Appliques, INSA de Rouen, France and Technical University of Bucharest, Romania (2007)
55. Apatean, A., Rogozan, A., Bensrhair, A.: Objects recognition in visible and infrared images from the road scene. In: IEEE International Conference on Automation, Quality and Testing, Robotics, vol. 3, pp. 327–332 (2008)
56. Apatean, A., Emerich, S.: Obstacle recognition by the use of different wavelet families in visible and infrared images. Carpathian J. Electron. Comput. Eng. **1**(1) (2008)
57. Apatean, A., Rogozan, A., Emerich, S., Bensrhair, A.: Wavelets as features for objects recognition. Acta Tehnica Napocensis -Electronics and Telecommunications, vol. 49, pp. 23–26 (2008)
58. Apatean, A., Emerich, S., Lupu, E., Rogozan, A., Bensrhair, A.: Wavelets and moments for obstacle classification. In: International Symposium on Communications, Control and Signal Processing, Malta, pp. 882–887 (2008)
59. http://www.csie.ntu.edu.tw/cjlin/libsvm/
60. http://www.cs.waikato.ac.nz/ml/weka/

Image Feature Extraction Acceleration

Jorge Fernández-Berni, Manuel Suárez, Ricardo Carmona-Galán,
Víctor M. Brea, Rocío del Río, Diego Cabello
and Ángel Rodríguez-Vázquez

Abstract Image feature extraction is instrumental for most of the best-performing algorithms in computer vision. However, it is also expensive in terms of computational and memory resources for embedded systems due to the need of dealing with individual pixels at the earliest processing levels. In this regard, conventional system architectures do not take advantage of potential exploitation of parallelism and distributed memory from the very beginning of the processing chain. Raw pixel values provided by the front-end image sensor are squeezed into a high-speed interface with the rest of system components. Only then, after deserializing this massive dataflow, parallelism, if any, is exploited. This chapter introduces a rather different approach from an architectural point of view. We present two Application-Specific Integrated Circuits (ASICs) where the 2-D array of photo-sensitive devices featured by regular imagers is combined with distributed memory supporting concurrent processing. Custom circuitry is added per pixel in order to accelerate image feature extraction right at the focal plane. Specifically, the proposed sensing-processing chips aim at the acceleration of two flagships algorithms within the computer vision community:

J. Fernández-Berni (✉) · R. Carmona-Galán · R. del Río · Á. Rodríguez-Vázquez
Institute of Microelectronics of Seville (CSIC - Universidad de Sevilla),
C/ Américo Vespucio s/n, 41092 Seville, Spain
e-mail: berni@imse-cnm.csic.es

R. Carmona-Galán
e-mail: rcarmona@imse-cnm.csic.es

R. del Río
e-mail: rocio@imse-cnm.csic.es

Á. Rodríguez-Vázquez
e-mail: angel@imse-cnm.csic.es

V.M. Brea · M. Suárez · D. Cabello
Centro de Investigación en Tecnoloxías da Información (CITIUS),
University of Santiago de Compostela, Santiago de Compostela, Spain
e-mail: victor.brea@usc.es

D. Cabello
e-mail: diego.cabello@usc.es

© Springer International Publishing Switzerland 2016
A.I. Awad and M. Hassaballah (eds.), *Image Feature Detectors and Descriptors*,
Studies in Computational Intelligence 630, DOI 10.1007/978-3-319-28854-3_5

the Viola-Jones face detection algorithm and the Scale Invariant Feature Transform (SIFT). Experimental results prove the feasibility and benefits of this architectural solution.

Keywords Image feature extraction · Focal-plane acceleration · Distributed memory · Parallel processing · Viola-Jones · SIFT · Vision chip

1 Introduction

1.1 Embedded Vision

Embedded vision market is forecast to experience a notable and sustained growth during the next few years [1]. The integration of hardware and software technologies is reaching the required maturity to support this growth. At hardware level, the ever-increasing computational power of Digital Signal Processors (DSPs), Field Programmable Gate Arrays (FPGAs), General-Purpose Graphics Processing Units (GP-GPUs) and vision-specific co-processors permit to address the challenging processing requirements usually demanded by embedded vision applications [2]. At software level, the development of standards like OpenCL [3] or OpenVX [4] as well as tools like OpenCV [5] or CUDA [6] allow for rapid prototyping and shorter time to market.

A noticeable trend within this ecosystem of technologies is hardware paralleliza-tion, commonly in terms of processing operations [7, 8]. However, improving perfor-mance is not only a matter of parallelizing computational tasks. Memory management and dataflow organization are crucial aspects to take into account [9, 10]. In the case of memory management, the limitation arises from the so-called memory gap [11], leading to a substantial amount of idle time for processing resources due to slow memory access. The influence of a well-designed dataflow organization on the sys-tem performance is intimately related to this limitation. The overall objective must be to avoid moving large amounts of information pieces back and forth between sys-tem components via intermediate memory modules [2]. Optimization on this point must be planned after a comprehensive analysis of the processing flow featured by the targeted algorithm [9]. Particularly, early vision involving pixel-level operations must be carefully considered as it normally constitutes the most demanding stage in terms of processing and memory resources.

1.2 Focal-Plane Sensing-Processing Architecture

When all these key factors shaping performance are closely examined from an archi-tectural point of view, a major disadvantage of conventional system architectures becomes evident. As can be observed in Fig. 1, vision systems typically consist of a

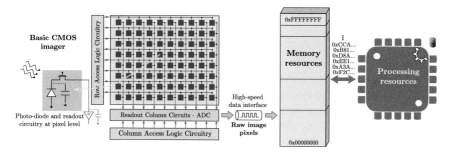

Fig. 1 Conventional architecture of embedded vision systems: image sensor, high-speed Analog-to-Digital Conversion (ADC), memory and processing resources (DSP, GPU etc.)

front-end imager delivering high-quality images at high speed to the rest of system components. This arrangement, by itself, generates a critical bottleneck associated with the huge amount of raw data rendered by the imager that must be subsequently stored and processed from scratch. But even more importantly, it precludes a first stage of processing acceleration from taking place just at the focal plane in a distributed and parallel way. Notice that the imager inevitably requires the physical realization of a 2-D array of photo-sensitive devices topographically assigned to their corresponding pixel values. This array can be exploited as distributed memory where the data are directly accessible for concurrent processing by including suitable circuitry at pixel level. As a result, the imager will be delivering pre-processed images, possibly in addition to the original raw information in case the algorithm needs it to superpose the processing outcome—e.g. highlighting the location of a tracked object. This architectural approach, referred in the literature as *focal-plane sensing-processing* [12] and represented in Fig. 2, presents two fundamental advantages when compared to that of Fig. 1. First of all, it enables a drastic reduction of memory accesses during low-level processing stages, where pixel-wise operations are common. Secondly, it permits to design ad-hoc circuitry to accelerate a vision

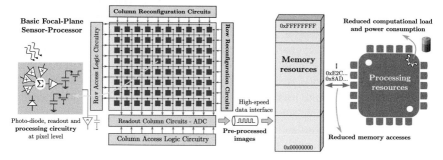

Fig. 2 Proposed architecture for focal-plane acceleration of image feature extraction. The pixel array is exploited as distributed memory including per-pixel circuitry for parallel processing

algorithm according to its specific characteristics. This circuitry can even be implemented in the analog domain for the sake of power and area efficiency since the pixel values at the focal plane have not been converted to digital yet. On the flip side, the incorporation of processing circuitry at pixel level reduces, for a prescribed pixel area, the sensitivity of the imager as less area is devoted to capture light. This drawback could be overcome by means of the so-called 3-D integration technologies [13, 14]. In this case, a sensor layer devoting most of its silicon area to capture light would be stacked and vertically interconnected onto one or more layers exclusively dedicated to processing. While not mature enough yet for reliable implementation of sensing-processing stacks, 3-D manufacturing processes will most surely boost the application frameworks of the research hereby presented.

All in all, this chapter introduces two full-custom focal-plane accelerator sensing-processing chips. They are our first prototypes aiming respectively at speeding up the image feature extraction of two flagships algorithms within the embedded vision field: the Viola-Jones face detection algorithm [15] and the Scale Invariant Feature Transform (SIFT) [16]. To the best of our knowledge, no prior attempts pointing to these algorithms have been reported for the proposed sensing-processing architectural solution. The chapter is organized as follows. After briefly describing both algorithms, we justify the operations targeted for implementation at the focal plane. We demonstrate that these operations feature a common underlying processing primitive, the Gaussian filtering, convenient for pixel-level circuitry. We then explain how this processing primitive has been implemented on both chips. Finally, we provide experimental results and discuss the guidelines of our future work on this subject matter.

2 Vision Algorithms

2.1 Viola-Jones Face Detection Algorithm

The Viola-Jones sliding window face detector [15] is considered a milestone in real-time generic object recognition. It requires a cumbersome previous training, demanding a large number of cropped frontal face samples. But once trained, the detection stage is fast thanks to the computation of the integral image, an intermediate image representation speeding up feature extraction, and to a cascade of classifiers of progressive complexity. A basic scheme of the Viola-Jones processing flow is depicted in Fig. 3. Despite its simplicity and detection effectiveness, the algorithm still requires a considerable amount of computational and memory resources in terms of embedded system affordability. Different approaches have been proposed in the literature in order to increase the implementation performance: by exploiting the highly parallel computation structure of GPUs [17, 18]; by making the most of the logic and memory capabilities of FPGAs [19, 20]; by custom design of specialized digital hardware [21] etc.

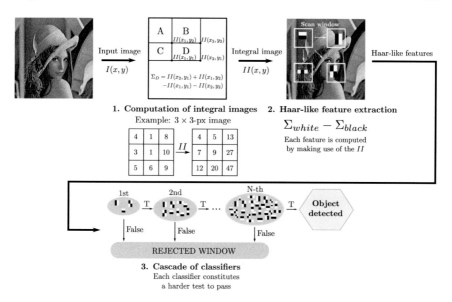

Fig. 3 Simplified scheme of the Viola-Jones processing flow

In order to evaluate the possibilities for focal-plane acceleration, our interest focuses on pixel-level operations. For the Viola-Jones algorithm, these operations take place during the computation of the integral image, defined as:

$$II(x, y) = \sum_{x'=1}^{x} \sum_{y'=1}^{y} I(x', y') \tag{1}$$

where $I(x, y)$ represents the input image. That is, each pixel composing $II(x, y)$ is equal to the sum of all the pixels above and to the left of the corresponding pixel at the input image. The first advantage of the integral image is that its calculation permits to compute the sum of any rectangular region of the input image by accessing only four pixels of the matrix $II(x, y)$. This is critical for real-time operation, given the potential large number of Haar-like features to be extracted—2135 in total for the OpenCV baseline implementation. The second advantage is that the computation of the integral image fits very well into a pipeline architecture—typically implemented in DSPs—by making use of the following pair of recurrences:

$$\begin{cases} r(x, y) = r(x, y - 1) + I(x, y) \\ II(x, y) = II(x - 1, y) + r(x, y) \end{cases} \tag{2}$$

with $r(x, 0) = 0$ and $II(0, y) = 0$. The matrix $II(x, y)$ can thus be obtained in one pass over the input image.

Despite these advantages, the purely sequential approach defined by Eq. (2) is still computationally expensive and memory access intensive [20, 22]. It usually accounts for a large fraction of the total execution time due to its linear dependence on the number of pixels of the input image [23]. Thus, its parallelization would boost the performance of the whole algorithm. In the next sections, we will propose an acceleration scheme that can clearly benefit from the concurrent operation and distributed memory provided by focal-plane architectures.

2.2 Scale Invariant Feature Transform (SIFT)

The SIFT algorithm constitutes a combination of keypoint detector and corresponding feature descriptor encoding [16]. It can be broken up into four main steps:

1. Scale-space extrema detection: generation of the Gaussian and subsequent Difference-of-Gaussian (DoG) pyramids, searching for the extrema points in the DoG pyramid.
2. Accurate keypoint location in the scale space.
3. Orientation assignment to the corresponding keypoint, searching for the main orientation or main component from the gradient in its neighborhood.
4. Keypoint descriptor: construction of a vector representative of the local characteristics of the keypoint in a wider neighborhood with orientation correction.

Numerous examples of SIFT implementations on different platforms have been reported: general-purpose CPU [24], GPU [25, 26], FPGA [27, 28], FPGA + DSP [29], specific digital co-processors [30] etc. As for the Viola-Jones, the lowest-level operation of the SIFT, namely the generation of the Gaussian pyramid, dominates the workload of the algorithm, reaching up to 90% of the whole process [31]. Figure 4

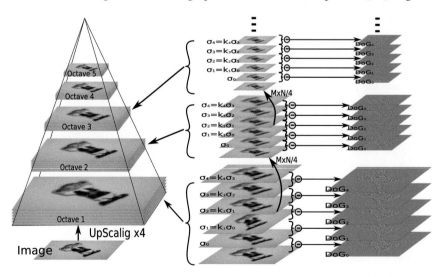

Fig. 4 Gaussian pyramid with its associated DoGs

shows an example of Gaussian pyramid with its associated DoGs. It is made up of sets of filtered images (scales). Every octave starts with a half-sized downscaling of the previous octave. The filter bandwidth, σ, applied for a new scale within each octave is the one applied in the previous scale multiplied by a constant factor k: $\sigma_n = k\sigma_{n-1}$. Each octave is originally divided into an integer number of scales, s, so $k = 2^{1/s}$. A total of $s + 3$ [16] images must be produced in the stack of blurred images for extrema detection to cover a complete octave. Once the Gaussian pyramid is built, the scales are subtracted from each other, obtaining the DoGs as an approximation to the Laplacian operator.

Our objective is therefore to accelerate the SIFT Gaussian pyramid generation by means of in-pixel circuitry performing concurrent processing. For the sake of relaxation on the hardware requirements, we carried out a preliminary study to determine the number of octaves and scales to be provided by our focal-plane sensor-processor. For this study, we used a publicly available version of SIFT in MATLAB [32]. Every octave is generated from a scale of the previous octave downsized by a 1/4 factor ($1/2 \times 1/2$), decreasing the pixels per octave. Therefore, the maximum potential keypoints decrease rapidly with the octaves $o = 0, 1, 2 \ldots$ as $M \times N/2^{o \times 2}$, with $M \times N$ being the size of the input image. Assuming a resolution of 320×240 pixels (QVGA), we obtained the keypoints for two images under many scales and rotation transformations. The reason of this moderate resolution is that the area to be allocated for in-pixel processing circuitry makes it difficult to reach larger resolutions in standard CMOS technologies with a reasonable chip size. The results for two of the applied transformations together with the test images are represented in Fig. 5. Clearly, the 3 first octaves render almost all the keypoints. Concerning scales, we have two opposite contributions. On the one hand, less scales per octave means more distance between scales, causing more pixels to exceed the threshold to be sorted out as keypoints. On the other hand, reducing scales also means to diminish the total number of potential keypoints. Both combined effects make it difficult to choose a specific value for scales as in the case of the octaves. The result of the scale analysis for the same respective test images and transformations as in Fig. 5 is depicted

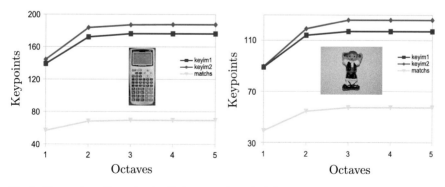

Fig. 5 The number of keypoints hardly increases from the 3 first octaves. This will be the reference value for our implementation

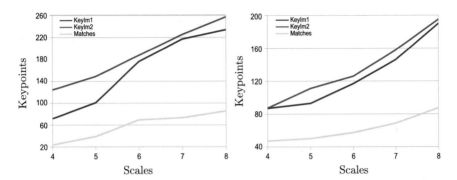

Fig. 6 The number of keypoints increases with the number of scales per octave. Trading this result for computational demand and hardware complexity leads to a targeted number of 6 scales. The same test images and transformations as in Fig. 5 are respectively used

in Fig. 6. It shows that the amount of keypoints increases monotonically with the scales. Nevertheless, increasing the scales per octave is not an option because of its corresponding computational demand and hardware complexity. Trading all these aspects, we conclude that 6 scales suffice for Gaussian pyramid generation at the focal plane. This figure coincides with the number of scales proposed in [16].

3   Gaussian Filtering

We demonstrate in this section that Gaussian filtering is the common underlying processing primitive for both the Viola-Jones and SIFT algorithms. While the role of Gaussian filtering is well defined for the latter, it is not obvious at all for the former. In order to understand the relation, we first need to establish a formal mathematical framework. Gaussian filtering is best illustrated in terms of a diffusion process. The concept of diffusion is widely applied in physics. It explains the equalization process undergone by an initially uneven concentration of a certain magnitude. A typical example is heat diffusion. Mathematically, a diffusion process can be defined by considering a function $V(\mathbf{x}, t)$ defined over a continuous space, in this case a plane, for every time instant. At each point $\mathbf{x} = (x_1, x_2)$, the linear diffusion of the function $V(.)$ is described by the following well-known partial differential equation [33]:

$$\frac{\partial V}{\partial t} = \nabla \cdot (D\nabla V) \tag{3}$$

where D is referred to as the diffusion coefficient. If D does not depend on the position:

$$\frac{\partial V}{\partial t} = D\nabla^2 V \tag{4}$$

and realizing the spatial Fourier transform of this equation, we obtain:

$$\frac{\partial \hat{V}(\mathbf{k})}{\partial t} = -4\pi^2 D |\mathbf{k}|^2 \hat{V}(\mathbf{k}) \tag{5}$$

where $\mathbf{k}$ represents the wave number vector in the continuous Fourier domain. Finally, by solving this equation we have:

$$\hat{V}(\mathbf{k}, t) = \hat{V}(\mathbf{k}, 0) e^{-4\pi^2 D t |\mathbf{k}|^2} \tag{6}$$

where $\hat{V}(\mathbf{k}, t)$ is the spatial Fourier transform of the function $V(.)$ at time instant t and $\hat{V}(\mathbf{k}, 0)$ is the spatial Fourier transform of the function $V(.)$ at time $t = 0$, that is, just before starting the diffusion. Equation (6) can be written as a transfer function:

$$\hat{G}(\mathbf{k}, t) = \frac{\hat{V}(\mathbf{k}, t)}{\hat{V}(\mathbf{k}, 0)} = e^{-4\pi^2 D t |\mathbf{k}|^2} \tag{7}$$

which, by defining $\sigma = \sqrt{2Dt}$, is transformed into:

$$\hat{G}(\mathbf{k}, \sigma) = e^{-2\pi^2 \sigma^2 |\mathbf{k}|^2} \tag{8}$$

This transfer function corresponds to the Fourier transform of a spatial Gaussian filter of the form:

$$G(\mathbf{x}, \sigma) = \frac{1}{2\pi \sigma^2} e^{-\frac{|\mathbf{x}|^2}{2\sigma^2}} \tag{9}$$

and therefore the diffusion process is equivalent to the convolution expressed by the following equation:

$$V(\mathbf{x}, t) = \frac{1}{2\pi \sigma^2} e^{-\frac{|\mathbf{x}|^2}{2\sigma^2}} * V(\mathbf{x}, 0) \tag{10}$$

We can see that a diffusion process intrinsically entails a spatial Gaussian filtering which takes place along time. The width of the filter is determined by the time the diffusion is permitted to evolve: the longer the diffusion time, t, the larger the width of the corresponding filter, σ. This means that, ideally, any width is possible provided that a sufficiently fine temporal control is available. From the point of view of the Fourier domain, we can define the diffusion as an isotropic lowpass filter whose bandwidth is controlled by t. The longer t, the narrower the bandwidth of the filter around the dc component (Fig. 7). Eventually, for $t \to \infty$, all the spatial frequencies but the dc component are removed. Furthermore, this dc component is completely unaffected by the diffusion, that is, $\hat{G}(\mathbf{0}, t) = 1$ $\forall t$. It is just this characteristic of the Gaussian filtering what constitutes the missing link with the computation of the integral image. When discretized and applied to a set of pixels, this property says that a progressive Gaussian filtering eventually leads to the average of the values the

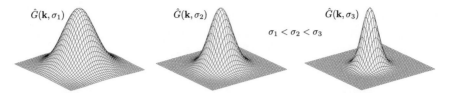

Fig. 7 Spatial Gaussian filters with increasing σ represented in the Fourier domain

pixels had before starting the filtering process. This average is a scaled version of
the sum of the original pixels, precisely the calculation required for each pixel of the
integral image. Furthermore, as we will see shortly, the averaging process inherent
to the Gaussian filtering is extremely helpful to cope at hardware level with the large
signal range demanded by the computation of the integral image.

4 Focal-Plane Implementation of Gaussian Filtering

The simple circuit depicted in Fig. 8a is our starting point to explain how we have
addressed the design of in-pixel circuitry capable of implementing Gaussian filtering.
Assuming that the initial conditions of the capacitors are V_{10} and V_{20}, the evolution
of the circuit dynamics is described by:

$$\begin{cases} C\frac{dV_1}{dt} = -\frac{V_1(t)-V_2(t)}{R} \\ C\frac{dV_2}{dt} = \frac{V_1(t)-V_2(t)}{R} \end{cases} \tag{11}$$

whose solution is:

$$\begin{bmatrix} V_1(t) \\ V_2(t) \end{bmatrix} = \frac{1}{2}(V_{10}+V_{20})\begin{bmatrix} 1 \\ 1 \end{bmatrix} + \frac{1}{2}(V_{10}-V_{20})\begin{bmatrix} 1 \\ -1 \end{bmatrix} e^{-2t/\tau} \tag{12}$$

where $\tau = RC$. Equation (12) physically represents a charge diffusion process—i.e.
Gaussian filtering—taking place along time between both capacitors at a pace deter-

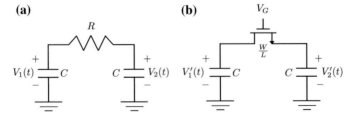

Fig. 8 2-node ideal diffusion circuit (**a**) and its transistor-based implementation (**b**)

mined by the time constant τ. For $t \to \infty$, both capacitors hold the same voltage, $(V_{10} + V_{20})/2$, that is, the average of their initial conditions. In order to achieve an area-efficient physical realization of this circuit, we can substitute the resistor by an MOS transistor (Fig. 8b) whose gate terminal additionally permits to control the activation-deactivation of the dynamics described by Eq. (12). Now suppose that V_{10} and V_{20} correspond respectively to two neighboring pixel values resulting from a photo-integration period previously set to capture a new image. If you are meant to compute the integral image from this new image, you will eventually want to add up both pixels as fast as possible. This can be accomplished by designing the proposed circuit with the minimum possible time constant τ in order to rapidly reach the steady state. Conversely, if you are meant to obtain the Gaussian pyramid, you will need fine control of the filtering process in order to increasingly blur the just captured image. In this case, the time constant τ cannot be arbitrarily small for the sake of making that fine control feasible. There are therefore conflicting design requirements depending on the specific task to be implemented by our basic circuit. In this scenario, we next present the particular realization of the diffusion process satisfying such requirements for both, the integral image computation and the Gaussian pyramid generation.

4.1 Focal-Plane Circuitry for Integral Image

A simplified scheme of how the charge diffusion process just described can be generalized for a complete image is depicted in Fig. 9. The MOS transistor in Fig. 8b has been substituted, to avoid clutter, for a simple switch for each connection between neighboring pixels in horizontal and vertical directions. This also highlights the fact that the MOS transistors are designed to have the minimum possible resistance when they are set ON, thus contributing to reduce the time constant τ. The state of these switches—ON or OFF—is controlled by the reconfiguration signals $EN_{S_{i,i+1_C}}$ and $EN_{S_{j,j+1_R}}$ for columns and rows respectively. The voltages $V_{px_{i,j}}$ represent the analog pixel values just after the photo-diode array has captured a new image. The integral image—really the averaged version provided by the diffusion process—is obtained by progressively establishing the adequate interconnection patterns in $EN_{S_{i,i+1_C}}$ and $EN_{S_{j,j+1_R}}$ according to the location of the pixel $II(x, y)$ being calculated at the moment. For example, we would need to activate $EN_{S_{1,2_C}}$ and $EN_{S_{1,2_R}}$, letting the remaining signals deactivated, in order to compute $II(2, 2)$. Each diffusion process producing an integral image pixel is followed by a stage of analog-to-digital conversion that takes place concurrently with the readjustment of the interconnection patterns for the next pixel to be computed. More details about the whole process and the additional circuitry required per pixel can be found in [34].

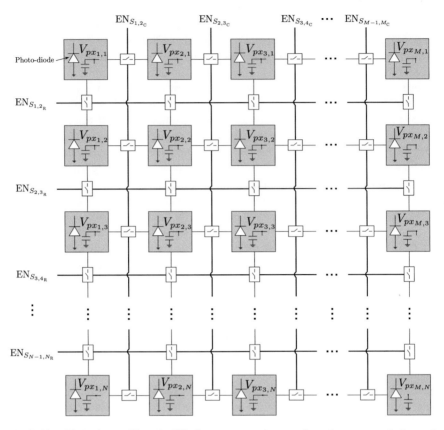

Fig. 9 Simplified scheme of how the diffusion process can be reconfigured to compute the integral image at the focal plane

4.2 Focal-Plane Circuitry for Gaussian Pyramid

As previously mentioned, the generation of the Gaussian pyramid requires an accurate control of the diffusion process in the circuit of Fig. 8b. A possible approach to achieve such control is to design specific on-chip circuitry providing precise timing over the gate signal of the MOS transistor [35]. Another possibility, featuring more linearity and even further diffusion control, is considered here. It is based on so-called Switched Capacitor (SC) circuits [36]. In this case, our reference circuit of Fig. 8a is transformed into that of Fig. 10. Two intermediate capacitors are introduced along with four switches enabling a gradual charge diffusion between the capacitors holding neighboring pixel values. Two non-overlapping clock phases driving the switches are used to carry out this progressive transfer of charge. It can be mathematically demonstrated [37] that this circuit configuration, called 'double Euler', is equivalent to apply a Gaussian filter with a width σ (see Eqs. (7) and (8)) given by:

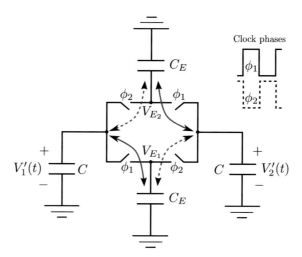

Fig. 10 In order to compute the Gaussian pyramid, two intermediate capacitors and four switches permit to gradually perform the charge diffusion between the capacitors holding neighboring pixel values

$$\sigma = \sqrt{\frac{2nC_E}{C}} \qquad (13)$$

where n is the number of cycles completed by the clock phases. We are assuming that $C_E \ll C$. A simulation example of a four-pixel diffusion featuring a diffusion cycle as short as 90 ns is shown in Fig. 11. In the final physical realization, this diffusion

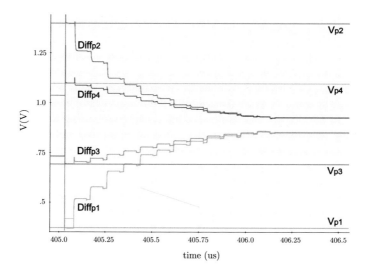

Fig. 11 Simulated temporal evolution of a four-pixel diffusion based on a double Euler SC circuit

cycle is adjusted in such a way that the targeted 3 octaves and 6 scales per octave can be attained. We therefore conclude that the discrete-time SC-based implementation of charge diffusion between capacitors provides the requested fine control of the underlying Gaussian filtering empowering the generation of the Gaussian pyramid at the focal plane.

5 Experimental Results

5.1 Viola-Jones Focal-Plane Accelerator Chip

The proposed prototype vision sensor presents the floorplan depicted in Fig. 12, featuring the elementary sensing-processing pixel shown in Fig. 13. The pixel array can be reconfigured block-wise by peripheral circuitry. The reconfiguration patterns are loaded serially into two shift registers that determine respectively which neighbor columns and rows can interact and which ones stay disconnected. There is also the possibility of loading in parallel up to six different patterns representing six successive image pixelation scales. This is achieved by means of control signals distributed regularly along the horizontal and vertical dimensions of the array [34]. The reconfiguration signals coming from the periphery map into the signals $\mathrm{EN}_{S_{i,i+1}}$, $\mathrm{EN}_{S_{j,j+1}}$, $\overline{\mathrm{EN}_{SQ_{i,i+1}}}$ and $\overline{\mathrm{EN}_{SQ_{j,j+1}}}$ at pixel level, where the coordinates (i, j) denote the location of the array cell considered. These signals control the activation of MOS switches for charge redistribution between the nMOS capacitors holding the voltages $V_{S_{ij}}$ and $V_{SQ_{ij}}$, respectively. Charge redistribution is the primary processing task that supports all the functionalities of the array, enabling low-power operation. Concerning A-to-D conversion, there are four 8-bit ADCs. These converters feature a tunable conversion range, including rail-to-rail, and a conversion time of 200 ns when clocked at 50 MHz. The column and row selection circuitry is also implemented by peripheral shift registers where a single logic '1' is shifted according to the location of the pixel to be converted.

The prototype chip together with the FPGA-based test system where it has been integrated can be seen in Fig. 14. An example of on-chip integral image computation is depicted in Fig. 15. As just explained, the sensing-processing array is capable of computing an averaged version of the actual integral image defined by Eq. (1), mathematically described as:

$$II_{av}(x, y) = \frac{1}{x \cdot y} \sum_{x'=1}^{x} \sum_{y'=1}^{y} I(x', y') \tag{14}$$

In Fig. 15, we can visualize the averaged integral image delivered by the chip and the integral image that can be directly derived from it. This integral image is compared with the ideal case obtained off-chip with MATLAB from the original image captured by the sensor, attaining an RMSE of 1.62 %. Notice that, in order

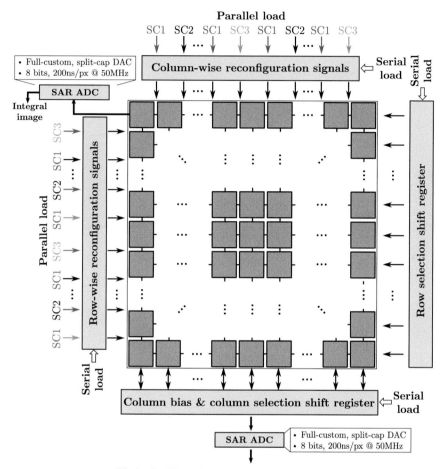

Fig. 12 Floorplan of the Viola-Jones focal-plane accelerator chip

to obtain $II(x, y)$, the only operation to be performed off-chip over $II_{av}(x, y)$ is to multiply each averaged pixel by its row and column number. No extra memory accesses are required for this task.

The chip has been manufactured in a standard 0.18 μm CMOS process. It features a resolution of 320 × 240 pixels and a power consumption of 55.2 mW when operating at 30 fps. This power consumption includes the image capture at that frame rate, the computation of the integral image for each captured image and the analog-to-digital conversion of the outcome for off-chip delivery. This figure is similar to that of state-of-the-art commercial image sensors, in this case with the add-on of focal-plane pre-processing alleviating the computational load of subsequent stages. The undesired effects of this add-on are reduced resolution and lower sensitivity. As mentioned in the introduction, these handicaps could be surmounted by 3-D integration.

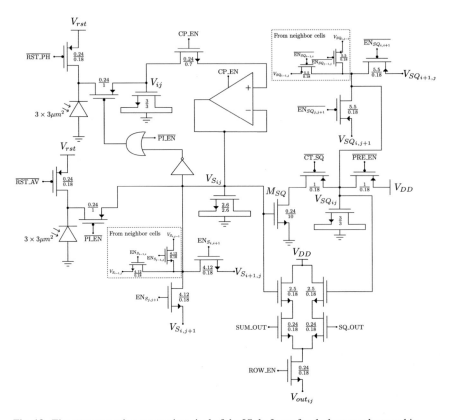

Fig. 13 Elementary sensing-processing pixel of the Viola-Jones focal-plane accelerator chip

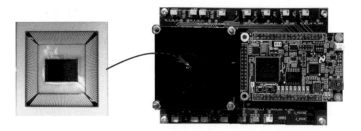

Fig. 14 Photograph of the Viola-Jones focal-plane accelerator chip and the FPGA-based system where it has been integrated

A direct transformation of the simplified scheme of Fig. 9 into a stacked structure is possible, as shown in Fig. 16. The top tier would exclusively include photo-diodes and some readout circuitry whereas the bottom tier would implement the reconfigurable diffusion network. The interconnection between both tiers would be carried out by the so-called Through-Silicon Vias (TSVs). This structure keeps maximum parallelism at processing while drastically increasing resolution and sensitivity.

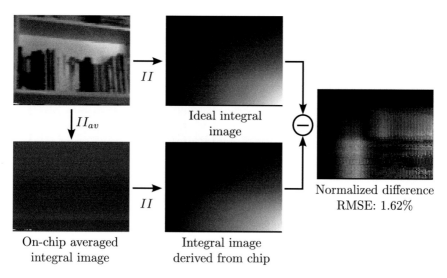

Fig. 15 Example of on-chip integral image computation and comparison with the ideal case

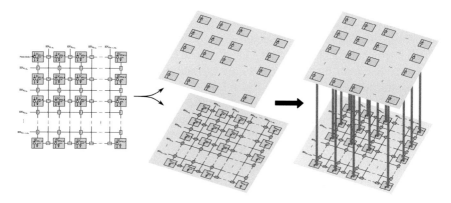

Fig. 16 Transformation of the simplified scheme of Fig. 9 into a stacked structure

5.2 SIFT Focal-Plane Accelerator Chip

The SIFT accelerator chip presents a similar floorplan to that of the Viola-Jones prototype. However, its elementary sensing-processing cell significantly differs. A simplified scheme is depicted in Fig. 17. The constituent blocks are mainly four photo-diodes, the local analog memories (LAMs), the comparator for A/D conversion and the switched capacitor network. During the acquisition stage, the photo-diodes, the capacitor C and the LAMs work together to implement a technique known as correlated double sampling [38] that improves the image quality. The LAMs jointly with the diffussion network carry out the Gaussian filtering. The capacitor C and the inverter make up the A/D comparator that would drive a register in the bottom

Fig. 17 Simplified scheme
of the elementary
sensing-processing cell
designed for the SIFT
focal-plane accelerator chip

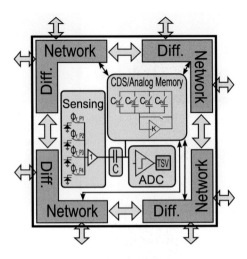

tier by a TSV on CMOS-3D technologies, or peripheral circuits on conventional
CMOS. Every cell is 4-connected to its closest neighbors in the North, South, East
and West directions. Given that every cell includes four photo-diodes, 4 internal and
8 peripheral interconnections are required.

Two microphotographs of the chip together with the different components of the
camera module built for test purposes are reproduced in Fig. 18. This prototype,
also manufactured in a standard 0.18 μm CMOS process, features a resolution of
176×120 pixels and can generate 120 Gaussian pyramids per second with a power
consumption of 70 mW. One of the operations required for Gaussian pyramid generation is downscaling. As previously commented, the 3 first octaves are the most
important ones in the performance of SIFT. This corresponds with downscaling at
ratios 4:1 and 16:1 for octaves 2 and 3, respectively. The chip includes the hardware required to implement this spatial resolution reduction. An example is shown
in Fig. 19. The images to the left are represented with the same sizes in order to visually highlight the effects of downscaling. Another example, in this case of on-chip
Gaussian filtering, is shown in Fig. 20. The upper left image constitutes the input
whereas the three remaining images, from left to right and top to down, correspond
to $\sigma = 1.77$, (clock cyles $n = 19$), $\sigma = 2.17$ ($n = 29$), and $\sigma = 2.51$ ($n = 39$). More
details about the performance of this chip can be found in [39].

This chip was conceived, from the very beginning, for implementation in 3-D integration technologies [40]. Unfortunately, these technologies are not mature enough
yet for reliable fabrication. Manufacturing costs of prototypes are also extremely
high for the time being, with long turnarounds, exceeding 1 year. In these circum-
stances, we were forced to redistribute the original two-tier circuit layout devised for
a CMOS 3-D stack in order to fit it into a conventional planar CMOS technology.
The result is depicted in Fig. 21.

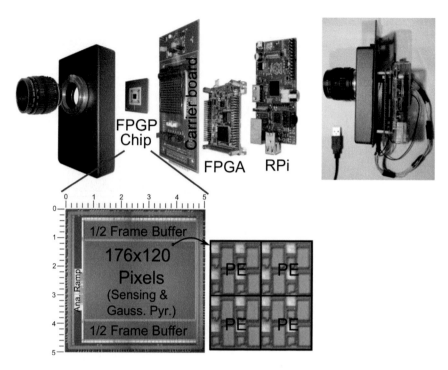

Fig. 18 Photograph of the SIFT focal-plane accelerator chip together with the camera module where it has been integrated

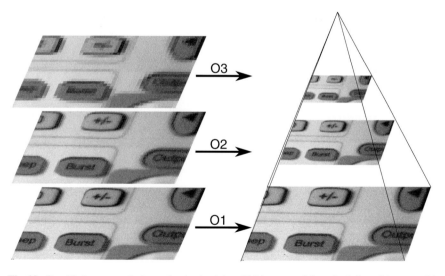

Fig. 19 On-chip image resolution reduction by 4:1 and 16:1 as part of the calculation of the pyramid octaves

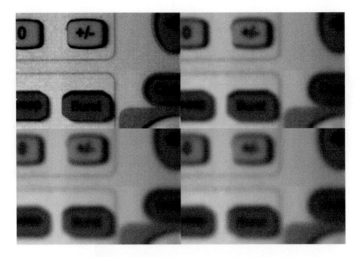

Fig. 20 Different snapshots of on-chip Gaussian pyramid

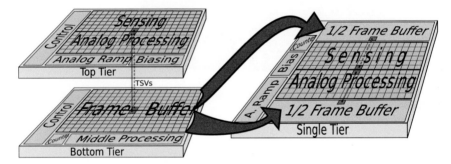

Fig. 21 Redistribution of circuits for Gaussian pyramid generation when mapping the original CMOS 3D-based architecture onto a conventional planar CMOS technology

5.3 Performance Comparison

Comparing the performance of the implemented prototypes with state-of-the-art focal-plane accelerator chips is not straightforward since every realization addresses a different functionality. As an example, we have included the most significant characteristics of our prototypes together with two recently reported focal-plane sensor-processor chips in Table 1. The Viola-Jones chip embeds extra functionalities in addition to the computation of the integral image [41] while featuring the largest resolution and the smallest pixel pitch, with a cost in terms of a reduced fill factor and increased energy consumption. Concerning the SIFT chip, one of the reasons of the energy overhead is the inherent high number of A/D conversions of the whole Gaussian pyramid plus the input scene, which amounts to 40 A/D conversions of the entire pixel array. Still, the acceleration at the focal plane provided by this chip

Table 1 Comparison of the implemented prototypes with state-of-the-art focal-plane sensor-processor chips

Reference	Ref. [42]	Ref. [43]	Viola-Jones chip	SIFT chip
Function	Edge filtering, tracking, HDR	2-D optic flow estimation	HDR, integral image, Gaussian filtering, programmable pixelation	Gaussian pyramid
Tech. (μm)	0.18	0.18	0.18	0.18
Supply (V)	0.5	3.3	1.8	1.8
Resolution	64×64	64×64	320×240	176×120
Pixel pitch (μm)	20	28.8	19.6	44
Fill factor (%)	32.4	18.32	5.4	10.25
Dyn. range (dB)	105	–	102	–
Power consumption (nW/px·frame)	1.25	0.89	23.9	26.5

pays off when comparing with more conventional solutions, as shown in Table 2. The power consumption of conventional CMOS imagers from Omnivision [44] featuring the image resolution tackled by the corresponding processor is incorporated in each of the entries related to conventional solutions. We have not accounted for accesses to external memories first because such costs would also be present if our chip were part of a complete hardware platform for a particular application; and

Table 2 Comparison of the SIFT focal-plane accelerator chip with conventional solutions

Hardware solution	Functionality	Energy/frame	Energy/pixel	Mpx/s
SIFT chip 180 nm CMOS	Gaussian pyramid	176×120 resol. 70 mW @ 8 ms 0.56 mJ/frame	26.5 nJ/px	2.64
Ref. [45] OV9655 + Core-i7	Gaussian pyramid	VGA resol. 90 mW @ 30 fps + 35 W @ 136 ms 4.8 J/frame	15.5 μJ/px	2.26
Ref. [46] OV9655 + Core-2-Duo	Gaussian pyramid	VGA resolution 90 mW + 35 W @ 2.1 s 73.7 J/frame	240 μJ/px	0.15
Ref. [47] OV6922 + Qualcomm Snapdragon S4	Gaussian pyramid	350×256 resol. 30 mW + 4 W @ 98.5 ms 0.4 J/frame	4.4 μJ/px	0.91

second because they are hardly predictable even with memory models. The energy cost of our chip outperforms that of an imager + conventional processor unit—even a low-power unit—in three orders of magnitude with similar processing speed. This leads to a combined speed-power figure of merit which makes our chip outperform conventional solutions in the range of three to six orders of magnitude.

6   Conclusions and Future Work

Focal-plane sensing-processing constitutes an architectural approach that can boost the performance of vision algorithms running on embedded systems. Specifically, early vision stages can greatly benefit from focal-plane acceleration by exploiting the distributed memory and concurrent processing in 2-D arrays of sensing-processing pixels. This chapter provides an overview of the fundamental concepts driving the design and implementation of two focal-plane accelerator chips tailored, respectively, for the Viola-Jones and the SIFT algorithms. These are the first steps within a long-term research framework aiming at achieving image sensors capable of simultaneously rendering high-resolution high-quality raw images and valuable pre-processing at ultra-low energy cost. The future work will be singularly biased by the availability of monolithic sensing-processing stacks. 3-D technologies will remove the tradeoff arising when it comes to allocating silicon area for sensors and processors on the same plane. High sensitivity and high processing parallelization will be compatible on the same chip. 3-D stacks will also foster alternative ways of making the most of vertical across-chip interconnections, from transistor level up to system architecture. In summary, 3-D integration technologies are the natural solution to develop feature extractors with low power budget without degrading image quality. Our prototypes on planar processes already consider future migration to these technologies, and this will continue to be a compulsory requirement of forthcoming designs.

Acknowledgments This work has been funded by: Spanish Government through projects TEC2012-38921-C02 MINECO (European Region Development Fund, ERDF/FEDER), IPT-2011-1625-430000 MINECO and IPC-20111009 CDTI (ERDF/FEDER); Junta de Andalucía through project TIC 2338-2013 CEICE; Xunta de Galicia through projects EM2013/038, AE CITIUS (CN2012/151, ERDF/FEDER), and GPC2013/040 ERDF/FEDER; Office of Naval Research (USA) through grant N000141410355.

References

1. Market analysis, embedded vision alliance. http://www.embedded-vision.com/industry-analysis/market-analysis
2. Kolsch, M., Butner, S.: Hardware considerations for embedded vision systems. In: Kisacanin, B., Bhattacharyya, S.S., Chai, S. (eds.) Embedded Computer Vision, Advances in Pattern Recognition Series, pp. 3–26. Springer, London (2009)
3. Open computing language. https://www.khronos.org/opencl/

4. OpenVX: portable, power-efficient vision processing. https://www.khronos.org/openvx/
5. Open source computer vision. http://opencv.org/
6. Compute unified device architecture. http://www.nvidia.com/object/cuda_home_new.html
7. Bailey, D.: Design for Embedded Image Processing on FPGAs. Wiley, Singapore (2011)
8. Kim, J., Rajkumar, R., Kato, S.: Towards adaptive GPU resource management for embedded real-time systems. ACM SIGBED Rev. **10**, 14–17 (2013)
9. Tusch, M.: Harnessing hardware accelerators to move from algorithms to embedded vision. In: Embedded Vision Summit. Embedded Vision Alliance, Boston (2012)
10. Horowitz, M.: Computing's energy problem (and what we can do about it). In: International Solid-State Circuits Conference (ISSCC), pp. 10–14. San Francisco (2014)
11. Wilkes, M.V.: The memory gap and the future of high performance memories. SIGARCH Comput. Archit. News **29**, 2–7 (2001)
12. Zárándy, A. (ed.): Focal-Plane Sensor-Processor Chips. Springer, New York (2011)
13. Campardo, G., Ripamonti, G., Micheloni, R.: Scanning the issue: 3-D integration technologies. Proc. IEEE **97**, 5–8 (2009)
14. Courtland, R.: ICs grow up. IEEE Spectr. **49**, 33–35 (2012)
15. Viola, P., Jones, M.: Robust real-time face detection. Int. J. Comput. Vis. **57**, 137–154 (2004)
16. Lowe, D.: Distinctive image features from scale-invariant keypoints. Int. J. Comput. Vis. **60**, 91–110 (2004)
17. Jia, H., Zhang, Y., Wang, W., Xu, J.: Accelerating Viola-Jones face detection algorithm on GPUs. In: IEEE International Conference on Embedded Software and Systems, pp. 396–403. Liverpool (2012)
18. Masek, J., Burget, R., Uher, V., Guney, S.: Speeding up Viola-Jones algorithm using multi-core GPU implementation. In: IEEE International Conference on Telecommunications and Signal Processing (TSP), pp. 808–812. Rome (2013)
19. Acasandrei, L., Barriga A.: FPGA implementation of an embedded face detection system based on LEON3. In: International Conference on Image Processing, Computer Vision, and Pattern Recognition. Las Vegas (2012)
20. Ouyang, P., Yin, S., Zhang, Y., Liu, L., Wei, S.: A fast integral image computing hardware architecture with high power and area efficiency. IEEE Trans. Circuits Syst. **II**(62), 75–79 (2015)
21. Kyrkou, C., Theocharides, T.: A flexible parallel hardware architecture for adaboost-based real-time object detection. IEEE Trans. Very Large Scale Integr. VLSI Syst. **19**, 1034–1047 (2011)
22. Gschwandtner, M., Uhl, A., Unterweger, A.: Speeding up object detection fast resizing in the integral image domain. Technical Report, University of Salzburg (2014)
23. de la Cruz, J.A.: Field-programmable gate array implementation of a scalable integral image architecture based on systolic arrays. Master Thesis, Utah State University (2011)
24. Kumar, G., Prasad, G., Mamatha, G.: Automatic object searching system based on real time SIFT algorithm. In: IEEE International Conference on Communication Control and Computing Technologies, pp. 617–622. Ramanathapuram (2010)
25. Cornelis, N., Van Gool, L.: Fast scale invariant feature detection and matching on programmable graphics hardware. In: IEEE Computer Vision and Pattern Recognition Workshops, pp. 1–8. Anchorage (2008)
26. Cohen, B., Byrne, J.: Inertial aided SIFT for time to collision estimation. In: IEEE International Conference on Robotics and Automation, pp. 1613–1614. Kobe (2009)
27. Cabani, C., MacLean, W.J.: A proposed pipelined-architecture for FPGA-based affine-invariant feature detectors. In: IEEE Computer Vision and Pattern Recognition Workshops, pp. 121. New York (2006)
28. Nobre, H., Kim, H.Y.: Automatic VHDL generation for solving rotation and scale-invariant template matching in FPGA. In: IEEE Southern Conference on Programmable Logic, pp. 21–26. Sao Carlos (2009)
29. Song, H., Xiao, H., He, W., Wen, F., Yuan, K.: A fast stereovision measurement algorithm based on SIFT keypoints for mobile robot. In: IEEE International Conference on Mechatronics and Automation (ICMA), pp. 1743–1748. Takamatsu (2013)

30. Gao, H., Yin, S., Ouyang, P., Liu, L., Wei, S.: Scale invariant feature transform algorithm based on a reconfigurable architecture system. In: 8th IEEE International Conference on Computing Technology and Information Management (ICCM), pp. 759–762. Seoul (2012)
31. Noguchi, H., Guangji H., Terachi, Y., Kamino, T., Kawaguchi, H., Yoshimoto, M.: Fast and low-memory-bandwidth architecture of SIFT descriptor generation with scalability on speed and accuracy for VGA video. In: IEEE International Conference on Field Programmable Logic and Applications (FPL), pp. 608–611. Milano (2010)
32. Andrea Vedaldi's implementation of the SIFT detector and descriptor. http://www.robots.ox.ac.uk/vedaldi/code/sift.html
33. Jahne, B.: Multiresolution signal representation. In: Jahne, B., Haubecker, H., Geibler, P. (eds.) Handbook of Computer Vision and Applications (volume 2). Academic Press, San Diego (1999)
34. Fernández-Berni, J., Carmona-Galán, R., del Río, R., Rodríguez-Vázquez, A.: Bottom-up performance analysis of focal-plane mixed-signal hardware for Viola-Jones early vision tasks. Int. J. Circuit Theory Appl. (2014). doi:10.1002/cta.1996
35. Fernández-Berni, J., Carmona-Galán, R., Carranza-González, L.: FLIP-Q: a QCIF resolution focal-plane array for low-power image processing. IEEE J. Solid-State Circuits **46**, 669–680 (2011)
36. Allen, P.E.: Switched Capacitor Circuits. Springer, New York (1984)
37. Suárez, M., Brea, V.M., Cabello, D., Pozas-Flores, F., Carmona-Galán, R., Rodríguez-Vázquez, A.: Switched-capacitor networks for scale-space generation. In: IEEE European Conference on Circuit Theory and Design (ECCTD), pp. 190–193. Linkoping (2011)
38. Enz, C.C., Temes, G.C.: Circuit techniques for reducing the effects of op-amp imperfections: autozeroing, correlated double sampling, and chopper stabilization. Proc. IEEE **84**, 1584–1614 (1996)
39. Suárez, M., Brea, V.M., Fernández-Berni, J., Carmona-Galán, R., Cabello, D., Rodríguez-Vázquez, A.: A 26.5 nJ/px 2.64 Mpx/s CMOS vision sensor for gaussian pyramid extraction. In: IEEE European Solid-State Circuits Conference (ESSCIRC), pp. 311–314. Venice (2014)
40. Suárez, M., Brea, V.M., Fernández-Berni, J., Carmona-Galán, R., Liñán, G., Cabello, D., Rodríguez-Vázquez, A.: CMOS-3-D smart imager architectures for feature detection. IEEE J. Emerg. Sel. Top. Circuits Syst. **2**, 723–736 (2012)
41. Fernández-Berni, J., Carmona-Galán, R., del Río, R., Kleihorst, R., Philips, W., R., Rodríguez-Vázquez, A.: Focal-plane sensing-processing: a power-efficient approach for the implementation of privacy-aware networked visual sensors. Sensors **14**, 15203–15226 (2014)
42. Yin, C., Hsieh, C.: A 0.5V 34.4μW 14.28kfps 105dB smart image sensor with array-level analog signal processing. In: IEEE Asian Solid-State Circuits Conference (ASSCC), pp. 97–100. Singapore (2013)
43. Park, S., Cho, J., Lee, K., Yoon, E.: 243.3pJ/pixel bio-inspired time-stamp-based 2D optic flow sensor for artificial compound eyes. In: IEEE International Solid-State Circuits Conference (ISSCC), pp. 126–127. San Francisco (2014)
44. Omnivision image sensors. http://www.ovt.com/products/
45. Murphy, M., Keutzer, K., Wang, P.: Image feature extraction for mobile processors. In: IEEE International Symposium on Workload Characterization (IISWC), pp. 138–147. Austin (2009)
46. Huang, F., Huang, S., Ker, J., Chen, Y.: High-performance SIFT hardware accelerator for real-time image feature extraction. IEEE Trans. Circuits Syst. Video Technol. **22**, 340–351 (2012)
47. Wang, G., Rister, B., Cavallaro, J.: Workload analysis and efficient openCL-based implementation of SIFT algorithm on a smartphone. In: IEEE Global Conference on Signal and Information Processing, pp. 759–762. Austin (2013)

Part II
Applications of Image Feature Detectors and Descriptors

Satellite Image Matching and Registration: A Comparative Study Using Invariant Local Features

Mohamed Tahoun, Abd El Rahman Shabayek, Hamed Nassar,
Marcello M. Giovenco, Ralf Reulke, Eid Emary and Aboul Ella Hassanien

Abstract The rapid increasing of remote sensing (RS) data in many applications ignites a spark of interest in the process of satellite image matching and registration. These data are collected through remote sensors then processed and interpreted by means of image processing algorithms. They are taken from different sensors, viewpoints, or times for many industrial and governmental applications covering agriculture, forestry, urban and regional planning, geology, water resources, and others. In this chapter, a feature-based registration of optical and radar images from same and different sensors using invariant local features is presented. The registration process starts with the feature extraction and matching stages which are considered as key issues when processing remote sensing data from single or multi-sensors. Then, the geometric transformation models are applied followed by the interpolation method in

M. Tahoun (✉) · A.E.R. Shabayek · H. Nassar
Computer Science Department, Faculty of Computers and Informatics, Suez Canal University, Ismailia, Egypt
e-mail: tahoun@ci.suez.edu.eg

A.E.R. Shabayek
e-mail: a.shabayek@ci.suez.edu.eg

H. Nassar
e-mail: nassar@ci.suez.edu.eg

M.M. Giovenco · R. Reulke
Institut Für Optische Sensorsysteme, DLR-Berlin, Berlin, Germany
e-mail: marcello.giovenco@dlr.de

R. Reulke
Computer Science Department, Humboldt University in Berlin, Berlin, Germany
e-mail: reulke@informatik.hu-berlin.de

E. Emary · A.E. Hassanien
Information Technology Department, Faculty of Computers and Information, Cairo University, Giza, Egypt
e-mail: e.emary@fci-cu.edu.eg

A.E. Hassanien
e-mail: aboitcairo@egyptscience.net

M. Tahoun · E. Emary · A.E. Hassanien
(SRGE Group (www.egyptscience.net)), Cairo, Egypt

© Springer International Publishing Switzerland 2016
A.I. Awad and M. Hassaballah (eds.), *Image Feature Detectors and Descriptors*,
Studies in Computational Intelligence 630, DOI 10.1007/978-3-319-28854-3_6

order to get a final registered version. As a pre-processing step, speckle noise removal is performed on radar images in order to reduce the number of false detections. In a similar fashion, optical images are also processed by sharpening and enhancing edges in order to get more accurate detections. Different blob, corner and scale based feature detectors are tested on both optical and radar images. The list of tested detectors includes: SIFT, SURF, FAST, MSER, Harris, GFTT, ORB, BRISK and Star. In this work, five of these detectors compute their own descriptors (SIFT, SURF, ORB, BRISK, and BRIEF), while others use the steps involved in SIFT descriptor to compute the feature vectors describing the detected keypoints. A filtering process is proposed in order to control the number of extracted keypoints from high resolution satellite images for a real time processing. In this step, the keypoints or the ground control points (GCPs) are sorted according to the response strength measured based on their cornerness. A threshold value is chosen to control the extracted keypoints and finalize the extraction phase. Then, the pairwise matches between the input images are calculated by matching the corresponding feature vectors. Once the list of tie points is calculated, a full registration process is followed by applying different geometric transformations to perform the warping phase. Finally and once the transformation model estimation is done, it is followed by blending and compositing the registered version. The results included in this chapter showed a good performance for invariant local feature detectors. For example, SIFT, SURF, Harris, FAST and GFTT achieve better performance on optical images while SIFT gives also better results on radar images which suffer from speckle noise. Furthermore, through measuring the inliers ratios, repeatability, and robustness against noise, variety of comparisons have been done using different local feature detectors and descriptors in addition to evaluating the whole registration process. The tested optical and radar images are from RapidEye, Pléiades, TET-1, ASTER, IKONOS-2, and TerraSAR-X satellite sensors in different spatial resolutions, covering some areas in Australia, Egypt, and Germany.

1 Introduction

The detection and matching of features from images taken from different sensors, viewpoints, or at different times are mandatory tasks in manipulating and processing data for many fields like remote sensing and medical imaging. Many local feature detection methods have been proposed and developed for this purpose [1, 2]. These two processes are basic tasks when registering or fusing remote sensing data for many applications. Huge satellite imagery data are now existing in high resolutions and different formats through variety of sensors and cameras for different purposes. Depending on the application of the remote sensing data, some achievements have been remarked but there is still a need for improving the detection and the matching processes for more accurate alignment and registration of images especially those with different modalities [3, 4]. A set of corresponding features or the matching points between two images is used later to know how they both are related to each other. This step is very important for any further tasks like image alignment and registration and image fusion [3, 5].

Classical image registration process includes the detection of features from images as the first step. These features are classified into local and global features and might be represented by points, edges, corners and contours or by other features [6]. For example, corners usually represent a point in which the directions of two edges have a clear change while blobs are known as regions of interest. Once the features are extracted from images, the second step or the matching process starts by comparing the corresponding feature descriptors of the extracted keypoints. A Final set of inliers or tie points should be determined in order to stitch the input images. The results can be enhanced by means of bundle adjustment in case of multiple images or panorama [7]. The third and the fourth steps of the registration process include the estimation of the transformation model and resampling and compositing the registered version.

In literature, automatic alignment and stitching can be divided into two categories: direct [8] and feature based [7, 9]. Direct methods use all the image data which may provide very accurate registration based on a close initialization [8]. Feature based methods do not need initialization. Using invariant features enables reliable matching despite of different possible transformations (e.g. rotation, zooming, etc.). The main steps of automatic stitching can be summarized as follows [5]: (1) Extract the invariant features from all input images, (2) find a number of nearest neighbours for each feature, (3) for each image, select some candidate matching images that have the most feature matches to this image, (4) geometrically, find consistent feature matches, (5) verify image matches by finding connected components of image matches, (6) an enhancement could be chosen by performing bundle adjustment to find intrinsic and extrinsic camera parameters for each connected component, and finally, (7) apply multi-band blending to obtain the output stitched image [7]. A local feature is considered as an image pattern which differs from its direct neighbourhoods and is usually represented by points, edges, corners or others as in [6]. Based on such representations, many feature detectors have been proposed, for example, Harris and FAST detectors work based on corner detection while SIFT and SURF work based on points or blobs. The last two detectors have been recently used in many applications and recorded a good performance against rotation, scaling and blurring [2].

Most of the current proposed methods for satellite image registration still have some limitations especially on very high and high resolution images taken from same or different sensors. Furthermore, applications that use image registration/fusion methods require more reliable, accurate, and robust techniques that have real time processing and less hardware requirements. The work presented here aims to enhance the registration process through combining invariant local feature methods with efficient geometric transformation models and image resampling techniques for achieving a generalized and enhanced registration scheme applicable for satellite images.

In this chapter, we will focus on feature-based methods using invariant local features. Section 2 will highlight some concepts and list different applications of image registration. Section 3 will discuss the steps involved in the registration process, in addition to the proposed registration scheme. Then, the experiments on different optical and radar images are followed with examples and a detailed discussion.

Finally, the chapter will be concluded by outlining the most important findings and future work suggestions.

2 Concepts and Applications

The role of a feature detector is to identify the interest points (i.e. regions of interest) while a feature descriptor is responsible for computing the feature vector describing these interest points (regions). The matching process determines correspondence between the computed descriptors. A clear example is the SIFT algorithm proposed by David Lowe 2004, where keypoints are extracted by the SIFT detector and the SIFT descriptor computes descriptors of the detected keypoints. It is a common practice to use SIFT detector and SIFT descriptor independently [10]. Keypoints can be extracted without descriptors (using SIFT detector only) or SIFT descriptor can be used to describe custom keypoints. Each detector (e.g. SIFT) is used together with its own corresponding descriptor. If the detector does not have a corresponding descriptor (e.g. STAR detector), a convenient feature descriptor is chosen according to the application as will be explained in Sect. 4.

Image registration is the process of determining a geometrical transformation that aligns points or pixels in one image with the corresponding points in the other image having the same scene [6]. A successful image registration is mandatory for image fusion and data integration. Generally, the main steps of image registration process are: Feature identification, feature matching, spatial transformation, and finally the resampling step as illustrated in Fig. 1. The feature identification includes the detection of the required features like keypoints or GCPs from the input images. Once the features are extracted from images, the matching process starts by comparing the feature descriptors based on the detected keypoints in order to get the set of correspondence between these descriptors.

Mean squared differences like Euclidean distance or Root Mean Squared Error (RMSE), and mutual information are examples of these similarity functions or

Fig. 1 The main steps involved in classical image registration

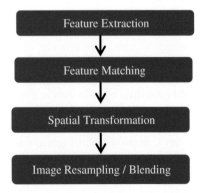

metrics. This step is followed by finding a geometrical transformation model that aligns the source image on the reference one which remains intact using the set of similarity points or matches between them.

The transformation model estimation is done based on this list of tie points. Different methods are existing for performing such transformations and can be classified as rigid and non-rigid image transformations [11]. In rigid transformations, the shape and the size of the objects do not change while they do change in non-rigid transformations like affine and similarity transformations. Based on the transformation model, the resampling techniques like linear interpolation are used to resize and composite the final registered image.

Good features extracted from images are supposed to be robust against rotation, scaling, illumination changes or invariant against any geometric changes. In [5], an evaluation of some invariant local features on satellite images is introduced with a focus on both the detection and the matching phases. In computer vision and image processing applications, these two processes are mandatory when manipulating images taken from the same or different sensors, different viewpoints or at different times.

There are different registration methods according to the type of features extracted from the input images and also on the transformation model that maps the extracted points from the input images. The registration is done once this mapping is determined and for some applications the two input images are combined together in one version by summing the intensity values in the two images. In general, image registration can be classified based on the applications as follows:

1. Multi-modal registration: where complementary information for multi-sensor images are integrated [12]. Images of the same scene are acquired by different sensors. The aim is to integrate the information obtained from different source streams to gain more complex and detailed scene representation. Examples of applications are fusion of remotely sensed information from sensors with different characteristics like panchromatic images, color/multispectral images with better spectral resolution, or radar images independent of cloud cover and solar illumination. Applications like land use or change detection make use of such type of registration where remote sensing and Geographic Information system (GIS) are combined together [13].
2. Multi-view registration: Images of the same scene are acquired from different view-points. The aim is to gain a larger 2D view or a 3D representation of the scanned scene. Examples of applications: remote sensing based mosaicking of images of the surveyed area and shape recovery from stereo cameras.
3. Temporal registration: Images of the same scene are acquired at different times, often on regular basis, and possibly under different conditions. The aim is to evaluate changes in the scene which appeared between the consecutive images acquisitions. Examples of applications: Remote sensing monitoring of global land usage, landscape planning, automatic change detection for security applications, detection for security monitoring, motion tracking, and medical imaging based monitoring of the healing therapy and monitoring of the tumor evolution.

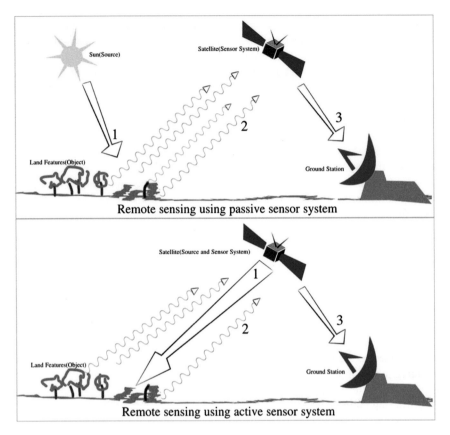

Fig. 2 Passive (*top*) and active (*down*) sensing processes (*Source* http://en.wikipedia.org/wiki/Remote_sensing#mediaviewer/File:Remote_Sensing_Illustration.jpg)

4. Scene to model registration: Images of a scene and a model of the scene are registered. The model can be a computer representation of the scene, for instance maps or digital elevation models (DEM) in GIS, another scene with similar content. The aim is to localize the acquired image in the scene/model and/or to compare them. Among the many different types of satellite images, two are mainly involved in our image analysis study that is radar and optical images (i.e. active and passive sensing respectively as in Fig. 2. In general, there are two categories of remote sensing systems: passive and active sensors. Passive sensors collect natural radiant reflected from targeted objects while active sensors transmit a signal then receive the reflected response. These two types of images provide the user with different and complementing information in order to achieve a better understanding of the analysed scene [14].

There are many applications for image registration in medical imaging and remote sensing and satellite imagery as well. In this work, we focus on satellite image registration and their applications on optical and radar images. These images are influenced by a number of effects based on the carrier frequency of the electromagnetic waves. It is well known that optical sensors are hindered by clouds to obtain information on the observed objects on Earth. Even the shadows of the clouds or the night transit over an area influence the interpretability of the imagery. On the other hand, Synthetic-Aperture Radar (SAR) suffers from severe terrain induced geometric distortions based on its side-looking geometry (layover, foreshortening, and shadow). To overcome these influences it is possible for example to combine different images acquired by the same or different instrument for getting one single image having information more than we get from a single sensor.

There are many available satellites with sensors in different formats, spectral bands and different ground resolution depending on the mission of each sensor. Famous satellite systems include: LANDSAT, SPOT, IKONOS, TerraSAR-X and RADARSAT. In our registration framework, we have tested different optical and radar data from TerrSAR-X, RapidEye, Pléiades, IKONOS-2, ASTER and TET-1 satellite sensors. More details about the tested images of this work in Sect. 4.

3 Image Registration

As illustrated in the previous section, Image registration is about to find the corresponding pixels or points between two or more images taken from same or different sensors or at different times or from different viewpoints [15]. In our work, we focus on registering multi-sensor data where images are taken from different sensors (optical and radar). The other two scenarios (multi-view and multi-temporal) are also considered. This section discusses in details the main steps involved in the registration process: feature extraction, feature matching, geometric transformation, and resampling phase. Then the proposed Algorithm is presented with more details.

3.1 Feature Extraction

The task of finding correspondences between two images of the same scene or object is part of many computer vision applications like object recognition, image indexing, structure from motion and visual localization—to name a few. In most images there are regions/points that can be detected with high repeatability since they possess some distinguishing, invariant and stable property. These regions/points represent the image local features. A local feature is an image pattern which differs from its immediate neighborhood. It is usually associated with a change of an image property or several properties simultaneously, although it is not necessarily localized exactly on this change. In the following subsections, some of the most recent local feature detectors found in literature will be briefly discussed. An evaluation of local feature detectors and descriptors is found in [13, 16].

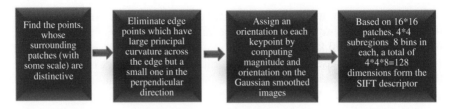

Fig. 3 The main steps of the SIFT feature detector and descriptor [17]

3.1.1 SIFT

SIFT (Scale Invariant Feature Transform) has been presented by Lowe in the year 2004 [17]. It has four major steps: scale-space extrema detection, keypoint localization, orientation assignment and finally building the keypoint descriptor. In the first step, points of interest are identified by scanning both the location and the scale of the image. The difference of Gaussian (DoG) is used to perform this step and then, the keypoint candidates are localized to sub-pixel accuracy.

Then the orientation is assigned to each keypoint in local image gradient directions to obtain invariance to rotation. In the last step a 128-keypoint descriptor or feature vector is built and ready for the matching process. SIFT gives good performance but still have some limitations against strong illumination changes and big rotation. Figure 3 shows the main steps of the SIFT algorithm.

3.1.2 SURF

SURF (Speeded-Up Robust Features) is a local invariant Interest point or blob detector [18]. It is partly inspired by the SIFT descriptor and is used too in static scene matching and retrieval. It is invariant to most of the image transformations like scale and illumination changes in addition to small changes in viewpoint. It uses Integral Images or an intermediate representation for the image and contains the sum of gray scale pixel values of image.

Then a Hessian-based interest point localization is obtained using Laplacian of Gaussian of the image. SURF is good at handling serious blurring and image rotation. However, it is poor at handling viewpoint and illumination changes. Figure 4 shows the main steps of the SURF algorithm.

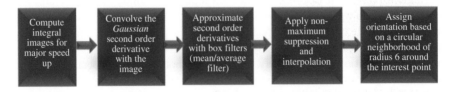

Fig. 4 The main steps of the SURF feature detector and descriptor [18]

Fig. 5 The main steps of the Harris corner and edge detector [19]

3.1.3 Harris

Harris is a combined corner and edge detector based on the local autocorrelation function [19]. It does not depend on rotation or shift or affine change of intensity. It extends the principle of Moravec's corner detector by considering the local autocorrelation energy [20].

Corners are usually good features to match especially with viewpoint changes. Figure 5 shows the main steps of the Harris corner and edge detector.

3.1.4 Star

Star Feature Detector is derived from CenSurE (Center Surrounded Extrema) detector [10, 21]. While CenSurE uses polygons such as Square, Hexagon and Octagons as a more computable alternative to circle, Star mimics the circle with 2 overlapping squares: one upright and one 45-degrees rotated. CenSurE determines large-scale features at all scales, and select the extrema across scale and location. It uses simplified bi-level kernels as center-surround filters. It focuses on finding kernels that are rotationally invariant. Figure 6 shows the main steps of the Star detector.

3.1.5 FAST

FAST (Features from Accelerated Segment Test) is a corner detection method [22]. Its importance lies in its computational efficiency as it is faster than many famous

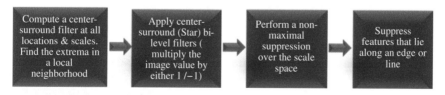

Fig. 6 The main steps of the Star corner detector (In our work, Star has been used as a feature detector without computing its own descriptor (instead, SIFT descriptor is used in the description phase))

Fig. 7 The main steps of the FAST corner detector [22]

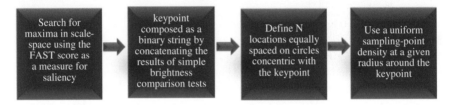

Fig. 8 The main steps of the BRISK feature detector and descriptor [23]

feature extraction methods (e.g. DoG, SIFT, Harris). It uses a Bresenham circle of radius 3 to find out whether a selected point is a corner. Each pixel in the circle is given a number from 1 to 16 clockwise. If a set of contiguous pixels inside the circle is brighter or darker than the candidate pixel then it is classified as a corner. FAST is considered as a high quality feature detector but still not robust to noise and depend on a threshold. Figure 7 shows the main steps of the FAST algorithm.

3.1.6 BRISK

BRISK (Binary Robust Invariant Scalable Keypoints) depends on easily configurable circular sampling pattern from which it computes brightness comparisons to form a binary descriptor string [23]. The authors claim it to be faster than SIFT and SURF. A further investigation is required to explore alternatives to the search of saliency scores scale space maxima to obtain higher repeatability while keeping speed. Figure 8 shows the algorithm main steps.

3.1.7 ORB

ORB (Oriented BRIEF-Binary Robust Independent Elementary Features) is a local feature detector and descriptor based on binary strings [10, 24]. It depends on a relatively small number of intensity difference tests to represent a patch of the image as a binary string. The construction and matching of this local feature is fast and performs well as long as invariant to large in-place rotations is not required. ORB investigates the variance under orientation which was critical in its construction that apparently improved the performance in nearest-neighbor applications. One of the

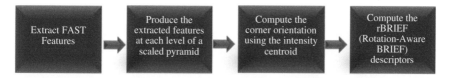

Fig. 9 The main steps of the ORB feature detector and descriptor [10]

Fig. 10 The main steps of the GFTT corner detector [10]

issues that was not properly addressed is scale invariance. Although they used a pyramid scheme, they did not explore per keypoint scale from depth cues or tuning the number of octaves [25]. Figure 9 shows the ORB main steps. For further details the reader is urged to refer to [25].

3.1.8 GFTT

GFTT (Good Features to Track) extracts the most prominent corners in the image as described in [26] where the corner quality measure at each pixel is calculated. Then a non-maximum suppression is applied. The corners with quality less than a certain threshold are rejected and the remaining corners are sorted by the quality measure in the descending order. Finally, each corner for which there is a stronger corner at a distance less than a threshold is thrown away. Their method shows that features with good texture properties can be found by optimizing the tracker accuracy. The GFTT are exactly those features that make the tracker work best. Two image motion models were found to be better than using one. Translation model gave more reliable results when the interframe camera translation was small. However, the affine model changes were mandatory to compare distant frames to determine dissimilarity. Figure 10 shows the GFTT main steps.

3.1.9 MSER

MSER (Maximally Stable Extremal Regions) is similar to SIFT (blob detection) as it extracts the co-variant regions from an image [27]. It was proposed to find correspondences between images from different viewpoints. Extremal regions are used which possess some important properties. The set is closed under continuous transformations of image coordinates and monotonic transformation of image intensities.

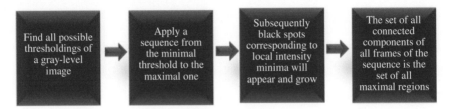

Fig. 11 The steps involved in MSER feature detector [27]

MSER has near linear complexity and practically fast to detect affinely invariant stable subset of extremal regions. Figure 11 shows, informally, the MSER main steps.

3.2 Feature Matching

Image similarity measures estimate the amount of correspondence between the image pairs or what is known as feature descriptors. Here, we give an overview on some of the most used similarity measures in the area of remote sensing:

3.2.1 Root Mean Squared Error (RMSE)

The correspondence can be measured as the sum-squared difference between the intensities of overlapping pixels to evaluate a given registration. This can be expressed as an error function where a value of zero represents a perfect match [28–31] see the below equation:

$$RMSE = \sqrt{\frac{1}{MN} \sum_{i=1}^{M} \sum_{j=1}^{N} (x\,(i,j) - y\,(i,j))^2} \tag{1}$$

where M and N are the image dimensions, x; y are the point list from reference and transformed image.

3.2.2 Peak Signal to Noise Ratio (PSNR)

It is a ratio of maximum possible power of signal to power of corrupting noise that affects the fidelity of representation. Its range is wide so it is expressed as logarithmic scale. Maximum value of PSNR indicates good match between the two images; can be expressed as [30].

$$PSNR = 10 \log_{10} \frac{(2^n - 1)^2}{MSE} \tag{2}$$

3.2.3 Normalized Cross Correlation (NCC)

NCC is a common similarity measure [11] and can be mathematically defined as:

$$NCC(A, B) = \frac{1}{N} \sum_{x,y} \frac{(A - \mu_A)(B - \mu_B)}{\sigma_A \sigma_B} \tag{3}$$

where N denotes the total number of pixels in image A and B, μ_A, μ_B are the mean of image A and B in order, and σ_A, σ_B denote the standard deviation of images A;B in order.

3.2.4 Mutual Information (MI)

Mutual information is a reliable and most used method based on the gray levels to measure the similarity metric between two images [30–32]. It measures the statistic correlations between two images based on the Shannon entropy. It is computed for an image using the distribution of the gray values within the image. If each pixel in an image is viewed as a random event, the information contained in the image can be measured by Shannon entropy. It can be viewed as a measure of uncertainty or how much information an image contains [32]. For two images A and B, mutual information MI (A, B) is computed as follows [33]:

$$MI(A, B) = H(A) + H(B) - H(A, B) \tag{4}$$

The interpretation of Eq. (4) form is that measures the distance between the joint distribution of image pixel values and the joint distribution in case of independence of the images A and B. H(A) and H(B) are the entropies of A and B respectively, and H(A, B) is the joint entropy of A with B. H(A) and H(A, B) can be calculated as:

$$H(A) = -\sum_{i=1}^{La} P_i(a) - \log P_i(a) \tag{5}$$

$$H(A, B) = \sum_{i=1}^{La} \sum_{j=1}^{Lb} P_{i,j}(a, b) \log P_{i,j}(a, b) \tag{6}$$

where La and Lb are the number of different colors in images A and B respectively, $P_i(a)$ is the probability of occurrence of color value i in image A, and $P_{ij}(a; b)$ is the probability of occurrence of color pair i and j in images A and B.

3.2.5 Structural Similarity Index Matrix (SSIM)

SSIM is a quality measure of one of the image being compared with another image having perfect quality. It is an improved version of the universal image quality index (UIQI). The mean SSIM is equal to one when two images are same. Unlike RMSE and PSNR, SSIM does not estimate the perceived errors. Conversely, SSIM considers image degradation as perceived change in structural data having strong interdependencies (important information about the structure of the object) specifically when they are spatially close. SSIM is calculated as follows:

$$SSIM = \frac{(2xy + C_1)(2\sigma xy + C_2)}{(\sigma x^2 + \sigma y^2 + C_2)(\bar{x}^2 + \bar{y}^2) + C_1} \tag{7}$$

where x(i, j) and y(i, j) are the reference and transformed images respectively, $\bar{x}, \bar{y}$ are the mean values for reference and sensed images respectively, σx^2 and σy^2 are the variance of x and y images, and C_1 and C_2 are the constants defined as:

$$C_1 = (k_1{}^*L)^2 \quad , \quad C_2 = (k_2{}^*L)^2 \tag{8}$$

where k_1 = 0:01 and k_2 = 0:03 and L = 2n–1

3.2.6 Ratio Image Uniformity (RIU)

Woods et al. proposed the ratio image uniformity (RIU) as the similarity measure in [34] for intra-modality alignment of medical images. The ration image uniformity is calculated as:

$$RIU = \frac{\sqrt{\frac{1}{N}\sum_x (R(x) - \bar{R})^2}}{\bar{R}} \tag{9}$$

where $\bar{R}$ is the mean value of R(x), R(x) = A(x)/B(x)

The registration strategy assumes that intensity ratio is maximally uniform across voxels if the two images are accurately registered. If σ is the standard deviation of R(x), and $\bar{R}$ is the mean value of R(x), this strategy uses σ/R as the similarity measure to evaluate how well the two images are registered.

3.2.7 Minkowski Similarity Distance

It is defined based on the Lp norm as follows:

$$D_p(S, R) = (\sum_{i=0}^{N-1} (S_i - R_i)^p)^{(1/p)} \tag{10}$$

where $D_p(S, R)$ is the distance between the two feature vectors $S = S_1, S_2, \ldots\ldots S_{N-1}$, $R = R_1, R_2, \ldots\ldots R_{N-1}$ representing the descriptors of the extracted keypoints from the input images [5]. Another distance measure is FLANN. It stands for Fast Library for Approximate Nearest Neighbors. FLANN matcher contains a collection of algorithms optimized for fast nearest neighbor search in large datasets and for high dimensional features. It works faster than other matchers [35].

3.3 Geometric Transformations

Pre-processing operations, sometimes known as image restoration and rectification. They are intended to correct specific radiometric and geometric distortions of data. Radiometric corrections may be necessary due to variations in scene illumination and viewing geometry, atmospheric conditions, and sensor noise and response. Each of these will vary depending on the specific sensor and platform used to acquire the data and the conditions during data acquisition. Also, it may be desirable to convert and/or calibrate the data to known (absolute) radiation or reflectance units to facilitate comparison between data [36]. Variations in illumination and viewing geometry between images (for optical sensors) can be corrected by modelling the geometric relationship and distance between the areas of the Earth's surface imaged, the sun and the sensor. This is often required to compare more readily images collected by different sensors at different dates or times, or to mosaic multiple images from a single sensor while maintaining uniform illumination conditions from scene to scene [36].

Scattering of radiation occurs as it passes through and interacts with the atmosphere. This scattering may reduce, or attenuate, some of the energy illuminating the surface. In addition, the atmosphere will further attenuate the signal propagating from the target to the sensor. Various methods of atmospheric correction can be applied ranging from detailed modelling of the atmospheric conditions during data acquisition, to simple calculations based solely on the image data.

All remote sensing imagery is inherently subject to geometric distortions (Fig. 12). These distortions may be due to several factors, including the perspective of the sensor optics, the motion of the scanning system, the motion of the platform, the platform altitude, attitude, and velocity, the terrain relief, and the curvature and rotation of the Earth. Geometric corrections are intended to compensate for these distortions so that the geometric representation of the imagery will be as close as possible to the real world. Many of these variations are systematic or predictable in nature and can be accounted for by accurate modelling of the sensor and platform motion and the geometric relationship of the platform with the Earth. Other unsystematic or random errors cannot be modelled and corrected in this way. Therefore, geometric registration of the imagery to a known ground coordinate system must be performed [36].

The geometric registration process involves identifying the image coordinates (i.e. row, column) of several clearly discernible points, (GCPs), in the distorted image

Fig. 12 Image-to-map
registration: *A* GCPs in the
distorted image. *B* True
ground coordinates
measured from a map
(*Source* Canada Center for
Remote Sensing. [Online]
http://crs.nrcan.gc.ca)

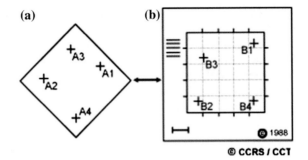

© CCRS / CCT

(A – A1 to A4), and matching them to their true positions in ground coordinates (e.g. latitude, longitude). The true ground coordinates are typically measured from a map (B – B1 to B4). This is known as image-to-map registration. Once several well-distributed GCP pairs have been identified, the coordinate information is processed by the computer to determine the proper transformation equations to apply to the original (row and column) image coordinates to map them into their new ground coordinates. Geometric registration may be also performed by registering one (or more) images to another image, instead of to geographic coordinates. This is called image-to-image registration and is often done prior to performing various image transformation procedures or for multi-temporal image comparison.

The transformation model determines which kind of geometrical transformation can be applied to the scene image to reach the model. This also controls which geometrical properties (e.g. size, shape, position, orientation, etc.) are preserved through the process [36].

Common models include rigid transform, which allows translations and rotations, similarity transform, which also admits scaling, and affine transformation, which can also represent shearing [37]. Figure 13 outlines the relation between the different transformation models. At the other end of the spectrum there are non-rigid (also called elastic) transformations, such as B-spline and thin-plate splines transformations, able to represent local deformations (warping) using hundreds or even

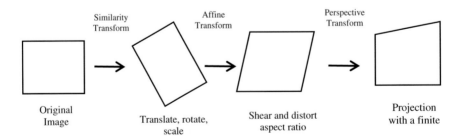

Fig. 13 Relation between different rigid transformations [37]

thousands of parameters. The following subsections briefly describe examples few transformation models.

3.3.1 Isometry Transformation

Isometry transform maps elements to the same or another metric space such that the distance between the image elements in the new metric space is equal to the distance between the elements in the original metric space. It considers only rotation and translation transformation and can be written in matrix form as [37]:

$$\begin{pmatrix} x' \\ y' \\ 1 \end{pmatrix} = \begin{pmatrix} \cos(\theta) & -\sin(\theta) & dx \\ \sin(\theta) & \cos(\theta) & dy \\ 0 & 0 & 1 \end{pmatrix} \begin{pmatrix} x \\ y \\ 1 \end{pmatrix}$$

where (x', y') is the transformed coordinate of (x, y), dx and dy are x-axis and y-axis translation respectively. The transformation matrix has the following properties: An orthogonal matrix, Euclidean distance is preserved, as three parameters; two for translation, and one for rotation

3.3.2 Similarity Transformation

This transformation considers rotation, translation and scaling transformation and can be written in matrix form as [37]:

$$\begin{pmatrix} x' \\ y' \\ 1 \end{pmatrix} = \begin{pmatrix} S_x \cos(\theta) & -\sin(\theta) & dx \\ \sin(\theta) & S_y \cos(\theta) & dy \\ 0 & 0 & 1 \end{pmatrix} \begin{pmatrix} x \\ y \\ 1 \end{pmatrix}$$

where (x', y') is the transformed coordinate of (x, y), dx and dy are x-axis and y-axis translation respectively, S and s is a scale factor. Minimum of two corresponding points in the images are required to determine the 4 parameters. Angles are preserved under the similarity transformation and it's useful when registering distant orthographic images of flat scenes. The transformation matrix has the following properties: An orthogonal matrix, similarity ratio (the ratio of two lengths) is preserved, it has four degrees of freedom; two for translation, one for rotation, and one for scaling.

3.3.3 Affine Transformation

An affine transform represents distortion of the aspect ratio and shearing of the image [37]. It is a combination between linear transformation and translation. It's considered as a simple transformation method, and also known as RST (Rotation, Scaling, and Translation). Although it is faster than the other transformations but it has some limitations and less accuracy than the polynomial model for example. Parallel lines remain parallel and straight lines remain straight. Affine transformation is represented in matrix form using homogeneous coordinates (with six unknown parameters) as:

$$\begin{pmatrix} x' \\ y' \\ 1 \end{pmatrix} = \begin{pmatrix} a_{11} & a_{12} & t_x \\ a_{21} & a_{22} & t_y \\ 0 & 0 & 1 \end{pmatrix} \begin{pmatrix} x \\ y \\ 1 \end{pmatrix}$$

where $x' = x + t_x$ and $y' = y + t_y$. In affine transformations, the relation between two corresponding points (x, y) and (X, Y) could be formulated as [38]:

$$x = m_1 + m_2 X + m_3 Y$$
$$y = n_1 + n_2 X + n_3 Y \tag{11}$$

where m_{1-3} and n_{1-3} are the transformation coefficients. The transformation matrix has the following properties: Ratio of lengths of parallel line segments is preserved, it has six unknown; two for translation, one for rotation, one for scaling, one for scaling direction, and one for scaling ratio

3.3.4 Polynomial Transformation

The first order polynomial warping includes XY interaction term which allows image shear as [38]:

$$x = m_1 + m_2 X + m_3 Y + m_4 XY$$
$$y = n_1 + n_2 X + n_3 Y + n_4 XY \tag{12}$$

where m_{1-4} and n_{1-4} are the transformation coefficients.

3.3.5 Projective Transformation

The perspective transform represents a full eight degrees of freedom in the homography. The perspective transform preserve the straightness of lines and planarity of surfaces, and the curved transformations, which do not. The projective transformations, which have the form, can be represented in homogeneous coordinates by [37]:

$$\begin{pmatrix} x' \\ y' \\ 1 \end{pmatrix} = \begin{pmatrix} a_{11} & a_{12} & a_{13} \\ a_{21} & a_{22} & a_{23} \\ a_{31} & a_{23} & a_{33} \end{pmatrix} \begin{pmatrix} x \\ y \\ 1 \end{pmatrix}$$

Fig. 14 Nearest Neighbour
Resampling (*Source* Canada
Center for Remote
Sensing. [Online] http://ccrs.
nrcan.gc.ca)

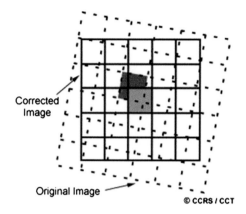

where (x', y') is the transformed coordinate of (x, y). The transformation matrix has
the following properties: Cross ratio preserved and it has 9 unknown parameters.

After getting the set of inliers, the warping phase starts using different non-rigid
geometric transformations including affine and polynomial transformation models.
We have also used a global homography transformation model as a warping method
with a linear blending resampling technique. In order to evaluate local feature detectors against different types of transformations, a global homography transformation
is used. An input image (S) is transformed to a new image (N) by applying the
following transformation [39]:

$$N(x', y') = s \left(\frac{h_{11}x + h_{12}y + h_{13}}{h_{31}x + h_{32}y + h_{33}}, \frac{h_{21}x + h_{22}y + h_{23}}{h_{31}x + h_{32}y + h_{33}} \right) \tag{13}$$

where h is a 3×3 transformation matrix. The homography matrix values are randomly generated within a uniform distribution.

3.4 Image Resampling

In order to actually geometrically correct the original distorted image, a procedure
called resampling is used to determine the digital values to place in the new pixel
locations of the corrected output image. The resampling process calculates the new
pixel values from the original digital pixel values in the uncorrected image. There
are three common methods for resampling: nearest neighbour, bilinear interpolation,
and cubic convolution.

Nearest-neighbor interpolation is the simplest but the least accurate interpolation
methods (Fig. 14). It uses the digital value from the pixel in the original image which
is nearest to the new pixel location in the corrected image. This is the simplest
method and does not alter the original values, but may result in some pixel values

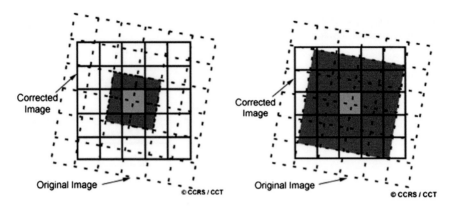

Fig. 15 Bilinear Interpolation Resampling (*left*), Cubic Convolution Resampling (*right*) (*Source* Canada Center for Remote Sensing. [Online] http://ccrs.nrcan.gc.ca)

being duplicated while others are lost. This method also tends to result in a disjointed or blocky image appearance.

Bilinear interpolation resampling takes a weighted average of four pixels in the original image nearest to the new pixel location (Fig. 15, left). The averaging process alters the original pixel values and creates entirely new digital values in the output image. This may be undesirable if further processing and analysis, such as classification based on spectral response, is to be done. If this is the case, resampling may best be done after the classification process [36].

Cubic convolution resampling goes even further to calculate a distance weighted average of a block of sixteen pixels from the original image which surround the new output pixel location (Fig. 15, right). As with bilinear interpolation, this method results in completely new pixel values. However, these two methods both produce images which have a much sharper appearance and avoid the blocky appearance of the nearest neighbour method. (http://ccrs.nrcan.gc.ca). As suggested in [39], preprocessing steps depend on the choice that is made for a suitable fusion level. In case of pixel based image fusion in fact, the geocoding is of vital importance and such details as geometric model, GCPs, DEM and resampling method need to be taken into consideration.

Trilinear Interpolation: it assumes the intensity varies linearly with the distance between the grid points along each direction. It considers all the contributions to interpolation point from the eight neighboring pixels. The distances are different between interpolation point and each of neighboring voxels. Therefore, the contribution (weight) from each voxel is different. Trilinear interpolation sums up the contribution (weight) from each neighboring voxel as the intensity value of interpolation point [36].

Partial Volume Interpolation: constructs the joint histogram without introducing new intensity values. Rather than interpolating the intensity value in the reference space, partial volume interpolation distributes the contribution from image intensity

S(s) over the neighboring eight voxels within the reference space. Eight entries of the joint histogram are updated by adding the corresponding weight at the same time. The calculation of weights from eight voxels is identical to the trilinear interpolation.

Triangulation Warping: A triangulation of set of points P= {p_1,...,p_n} is a maximal planar subdivision with vertex set P. Delaunay triangulation warping fits triangles to the irregularly spaced GCPs and interpolates values to the output grid.

In order to perform the resampling step between the sensed and the reference images, the following methods have been tested: Nearest Neighbor, linear/bilinear, and cubic interpolation. The nearest neighbor method uses the nearest pixel without any interpolation to create the warped image while the bilinear one performs a linear interpolation using four pixels to resample the warped image. In the cubic interpolation, 16 pixels (4 × 4) are used to approximate the since function using cubic polynomials to resample the image. It takes usually longer time compared to the linear or the linear methods so it is used when the speed is not the issue. It's smoother and less in distortion than other methods. In short, the registration process depends on the quality or the tie points and on the type of the transformation model used in the warping phase in addition to the resampling method.

3.5 The Proposed Registration Framework

The proposed registration framework (Fig. 16) starts by applying a speckle noise removal on the TerraSAR-X radar images. Such step increases the quality of SAR and active radar images against any possible degradation. This includes applying different adaptive filters like Lee or Frost (available in GIS software like ENVI with the possibility to change or use different noise models (multiplicative or additive). Also, optical images may need a pre-processing step in order to enhance the edges and lead to more accurate detection.

The next step is to detect and extract the GCPs from both optical and radar images. The detailed procedure for extracting and matching the GCPs from the input images is summarized as follows [38]:

(1) Extract the GCPs from the input images and compute their descriptors (CP1 and CP2 if there are two input images). If the number of the detected GCPs exceeds a predefined number of the key points, then the detected GCPs are filtered by sorting them according to their response.

(2) Once the GCP descriptors are built, the similarity measurements can be used to measure how the two sets of CPs are similar. In this step, the nearest neighbor is considered as the GCP with minimum Euclidean distance for the invariant descriptor. Different distance similarity measures have been tested in the experiments including Euclidean distance, Manhattan, and FLANN distance measures).

(3) The outlier removal is done using RANSAC (RANdom SAmple consensus) to exclude the inconsistent matches and help getting the list of tie points repre-

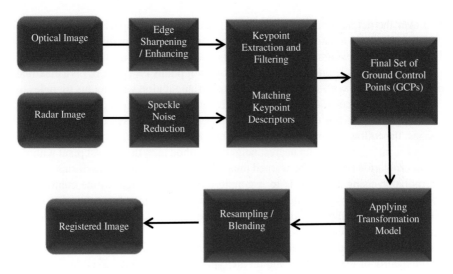

Fig. 16 The proposed registration scheme of optical and radar images

senting the actual matches between the two input images (known as the set of inliers).

Once we get the set of inliers, a suitable transformation model is chosen to map the points of the sensed image to the corresponding ones in the reference images (or vice versa). The resampling or the interpolation method is then chosen to warp the sensed image on the reference one or bring the geometry of one image to the other in order to get the registered version. Although it depends on the application, but the implemented system allows the user to get a mosaic version to subjectively judge how accurate is the registration process?.

The experiments have been also applied on same sensor registration which makes it available to register optical images together and radar images too as will be explained in the next section.

4 Experiments and Discussion

In this section, variety of experiments have been run in order to enhance the entire registration process. This include experiments on the same and different satellite sensors. In our case, we have tested RapidEye, Pléiades, ASTER, IKONOS-2, and TET-1 optical images in addition to and TerraSAR-X radar images covering different areas in Egypt, Germany and Australia (Fig. 17 shows different samples from the tested images). They are available in different spatial and ground resolution as follows [40]:

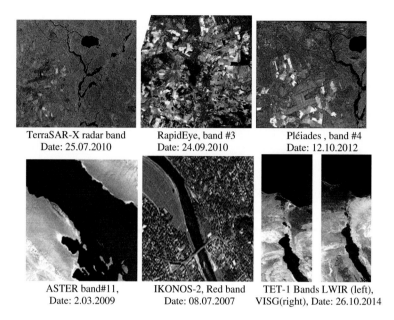

TerraSAR-X radar band Date: 25.07.2010	RapidEye, band #3 Date: 24.09.2010	Pléiades , band #4 Date: 12.10.2012
ASTER band#11, Date: 2.03.2009	IKONOS-2, Red band Date: 08.07.2007	TET-1 Bands LWIR (left), VISG(right), Date: 26.10.2014

Fig. 17 Samples from tested satellite images

TerraSAR-X. TerraSAR-X is a German Earth-observation satellite. Its primary payload is an X-band radar sensor with a range of different modes of operation, allowing it to record images with different swath widths, resolutions and polarisations. It offers space-based observation capabilities that were not available before. TerraSAR-X provides a value-added SAR data in the X-band, for research and development purposes as well as scientific and commercial applications. One of the things that makes TerraSAR-X stand out is its high spatial resolution using civilian radar systems. This enables scientists to examine detailed ground features, for instance the differentiation between different crops, in order to arrive at an improved classification of ground use (Table 1).

RapidEye. On August 29, 2008, a cluster of five identical medium-resolution satellites known as the RapidEye constellation was launched from Baikonur Cosmodrome, Kazakhstan, and reached an orbital height of 630 km. The constellation features 6.5-m resolution, 5-band multispectral (or blue, green, red, red edge and near-infrared/NIR) imagery. With a wide footprint, five satellites and daily revisits, the RapidEye constellation has amassed more than 5 billion km^2 of medium resolution imagery to date (Table 1).

Pléiades. Launched at the end of 2011 from Guiana Space Centre, Kourou, French Guiana, Pléiades 1 A was the first high resolution satellite in the Airbus Defense and Space constellation. It is available in 50 cm (centimetre) resolution -panchromatic and 2 m (meter) 4-band multispectral (i.e. blue, green, red and near-infrared/NIR)

Table 1 Samples from RapidEye, Pléiades, and TerraSAR-X satellite data

Optical/radar images	RapidEye		Pléiades		TerraSAR-X
Original dimension	5000 × 5000		37197 × 30258		25468 × 45065
Ground resolution	5 m-5 spectral bands		0.5 m-4 spectral bands		2.75 m
Bands	Blue	440–510 nm	Blue	430–550 nm	Active X-band microwave
	Green	520–590 nm	Green	490–610 nm	
	Red	630–685 nm	Red	600–720 nm	
	Red E.	690–730 nm	NIR	750–950 nm	
	NIR	760–850 nm			

products with the widest footprint of any high resolution satellites at 20 km (kilometres) (Table 1).

TET-1. TET-1 (Technologie Erprobungs Träger-1) is a German technology demonstration microsatellite of DLR (German Aerospace Center) within its OOV (On-Orbit Verification) program [41]. The overall objective is to provide industry and research institutes with adequate means for the in-flight validation of space technology. Certain programmatic rules were established for the space segment and the ground segment to realize TET-1 as a low-cost mission within a relatively short timeframe under the leadership of an industrial space company as prime contractor It covers areas with high temperatures like forest fire [40].

ASTER. ASTER (Advanced Spaceborne Thermal Emission and Reflection Radiometer) satellite is one of the five state-of-the-art instrument sensor systems on-board Terra, (December 1999), at Vandenberg Air Force Base, California, USA [GeoImage2013]. ASTER is a high-resolution sensor produces stereo imagery for creating detailed digital terrain models (DTMs). The resolution of images is between 15 to 90 m, and is used to create detailed maps of surface temperature of land, reflectance, and elevation.

IKONOS-2. IKONOS is the first commercially available high resolution satellite with imagery exceeding 1 m resolution (September, 1999) at Vandenberg Air Force Base, California, USA. Its capabilities include capturing a 3.2 m multispectral, Near-Infrared 0.82 m panchromatic resolution at Nadir with 681 orbit altitude. IKONOS-2 bands are: Panchromatic (1 m), blue, green, red, near IR (4 m multi-spectral).

Binary-valued feature descriptors (like BRIEF, BRISK and ORB) and also vector-based feature descriptors (like SIFT and SURF) have been also tested. In general, binary feature descriptors have some advantages over vector-based features in terms of less computation time and they are compact to be stored, in addition that they are efficient to be compared. Using SURF detection method, a comparison among these descriptors as in Table 2. The results on RapidEye and TerraSAR-X have showed that SIFT descriptor takes longer time to be built compared to other descriptors. Also,

Satellite Image Matching and Registration … 159

Table 2 The tested descriptors and their performance on RapidEye and TerraSAR-X images

Item/descript.	Optical/radar	BRIEF	BRISK	ORB	SIFT	SURF
# of detected Keypoints (KPs)	RapidEye	# of detected KPs using SURF detector is 271747, then these KPs are filtered later to n KPs, finally build the descriptor				
	TerraSAR-X					
KPs detection time (in seconds)	RapidEye	13.947	13.639	13.940	11.510	15.515
	TerraSAR-X	21.846	18.248	18.578	18.453	22.768
KP filtering time (in seconds)	RapidEye	0.035	0.86	0.025	0.028	0.034
	TerraSAR-X	0.058	0.050	0.060	0.052	0.060
Descriptor-time (in seconds)	RapidEye	0.183	0.213	0.372	4.415	0.540
	TerraSAR-X	0.151	0.169	0.498	4.566	0.367
Matching time using Euclidean distance (in seconds)		0.074	0.133	0.113	0.090	0.068

the matching time of SIFT descriptor using Euclidean distance is also greater than other descriptors but it still giving the most robust and stable performance.

4.1 Same Sensor Experiments

Same sensors images are usually taken at different times or from different viewpoints. In this part of the experiments we have tested both optical and radar images taken from different viewpoints and at different times respectively. Figure 18 shows the inliers between a RapidEye image and its transformed version using random homography.

A linear blending method is also tested in our work where Dyadic (two-input) operator is the linear blend operator:

$$g(x) = (1 - \alpha) f_0(x) + \alpha f_1(x) \qquad (14)$$

where $f_0(x)$ and $f_1(x)$ are the two source images of the same type and size. By varying α (weight of the first image) from 0 to 1 this operator can be used to perform a temporal cross-dissolve between the two images. The $g(x)$ will generate an image, we consider $\beta = (1 - \alpha)$ as the weight of the second image. Then we calculate the weighted some of the two arrays (with 6 parameters, $f_0(x), f_1(x), \alpha, \beta, \gamma,$ and dst) where $dst = \alpha f_0(x) + \beta f_1(x) + \gamma$. (Figure 19 shows an example of linear blending on Tet-1 images for an area near Canberra in Australia).

Table 3 presents a comparison between a full registration of two TET-1 bands using ORB and SURF detectors. It shows that ORB runs faster than SURF as it finds less number of keypoints that's why it takes less extraction and pairwise matching time. Here ORB and SURF detectors compute their own descriptors.

Fig. 18 The inliers between an original RapidEye image (*left*), and its transformed version using random homography (*right*)

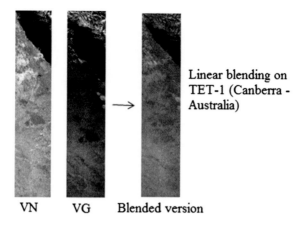

**Linear blending on
TET-1 (Canberra -
Australia)**

VN   VG   Blended version

Fig. 19 A linear blending between two TET-1 bands (VN and VG) and their blended version (*right*)
(*α = 50*)

Table 3 A Comparison between SURF and ORB detectors of registering two satellite bands from
TET-1 satellite images covering the area of Suez Canal University.)

	Item/Descript.	ORB	SURF
TET-1 VISG band	No. of extracted keypoints	1530	7789
TET-1 VISN Band	No. of extracted keypoints	1530	3989
	Extraction time (in seconds)	0.323	2.459
	Pairwise matches time (in seconds)	0.212	0.539
	Warping time (in seconds)	0.005	0.005
	Compositing time (in seconds)	0.732	0.504
	Total registration time (in seconds)	1.628	4.105

In Fig. 20a, two TET-1 bands covering Suez Canal area in Egypt are registered together using SURF detector. While in Fig. 20b, another same sensor registration but for two TerraSAR-X images in one area in Dresden in Germany, using ORB detector.

Figure 21 shows the general performance of different feature detectors on optical and radar images. SIFT, SURF, FAST, GFTT, and Harris have a good performance on different optical images while SIFT has performed better on radar images compared to the other detectors. SIFT descriptor has been used as a default description method in this evaluation.

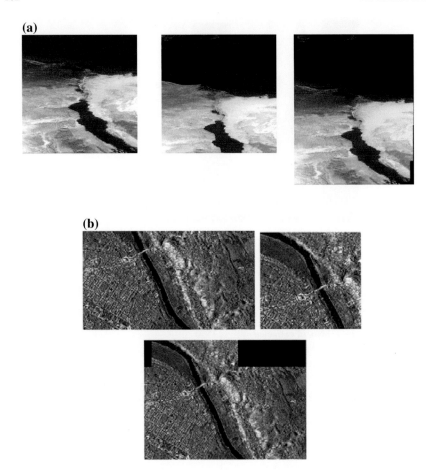

Fig. 20 Two examples of same sensor registration. **a** VISG (*left*) and VISN (*right*) TET-1 bands of the Suez Canal Area in Egypt, and their registered version using SURF (*right*), **b** two TerraSAR-X radar images covering an area in Dresden in Germany (*top left* and *right*), and their registered version using ORB (*down*)

4.2 Different Sensor Experiments

Image distortions may occur due to several factors. They may include: the perspective of the sensor optics, the motion of the scanning system or the platform, the platform altitude, attitude, and velocity, the terrain relief, and the curvature and rotation of the Earth [36]. Both of these effects should be corrected before further enhancement or classification is performed. Both Optical and radar images have different illumination characteristics. Furthermore, radar image suffer from speckle noise which can lead to false GCPs detections. Speckle noise is generally found in satellite and

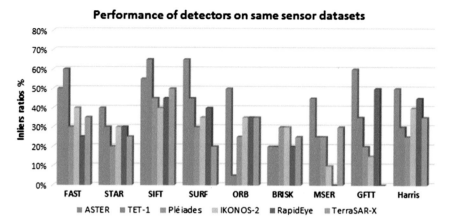

Fig. 21 The general inliers ratios (%) of the tested detectors (using SIFT descriptor) on different optical and radar satellite Images

medical images. They are usually degraded by noise during the acquisition and the transmission processes. Any speckle noise removal aims to remove speckle noise by retaining the important features within the images.

Methods used for noise reduction include adaptive and non-adaptive methods. Adaptive speckle filters are supposed to be better as they preserve both edges and details in high-texture areas like forests and urbans. On the other hand, non-adaptive filters are simpler to implement, and requires less computational power. Lee filters are used to smooth noisy (speckled) data that have an intensity related to the image scene and have also an additive and/or multiplicative component. Lee filter is a standard deviation based filter that filters data based on statistics calculated within individual filter windows (e.g. 3×3 or 5×5). The main idea is that the pixels are filtered and replaced by a value calculated using the surrounding pixels. Within each windows, the local mean and variance are estimated. The filter outputs the local mean when no signal is detected while it passes the original signal unchanged. Optical and radar images are representing two different ways of sensing named active and passive sensing. In this part of the experiments, we have tested multi-sensor registration between RapidEye/Pléiades optical images and TerraSAR-X radar image of the area Berlin Brandenburg airport area. On the other hand, optical images are also processed before the feature detection process starts (if required). They have different radiometric correlation than radar images. For this reason, applying edge detection methods aims to overcome this problem. Edges do not require such radiometric correlations as they are defined as a boundary between any two radiometric features. Canny edge detection has been applied on optical images in the multi-sensor framework. It detects edges with noise suppressed at the same time. The steps for applying Canny edge detection are: smooth the image by Gaussian convolution, highlight regions of the image with high first spatial derivatives.

Edges give rise to ridges in the gradient magnitude image, track along the top of these ridges and sets to zero all pixels that are not actually on the ridge top so as to give a thin line in the output. This process is known as non-maximal suppression. Then, threshold edges in the images with the hysteresis process to eliminate false response. Two thresholds t_1 and t_2 are used where $t_1 > t_2$ as in Fig. 22a. Tracking can only begin at a point on a ridge higher than t_1. Tracking then continues in both directions out from that point until the height of the ridge falls below t_2. Also, in Fig. 22b, an example of an original TerraSAR-X after apply the speckle noise removal. Then the enhanced versions are used as sensed or reference images in the registration process. The quality of the registration process between RapidEye and Pléiades with the de-noising version from TerraSAR-X is better than using the original used band.

In Figs. 23 and 24, RapidEye and Pléiades optical images of the area of Berlin Brandenburg airport are registered to TerraSAR-X radar image using SIFT detector and global homography transformation.

A proposed filtering process to control and number of detected GCPs with the aim to use only robust and strong keypoints. The aim of this study is to develop a generalized scheme for the registration process based on the same or different sensors. The registration algorithm presented in this chapter has the following characteristics: an automatic detection and extraction of local features from images.

An evaluation of polynomial, affine, and global homography transformations with different interpolation techniques on RapidEye and TerraSAR-X images is presented in Fig. 25. This comparison has been done on different samples from RapidEye and TerraSAR- X images covering the whole and some parts from Berlin Brandenburg airport area. This will help to minimize the running time and hardware requirements when manipulating such huge high resolution satellite images. Different feature matching methods in order to enhance the overall performance of the extraction and the matching phase.

Few transformation models and resampling techniques in order to warp and composite the registered version with different subjective and objective evaluation, furthermore, a synthesize test with other evaluation methods to measure how robust the tested detectors to rotation, translation and scaling and their performance against noise as well. A full registration process on optical and radar images from the same and different sensors covering different viewpoints and times in addition to different sensors. The quality of the registration process depends on how accurate is the correspondence between pixels from the input images which emphasizes the importance of the detection and the matching steps. Here, the role of both local feature detectors and descriptors in the extraction and the matching of features from satellite images is increasing and providing a faster and more accurate registration of remotely sensed images.

(a)

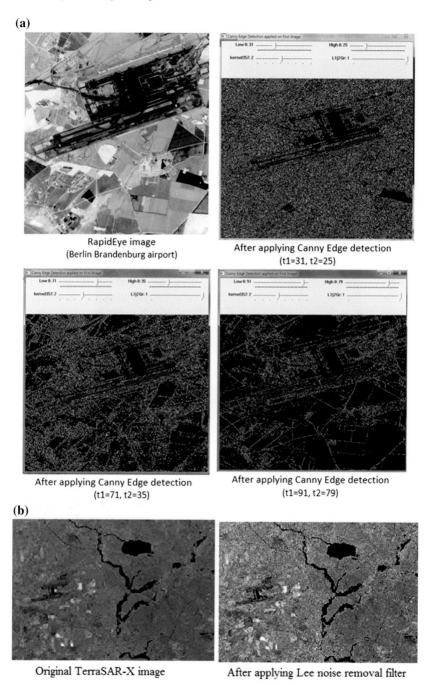

(b)

Fig. 22 Samples from the pre-processing steps on optical (Canny Edge detection as in (**a**)) and radar (Lee speckle noise removal as in (**b**))

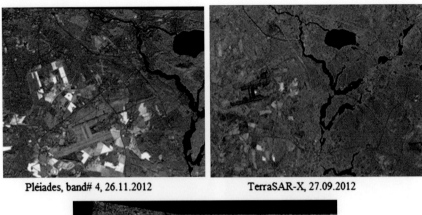

Pléiades, band# 4, 26.11.2012   TerraSAR-X, 27.09.2012

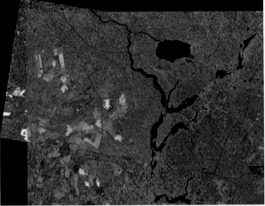

Fig. 23 TerraSAR-X and Pléiades image registration using SIFT and global homography and linear blending (SIFT detector is usually used with its own descriptor in the whole experiments)

5 Conclusions and Future Work

5.1 Conclusions

Satellite image matching and registration are two basic operations when manipulating remote sensing data for many applications. In this chapter, different local feature detectors and their corresponding descriptors have been tested and evaluated on different optical and satellite images. It studies the behaviour of several invariant local features on same and different sensor image registration. It aims to investigate and enhance the whole registration process including: feature extraction and matching, geometric transformations and image resampling. Furthermore, a full registration between optical and radar images with variety of options is presented.

The suggested framework filters the detected keypoints (in case of high resolution images) in order to reduce the running time during the extraction and the matching

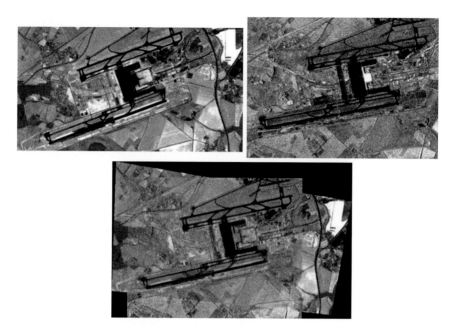

Fig. 24 Two RapidEye and TerraSAR-X Images (*top left* and *right* respectively), and their registered version using SIFT and global homography transformation

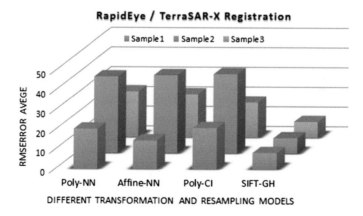

Fig. 25 A comparison among different geometric transformations with resampling methods on three optical/radar samples

stages. RANSAC is used as an outlier removal for excluding the false matches and get an enhanced tie points' list between the sensed and the reference images. The results showed a good performance of SIFT on TerraSAR-X radar images while other detectors like SURF and Harris have a good performance on RapidEye/Pléiades optical images. A comparison among different non-rigid geometric transformation methods is also presented. A full registration of Rapid-Eye and Pléiades optical, and TerraSAR-X radar images is done using SIFT, and global homography transformation with linear blending.

We aim to develop a robust and generalized scheme for registering both optical and radar images for supporting multi-modal remote sensing applications. The role of local features detectors and descriptors in this development is important as they offer a faster and an automatic way of the matching and registration of same and different sensor satellite imagery.

5.2 Future Work

Any future work ideas will depend on the ability of developing the extraction and the matching processes for specific applications. The work presented here aims to developing a generalized and robust scheme for registering single and multisensor remote sensing data. Several enhancements to the automatic registration process can be developed. This can include:

- Testing additional very high resolution optical and radar images for different applications using feature-based image extraction.
- Studying the role of domain experts in order to switch to intelligent and optimized registration schemes. This can handle some of the current obstacles in some applications.
- Investigating a possible enhancement via combining correlation and feature-based methods. This aims to benefit from the advantages of using the two approaches. Dissimilarity measures can play a critical role in such combination together with the estimation of the transformation model.
- Enhancing the registration of remote sensing data so that specific applications can be done with a real-time processing and good performance.
- Feature-based precision corrections of image registration is still a challenge on remote sensing data. Geolocation accuracy has to be refined by selecting optimal or correct control points.

6 Key Terms and Definitions

Feature Extraction: Is the process of detecting features like control points or corner or edges from images and building descriptors that contain such features.

Feature detector: identifies the interest points (i.e. regions of interest) within images.

Feature descriptor: computes the feature vector describing the detected interest points (regions)

Target image: is the reference image or the image which does not change. Other images are aligned or warped to it.

Source image: is the sensed image and it has to be aligned or warped on the reference one.

Invariant Local Features: they are features within an images which are not changing against certain transformations and could be used as a robust image representations.

Image Matching: is the process of matching saved descriptors or feature vectors in order to find how far the matched images are similar to each other. Recent advances in image matching includes multimodal image matching where different kind of features are extracted and tested for increasing the accuracy and reliability of the matching.

Similarity Measurements: is a real-valued function that measures the similarity between two feature vectors or descriptors. Euclidean distance, cross correlation, mutual information are such examples of famous similarity measurements or matchers.

Image registration: is the process of determining a geometrical transformation that aligns points or pixels in one image with the corresponding points in the other image having the same scene.

Remote sensing: is the acquisition of information about objects without having a direct contact with the objects. This term is mainly about the use of aerial sensor technologies to detect and classify objects on Earth. Change detection, land use, weather forecasting.

GIS: Geographic Information System: is the optimum tool to handle and integrate large amount of spatially referenced data including remotely sensed data.

A spatial transformation: defines a geometric relationship between each point in the input and output images.

References

1. Tuytelaars, T., Mikolajczyk, K.: Local invariant feature detectors: a survey, foundations and trends. Comput. Graph. Vis. **3**(3), 177–280 (2007)
2. Juan, L., Gwon, O.: A comparison of sift, PCA-SIFT and SURF. Int. J. Image Process. **3**(4), 143–152 (2009)
3. Hong, T.D., Schowengerdt, R.A.: A Robust technique for precise registration of radar and optical satellite images. Photogram. Eng. Remote Sens. **71**(5), 585–594 (2005)
4. Khan, N.Y., McCane, B., Wyvill, G.: SIFT and SURF performance evaluation against various image deformations on benchmark dataset. In: Proceedings of the 2011 International Conference on Digital Image Computing: Techniques and Applications (DICTA '11), pp. 501–506 (2011)
5. Tahoun, M., Shabayek, A., Hassanien, A., Reulke, R.: An Evaluation of Local Features on Satellite Images. In: Proceedings of the 37th International Conference on Telecommunications and Signal Processing (TSP), pp. 695–700. Berlin, Germany (2014)
6. Bouchiha, R., Besbes, K.: Automatic remote sensing image registration using SURF. Int. J. Comput. Theory Eng. **5**(1), 88–92 (2013)

7. Brown, M., Lowe, D.: Automatic panoramic image stitching using invariant features. Int. J. Comput. Vis. **74**(1), 59–73 (2007)
8. Shum, H.Y., Szeliski, R.: Construction of panoramic image mosaics with global and local alignment. Panoramic Vision, pp. 227–268. Springer, New York (2001)
9. McLauchlan, P., Jaenicke, A.: Image Mosaicing using sequential bundle adjustment. Image Vis. Comput. **20**, 751–759 (2002)
10. Bradski, G.: OpenCV Library. Dr. Dobb's journal of software tools (2000). http://opencv.org/
11. Zhang, Y., Wu, L.: Rigid image registration based on normalized cross correlation and chaotic firey algorithm. Int. J. Digital Content Technol. Appl. **6**(22), 129–138 (2012)
12. Pohl, C., Van Genderen, J.L.: Multisensor image fusion in remote sensing: concepts, methods and applications. Int. J. Remote Sens. **19**(5), 823–854 (1998)
13. Schmidt, A., Kraft, M., Kasinski, A.J.: An evaluation of image feature detectors and descriptors for robot navigation. In: ICCVG (2)'10, pp. 251–259 (2010)
14. Murthy, V.V.S., Mounica, Y., Vakula, K., Krishna, K.V., Ravichandra, S.: Review of performance estimation for image registration using optimization technique. Int. J. Syst. Technol. **7**(1), 7–12 (2014)
15. Deshmukh, M., Bhosle, U.: A survey of image registration. Int. J. Image Process. **5**, 245–269 (2011)
16. Mikolajczyk, K., Tuytelaars, T., Schmid, C., Zisserman, A., Matas, J., Schaffalitzky, F., Kadir, T., Gool, L.: A comparison of affine region detectors. Int. J. Comput. Vis. **65**(1–2), 43–72 (2005)
17. Lowe, D.G.: Distinctive image features from scale-invariant keypoints. Int. J. Comput. Vis. **60**(2), 91–110 (2004)
18. Bay, H., Tuytelaars, T., Gool, L.V.: Surf: speeded-up robust features. Computer Vision—ECCV 2006. Lecture Notes in Computer Science, vol. 3951, pp. 404–417 (2006)
19. Harris, C., Stephens, M.: A combined corner and edge detector. In: Proceedings of the 4th Alvey Vision Conference (AVC), pp. 147–151 (1988)
20. Moravec, H.: Obstacle avoidance and navigation in the real world by a seeing robot rover. Technical Report CMU-RI-TR-3 Carnegie-Mellon University, Robotics Institute (1980)
21. Agrawal, M., Konolige, K., Blas, M.R.: Censure: center surround extremas for realtime feature detection and matching. In: European Conference on Computer Vision (ECCV), pp. 102–115. Springer, Berlin (2008)
22. Rosten, E., Drummond, T.: Machine learning for high speed corner detection. In: The 9th European Conference on Computer Vision, vol. 1, pp. 430–443 (2006)
23. Leutenegger, S., Chli, M., Siegwart, R.: BRISK: binary robust invariant scalable keypoints. In: Proceedings of the IEEE International Conference on Computer Vision (ICCV), pp. 2548–2555 (2011)
24. Calonder, M., Lepetit, V., Strecha, C., Fua, P.: Brief: binary robust independent elementary features. In: 11th European Conference on Computer Vision (ECCV). Computer Vision—ECCV 2010, Lecture Notes in Computer Science, vol. 6314, pp. 778–792 (2010)
25. Rublee, E., Rabaud, V., Kurt, K., Bradski, G.: An efficient alternative to SIFT or SURF. In: Proceedings of the 2011 International Conference on Computer Vision (ICCV '11), pp. 2564–2571. IEEE Computer Society, Washington, DC, USA (2011)
26. Shi, J., Tomasi, C.: Good features to track. In: Proceedings of the IEEE Conference on Computer Vision and Pattern Recognition, pp. 593–600 (1994)
27. Matas, J., Chum, O., Urban, O., Pajdla, T.: Robust wide baseline stereo from maximally stable extremal regions. In: Proceedings of British Machine Vision Conference, pp. 384–396 (2002)
28. Machowski, L.A., Marwala, T.: Evolutionary optimization methods for template based Image Registration, School of Electrical and Information Engineering (2004). arXiv:0705.1674
29. Zhang, X., Zhang, C.: Satellite cloud image registration by combining curvature shape representation with particle swarm optimization. J. Softw. **6**(3), 483–489 (2011)
30. Kher, H.R.: Implementation of image registration for satellite images using mutual information and particle swarm optimization techniques. Int. J. Comput. Appl. **97**(1), 7–14 (2014)

31. Maddaiah, P.N., Pournami, P.N., Govindan, V.K.: Optimization of image registration for medical image analysis. Int. J. Comput. Sci. Inf. Technol. **5**(3), 3394–3398 (2014)
32. Ayatollahi, F., Shokouhi, S.B., Ayatollahi, A.: A new hybrid particle swarm optimization for mutlimodal brain image registration. J. Biomed. Sci. Eng. **5**, 153–161 (2012)
33. Zhang, X., Shen, Y., Li, S., Ji, Y.: Medical image registration using a real coded genetic algorithm. J. Comput. Inf. Syst. **8**(12), 5119–5128 (2012)
34. Woods, R.P., Cherry, S.R., Mazziotta, J.C.: Rapid automated algorithm for aligning and reslicing PET images. J. Comput. Assist. Tomogr. **16**, 620–633 (1994)
35. Muja, M., Lowe, D.G.: Fast approximate nearest neighbors with automatic algorithm configuration. In: Proceedings of the International Conference on Computer Vision Theory and Application (VISSAPP), pp. 331–340 (2009)
36. Canada Center for Remote Sensing—CCRS (2007). Fundamentals of remote sensing. http://www.ccrs.nrcan.gc.ca
37. Lim, D.: Achieving accurate image registration as the basis for super resolution. PhD Thesis. School of Computer Science and Software Engineering, University of Western Australia (2003)
38. Tahoun, M., Hassanien, A., Reulke, R.: Registration of Optical and Radar Satellite Images Using Local Features and Non-rigid Geometric Transformations. Lecture Notes in Geoinformation and Cartography Series, pp. 249–261. Springer (2015)
39. Zaragoza, J., Chin, T., Brown, M., Suter, D.: As-projective-as-possible image stitching with moving DLT. In: Proceedings of Computer Vision and Pattern Recognition (CVPR), vol. 36, pp. 2339–2344 (2013)
40. GeoImage: GeoImage satellite overview (2013). http://www.geoimage.com.au/satellites/satellite-overview
41. Small satellite TET-1, Astrofein (2010). http://www.astrofein.com/2728/dwnld/admin/AstroFein_TET_EN.pdf

Redundancy Elimination in Video Summarization

**Hrishikesh Bhaumik, Siddhartha Bhattacharyya
and Susanta Chakraborty**

Abstract Video summarization is a task which aims at presenting the contents of
a video to the user in a succinct manner so as to reduce the retrieval and browsing
time. At the same time sufficient coverage of the contents is to be ensured. A trade-
off between conciseness and coverage has to be reached as these properties are
conflicting to each other. Various feature descriptors have been developed which can
be used for redundancy removal in the spatial and temporal domains. This chapter
takes an insight into the various strategies for redundancy removal. A method for
intra-shot and inter-shot redundancy removal for static video summarization is also
presented. High values of precision and recall illustrate the efficacy of the proposed
method on a dataset consisting of videos with varied characteristics.

Keywords Video summarization · Redundancy removal · Feature descriptors ·
Metrics for video summary evaluation · Three-sigma rule

1 Introduction

The ever growing size of online video repositories like DailyMotion, YouTube,
MyVideo etc. have propelled the need for efficient Content Based Video Retrieval
Systems. This has augmented research in several related fields such as, feature extrac-
tion, similarity/dissimilarity measures, video segmentation (temporal and semantic),

H. Bhaumik (✉) · S. Bhattacharyya
Department of Information Technology, RCC Institute of Information Technology,
Kolkata, India
e-mail: hbhaumik@gmail.com

S. Bhattacharyya
e-mail: dr.siddhartha.bhattacharyya@gmail.com

S. Chakraborty
Department of Computer Science and Technology,
Indian Institute of Engineering Science and Technology, Shibpur, Howrah, India
e-mail: susanta.chak@gmail.com

© Springer International Publishing Switzerland 2016
A.I. Awad and M. Hassaballah (eds.), *Image Feature Detectors and Descriptors*,
Studies in Computational Intelligence 630, DOI 10.1007/978-3-319-28854-3_7

key-frame extraction, indexing, annotation, classification and retrieval of videos. Video Summarization is a task which stands at the intersection of these research issues. It reduces to a great extent the demand for computing resources in processing the huge volumes of video data. The basic objective of the video summarization process is to provide a concise, yet meaningful representation of the video to the user. The efficacy of a video summarization system depends on maximizing two conflicting features-coverage and succinctness. The important application areas of video summarization include content-based video retrieval [1–3], semantic indexing [4], Copied Video Detection (CVD) [5], video surveillance [6], generation of highlights for sports [7–10], movies and drama [11–14], bandwidth-constrained video processing applications [15] etc. The hierarchical levels of the composing units in a video may consist of scenes, shots or frames depending on the granularity intended as depicted in Fig. 1. The composing units share a temporal relationship with each other. Distortion of semantic content occurs if these temporally sequenced units are disordered. A video may be represented as $V = u_1 \otimes u_2 \otimes u_3 \ldots \otimes u_n$, where u_i is the ith composing unit. Depending on the summarization approach used, a Boolean decision is made for each constituent unit of the video as to whether it will be a part of the generated summary or not. The mechanisms for selecting these units determine the efficacy of the approach used. Video summarization can also be viewed as a task of amalgamating those video units which have the maximum entropy. The system generated summary (SGS) consists of a subset of V (extraction type) or a transformed set of the elements in V (abstraction type). In both cases the duration of SGS is far smaller than V. Static video summarization falls under extraction type where a set of key-frames are chosen to represent the video. This is particularly helpful in bandwidth constrained scenarios, where the user needs to get an overview of the contents of a video. On the other hand, a dynamic video summary may be produced by coalescing together the units which have greater significance. In such case, the summary generated may be either extraction based (e.g. sports highlights package) where the chosen units have the same time-sequence in which they occur in the original video or abstraction based (e.g. movie trailer) where the selected units may be intermingled in a manner so as to produce a meaningful abstract of the given

Fig. 1 Hierarchical representation of video units

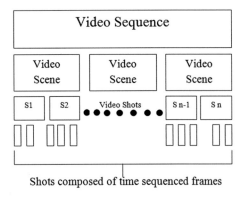

Shots composed of time sequenced frames

video. The challenges for designing a video summarization system arise from the fact that the summary generated must be represented by the most significant composing units. The ranking of video units according to significance is the crux of the problem besetting researchers in this field. An equally considerable problem is to remove redundancy without diminishing the coverage made by the representative units. Elimination of visual redundancy is possible by extracting mid-level features such as interest points. However, removal of semantic redundancy is a problem of a different dimension. The task of semantic redundancy elimination is more complex as it encompasses fields like object recognition, tracking, gesture/action identification, event detection etc. to name a few. Hence, such a task requires extraction of high level features. In this chapter, we focus on approaches taken for reduction of visual redundancy.

The rest of the chapter is organized as follows. Section 2 details the related works in video summarization using redundancy elimination approaches. Section 3 presents an insight into the redundancy elimination problem in video summarization and why it needs to be tackled. Section 4 enumerates the role of interest points detection in video summarization. This section also elaborates the various feature descriptors in use. The proposed method is presented in Sect. 5. In Sect. 6 some widely used metrics for measuring the quality of summary is presented. The results of the summarization process using the proposed method along with the details of the dataset used is elaborated in Sect. 7. Some concluding remarks are presented in Sect. 8.

2 Survey of Related Works

Redundancy occurs due to the appearance of similar visual content at several points in a video. This invariably increases the size of the resulting summary as it is formed by coalescing together several video units taken from different points in the video. The task of redundancy removal refers to the elimination of the repeated content which conflicts with the objective of producing a concise summary. Several techniques for redundancy removal have been devised over the years. These methods can be broadly categorized into two groups:

1. Techniques using feature descriptors
2. Other redundancy elimination techniques.

2.1 Methods of Redundancy Elimination Using Feature Descriptors

Similarity in visual content may be aptly captured by using feature descriptors. The feature descriptors capture medium-level semantic content. This is intermediate between low-level characteristics such as histogram comparisons [16, 17], statistical differences [18, 19], standard deviation of pixel intensities [20], frame-to-frame pixel intensity difference [21], gray level histograms of successive frames [17, 22], statistical information like mutual and joint entropy [19], mean and variance of pixel values, luminance etc. and high-level features like shapes of objects, edges in the frames, optical flow [23], motion vectors [24], event modeling [25], etc. The high-level features which are connected to the content of a video such as scenes, objects etc. are more natural to humans than the low-level features. As such the features which capture points on the objects rather than the whole semantic meaning come under mid-level features. The mid-level features are useful for detection and recognition of objects which have consistent low-level characteristics. However, these mid-level characteristics may not be useful for semantic analysis of the content in a video. Feature descriptors have been used in several video analysis problems. These include Shot Boundary Detection [18], Video Summarization [26], Object tracking [27] etc.

The various approaches related to Shot Boundary Detection aim at extracting feature descriptors from the time sequenced frames of a video. Depending on the feature descriptors extracted from the frames the similarity between consecutive frames are computed. The number of matched features is tracked to find abrupt discontinuities. The points of discontinuities are the shot boundaries in the video sequence. Gradual discontinuity patterns indicate smooth transition from one shot to another. These are achieved through fades, wipes and dissolves. Apart from detecting shot boundaries, the matched feature descriptors also serve as indicators to the amount of content match between two shots. This aspect is exploited by researchers in tasks related to summarization of a video. In [28], key-points are recognized for all the frames of each shot in a video. A set of unique key-points is built for the shots. The set of feature descriptors corresponding to the key-points are extracted. The representative set of key-frames is constructed such that minimum number of frames covers the entire pool of key-points. This ensures maximizing the coverage and minimizing the redundancy [29]. In [30], an approach for static video summarization using semantic information and video temporal segmentation is taken. The performance and robustness of local descriptors are also evaluated as compared to global descriptors. The work also investigates, as to whether descriptors using color information contribute to better video summarization than those which do not use it. Also the importance of temporal video segmentation is brought out in the work. The summarization process uses a bag of visual words concept where the feature descriptors are used to describe a frame. Visual word vectors are formed to cluster similar frames and finally filter out the representative frames. The performance of various feature detectors and descriptors in terms of tracking speed and effectiveness were evaluated in [31]. The work pertains to evaluation of these feature descriptors for face detection in real-time videos. Change in structure for non-rigid objects, sudden changes in object motion resulting in varied optical flow, change in manifestation of objects, occlusions in the scene and camera motion are some of the inherent challenges which have to be overcome for accurate tracking of objects. A proper amalgamation of these feature descriptors may serve to improve the overall tracking precision. The work concludes that a single feature detector may not provide enough accuracy for object tracking.

An event detection system has been proposed in [32] for field sport videos. The system runs two parallel threads for detecting text and scene in the video stream. The output of a text detector is provided to a scoreboard analyzer for notifying the user of an interesting event. The scene analyzer which runs parallel to the text analyzer takes input from the scene detector and provides output to the event notification system to tag the interesting sequences. Since the approach is designed for real-time videos, the authors stress the need for using feature detectors which are inherently fast and have an acceptable recognition rate.

In sports videos, the event is usually covered by a fixed set of cameras on stands. As such the coverage is free from rotational variance. Also there are long shots and close-ups which need to be distinguished. The algorithms designed for such purposes may therefore ignore scale and rotation invariance strategies. The BRIEF descriptor was chosen for this work as it satisfies these considerations and is computationally efficient. BRIEF has been reported to be almost sixty times faster than SURF, while ensuring an acceptable recognition rate [33]. In [34] an elaborate comparison of the various descriptor extraction techniques is presented. The work reviews techniques like SIFT, DIFT, DURF and DAISY in terms of speed and accuracy for real-time visual concept classification. A number of high speed options have been presented for each of the components of the Bag-of-Words approach. The experiments consist of three phases i.e. descriptor extraction, word assignment to visual vocabulary and classification. The outcome of this work can be extended for designing robust methods for redundancy elimination based on visual concept classification.

Li [35] employs SIFT as the basis for computing content complexity and frame dissimilarity. This allows detection of video segments and merging of the shots based on similarity. Key-frames are then extracted from these merged shots. In [36], Compact Composite Descriptors (CCDs) [37] and the visual word histogram are extracted for each image. The object descriptor used is based on SURF. The CCD consists of four descriptors i.e. the Color and Edge Directivity Descriptor (CEDD) [38], the Fuzzy Color and Texture Histogram (FCTH) [39], the Brightness and Texture Directionality Histogram (BTDH) [40] and the Spatial Color Distribution Descriptor (SpCD) [41]. A Self-Growing and Self-Organized Neural Gas (SGONG) network is used for frame clustering. The main aspect of this method is the ability to determine the appropriate number of clusters. As in some of the other methods, the cluster centers are chosen to generate the summary. Redundancy elimination is carried out in [42] by extracting the SURF and GIST features from the representative frames obtained by generating a Minimal Spanning Tree for each shot. The duplicate frames in the representative set are eliminated using a threshold based on the three-sigma rule in accordance with the number of descriptor matches for each pair of frames in the representative set. A comparison of the summaries after redundancy elimination using SURF and GIST are also elaborated.

2.2   Other Methods of Redundancy Elimination

Apart from using mid-level features in the form of interest points for removal of content duplication, several other methods have been proposed by the researchers. A few of the important approaches are presented in this section.

The approaches using key-frame selection for static video summarization aim to summarize the video by selecting a subset of frames from the original set of decomposed frames. In order to remove redundancy from the set of selected frames, clustering is applied on the set of selected key-frames by extracting features. One such method is presented in [43] which includes a feature extraction phase required for clustering. Duplicates from the selected key-frames are removed using a combination of local and global information. In [44] an exploration framework for video summarization is proposed. Key-frames are selected from each shot based on the method described in [45]. The redundant frames are eliminated using a self-organizing map. The redundancy eliminated set of key-frames are connected in a network structure to allow the users to browse through the video collection. The power and simplicity of color histograms have been exploited in several works for finding the similarity between frames and thereby remove duplication. In [46] the main low-level feature used is a color histogram. The given video is first segmented into shots and clustering is performed on the set of frames based on color histogram extracted from each frame. The frame at the centroid of each cluster forms a part of the final key-frame set. Although color histogram is a very elegant low-level feature, however, the computational complexity involved for extraction and comparison is high as it represents a vector of high dimensionality. In order to eliminate the components having lower discrimination power, singular value decomposition (SVD) is used in [47]. In [48] principal component analysis (PCA) is applied on the color histogram to reduce the dimensionality of the feature vector. Delaunay clustering is used to group the frames using the reduced feature vector. The center of each cluster represents a key-frame of the storyboard. PCA has also been used in [49, 50] to reduce the elements in a histogram. Further, in [49], Fuzzy C-means and frame difference measures are used to detect shot boundaries in the video. The use of Fuzzy-ART and Fuzzy C-Means is also proposed in [50] to extract shots from the given video by identifying the number of clusters without any *apriori* information. A cost-benefit analysis of using PCA has not yet been done.

Furini et al. proposed a tool called STIMO (Still and Moving Video Storyboard) in [51] which was capable of generating still and moving storyboards on the fly. The tool also enabled users to specify the length of summary and waiting time for summary generation. A clustering algorithm is executed on the HSV color descriptors extracted from each of the frames. A representative frame is selected from each cluster to produce the static storyboard. A similar approach is used to cluster the shots and choose sequences from the clusters to produce a moving storyboard. A similar approach is used in [52] where the K-means clustering algorithm is used on the HSV color features. The final storyboard is formed by choosing a frame from each cluster. In [53] an approach for summarization of news videos is discussed. The extracted key-frames are clustered together using affinity propagation. A vector space model approach is then used to select shots having high information content. This ensures that the key-frames having discrimination power are retained and visual redundancy is removed. Liu et al. [54] and Ren et al. [55] are works which aim to summarize the rushes video [56]. Liu et al. [54] implements a multi-stage clustering algorithm to remove redundant shots. A value for frame significance is computed based on change of content with time and spatial image salience. The most important parts of the video are extracted based on the frame significance value. Using formal language technique, [55] introduces a hierarchical model to remove unimportant frames. An adaptive clustering is used to remove redundancy while summarizing the rushes video. In [57], a pair of clips is modeled as a weighted bipartite graph. The similarity between the clips of a video is computed based on max-weighted bipartite matching algorithm. The clustering process is based on affinity propagation algorithm and serves to remove redundancy. In [58] a method for video object segmentation is presented which removes redundancy from the spatial, temporal and content domains. A 3D graph-based algorithm is used to extract video objects. These objects are clustered based on shapes using the K-means algorithm. Key objects are identified by selecting objects from the clusters for obtaining intended summarization. A joint method for shot boundary detection and key frame extraction is presented in [59] wherein a method based on three probabilistic components is considered. These are the prior of the key frames, the conditional probability of shot boundaries and the conditional probability of each video frame. Gibbs sampling algorithm [60] is used for key frame extraction and generation of the storyboard. This also ensures that duplication is removed from the final summary.

3 Redundancy Elimination in Video Summarization

Duplication in video content occurs when the same scene or objects are covered by a set of multiple cameras. This duplication of visual content may occur within a given shot (intra-shot level) or between several shots (inter-shot level). Removal or retention of such redundant content is contextual and depends on the genre of the video. Duplication of content holds a different perspective for a sports video like soccer than for news video or a documentary. It is still different for video surveillance applications where only the frames containing some event or activity might be of interest. This emphasizes the point that different approaches to redundancy removal are required in different situations and the same algorithm may not work in all cases. The basic objective of the video summarization task is to provide maximum coverage of the contents while attempting to select the minimum number of video units possible. It can be easily perceived that the two objectives are inversely proportional and conflicting to each other. Redundancy elimination aims to achieve the later objective without affecting the former. Hence it is seen as one of the most important steps in the summarization task. Since redundancy removal is a phase where an attempt is made to eliminate visually redundant units of the video, it assumes vital importance in bandwidth constrained scenarios where the user perception has to be maximized with minimum amount of data transmission between the source and destination. Redundancy elimination thus helps the user to get an insight into the contents of the video in least possible time. This is also important for video indexing applications where the non-redundant frames can be viewed as the features of the video. This characterization helps to symbolize the video in order to facilitate content based video retrieval. Visual redundancy is removed through the use of one or more members from the family of feature descriptors like SIFT, SURF, DAISY, GIST, BRIEF, ORB etc. (described in later sections). The interest points extracted by using these feature descriptors serve as mid-level features necessary for finding the overlap in visual content between the composing units of the video. Setting a threshold for permitting overlap is another important task in this process. A stringent threshold ensures that there is almost no overlap in visual content. This is sometimes necessary for a storyboard representation of the video. For duplication removal in video skimming applications, the amount of similarity between shots may be computed from the number of matching interest points in the frames composing the shots. A decision on elimination is taken on a threshold computed on the similarity values of these features. Figure 2 depicts removal of duplicate frames in a video. An elaboration on the various feature descriptors used for redundancy control in video summarization tasks is presented in the next section.

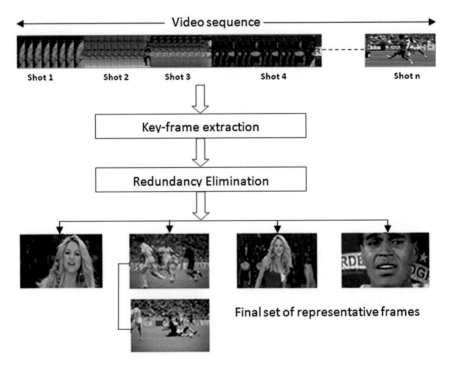

Fig. 2 Redundancy removal

4 Role of Interest Point Detection in Video Summarization

Interest point detection is a field of computer vision that refers to the identification of points which serve as features to the contents in an image. Interest point generation is characterized by a definite position in the spatial domain of an image and is defined by a strong mathematical model. This ensures a high degree of reproducibility in images under different transformations. An interest point descriptor is used to describe the texture around the point. Detection of image features has been the focal point of research in the field of computer vision over the last few decades. Image features include edges, corners, ridges, blobs, textures, interest points etc. The application areas of image feature extraction encompass object identification and tracking [61, 62], video surveillance [6, 63], image similarity/dissimilarity metrics [64], content-based image and video retrieval [1–3, 65, 66], image and video mosaicing [67, 68], video stabilization [69], 3D image construction [70], video summarization [71] etc., to name a few. The matching of a pair of images using feature points involves three stages i.e. detection of the feature points, description of these points using an n-dimensional vector and matching these feature vectors. In this chapter, we focus mainly on interest point detectors and descriptors which can be used for elimination of redundant frames in a video summarization task.

Initially, interest point detectors were developed with the motivation of extracting robust and stable features which could reliably represent salient points in the image and serve as identifiers to it. As research progressed in this field, the focus was on developing algorithms which extracted feature points immune to variations in light intensity, scale, rotation etc. Further advances in the field centered on development of methods which could reliably extract feature points in lesser time by eliminating information around the chosen points which would not degrade the performance of the interest point detector. Interest point detectors are based on well-substantiated mathematical models. Interest points are illustrated by a distinct position in the image space and are usually represented by a multi-dimensional representative vector. These vectors encompass local information content of that point which would help to discriminate it from other points and would also distinctly identify that point in a perturbed image. It is important to note that the change in relative position of the selected interest points can be used to estimate the amount of geometric transform in the objects of a given set of images. The noise points or outliers are detected by tracking huge change in the estimated transform for the objects in a given image with respect to the original scene. The interest points corresponding to an object in an image have mid-level semantic features for describing and identifying it. A majority of the interest points detected lie on the high frequency regions of the image.

Feature point descriptors have been used by researchers to boost the algorithms designed for video summarization. This is in contrast to summarization methods which use visual descriptors [72, 73]. Computer vision algorithms aim to extract the semantic meaning of images composing the video. This augments the target of video summarization algorithms to provide a content revealing summary through a concise representation. Semantic understanding of videos is still a far-fetched reality. In the further sections, the various interest point detectors and descriptors are presented which have been used for the video summarization task in various ways.

4.1   Scale Invariant Feature Transform (SIFT)

SIFT is a feature point detector and descriptor method, proposed by Lowe [74] in 1999. The goal here is to extract certain key-points corresponding to objects in an image. These key-points are represented by a set of low-level features, necessary for identification of the objects in an experimental image containing other objects. The feature points so taken are immune to various translations (such as rotation and scaling) and also to changes in light intensity. The points are chosen from high contrast regions, rendering them to be detected under several types of perturbations. The four steps involved in this method include:-

1. Scale-space Extrema Detection
2. Key-point Localization
3. Orientation Assignment
4. Key-point Description

The SIFT detector and descriptor is designed to be fully immune to changes in scale and orientation. It is also partially immune to affine distortion and changes in light. It can be used to identify an object from a group of objects in a given scene. SIFT feature points are described by feature vectors having 128 elements. Given a pair of images, the feature points are first detected from both images and the corresponding descriptors are computed. Euclidean distance between the two set of feature vectors is then calculated to find the initial set of candidate matches. A subset of the feature point matches for an object is taken which agree on the scale, orientation and location is taken to separate out the superior matches. A hash table based on the generalized Hough transform is used to find the consistent clusters. A cluster must contain at least three feature points to be considered for the next stage of model verification. The probability for presence of an object is computed based on the set of features given. The matches that pass these checks are recognized as true matches with high confidence. Figure 3 depicts the detection of SIFT interest points on an image. A number of variants for SIFT such as PCA-SIFT (based on Principal Component Analysis) [75], Harris-SIFT (based on Harris interest points) [76] etc. with different characteristics have been designed for various uses. Various approaches to video summarization use SIFT or its variants. In [77] a video summarization method is presented where web images are used as prior input for summarizing videos containing similar set of objects. SIFT Flow [78] is used to define frame distance in order to determine the similarity of one frame with another frame. As mentioned previously, the SIFT descriptor has been used vastly in computer vision for its robustness and ability to handle intensity, rotation, scale variations despite its high computational cost. In [79], SIFT and SURF (described in the next section) feature descriptors have been used to detect forgery in images where copy-move

Fig. 3 An image marked with SIFT interest points

technique has been used. An approach to video summarization where the semantic content is preserved has been presented in [80]. The video is segmented into shots and SIFT features for each shot are extracted. The latent concepts are detected by spectral clustering of bag-of-words features to produce a visual word dictionary.

4.2 Speeded-Up Robust Features (SURF)

Speeded-Up Robust Features (SURF) [81] was proposed by Herbert Bay et. al. It was inspired from SIFT. The main advantage of SURF over SIFT is its low execution speed and computational complexity over the latter. It is claimed to be more robust than SIFT for different image transformations. It provides reliable matching of the detected interest points by generating a 64 element vector to describe the texture around each point of interest. The generated vector for each interest point are designed to be immune to noise, scaling and rotation. SURF has been used widely in object detection and tracking. Determinant of the Hessian blob detector is used for the detection of interest points. To detect scale-invariant features, a scale-normalized second order derivative on the scale space representation is used. SURF approximates this representation using a scale-normalized determinant of the Hessian (DoH) operator. The feature descriptor is computed from the sum of the Haar wavelet [82] response around the point of interest. To find the similarity between a pair of images, the interest points detected are matched. The amount of similarity between the images is the ratio of descriptor matches to the total number of interest points detected. Figure 4 illustrates the SURF correspondences on two similar video frames. Research article [83] deals with identifying faces in CCTV cameras installed for surveillance purposes. A database of human faces is created as new faces appear in front of the camera. A Haar classifier is used for recognizing human faces in images. SURF descriptors provide a match between the detected face and existing faces in the database. In case the faces in the database do not match, the new

Fig. 4 SURF correspondences between two similar images

face is updated in the database. Bhaumik et al. [84] presents a technique for static
video summarization in which key-frames detected in the first phase from each shot
are subjected to redundancy elimination at the intra-shot and inter-shot levels. For
removal of redundant frames, SURF and GIST feature descriptors were extracted for
computing the similarity between the frames. The work also compares the quality of
summary obtained by using SURF and GIST descriptors in terms of precision and
recall.

4.3 DAISY

The DAISY feature descriptor was proposed by Tola et al. [85]. It was inspired by
the SIFT and GLOH [86] feature descriptors and is equally robust. DAISY forms
25 sub-regions of 8 orientation gradients, resulting in a 200 dimensional vector. The
sub-regions are circular in nature and can be computed for all pixels in an image.
A Gaussian kernel is used in DAISY as opposed to a triangular kernel for SIFT
and GLOH. In this descriptor, several Gaussian filters are used on the convolution
of the gradients in definite directions. This is in contrast to the weighted sums of
gradient norms used in SIFT and GLOH. DAISY provides very fast computation of
feature descriptors in all directions and is therefore appropriate for dense-matching.
According to [34], DAISY is 2000 % faster than SIFT 4 × 4, when sampling each
pixel.

4.4 GIST

GIST [87] feature descriptor was proposed by Oliva et al. in 2001 to represent the
dominant spatial structure of a scene. This low-level representation is done using a
set of five perceptual dimensions i.e. naturalness, openness, roughness, expansion
and ruggedness. The spectral components at different spatial locations of the spatial
envelope is computed by using a function called the Windowed Discriminant Spectral

Input Image   **GIST Descriptor**

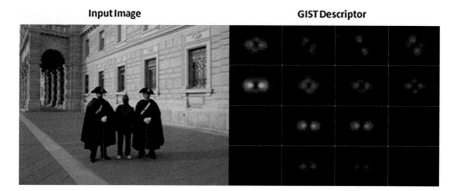

Fig. 5 GIST descriptors for an image

Template (WDST). The perceptual dimensions can be reliably computed by using spectral and coarsely localized information. GIST has been used by researchers for various applications such as finding similarity in images for redundancy removal [42, 84], similar image retrieval [88] and 3D modeling [89], scene classification [90], image completion [91] etc. Different approaches have been developed by Torralba et al. [92, 93] to compress the GIST descriptor. Figure 5 depicts GIST descriptors for an image.

4.5 Binary Robust Independent Elementary Features (BRIEF)

BRIEF [94] was proposed by Calonder et al. in 2010. It is a feature point descriptor which can be used with any available feature detector. Commonly used feature detectors like SIFT and SURF generate long vectors of 128 and 64 dimensions respectively. Generation of such features for a large number of points not only takes a fair amount of computation time but also consumes a lot of memory. A minimum of 512 and 256 bytes are reserved for storing a feature point in SIFT and SURF respectively. This is because of using floating point numbers to store the dimension values. As a result an appreciable time is taken to match the feature descriptors due to large number of elements in the descriptor vectors. Since all the elements are not required for matching, methods like PCA or LDA may be used to find the more important dimensions. Local Sensitive Hashing (LSH) may be used to convert the floating point numbers to string of binary values. Hamming distance between the binary strings is used to compute the distance by performing the XOR operation and finding the number of ones in the result. BRIEF provides a shorter way to find the binary strings related to an interest point without finding the descriptor vectors. As BRIEF is a feature descriptor, feature detectors like SIFT, SURF etc. have to be used to find the interest points. BRIEF is thus a quicker method for computing the feature descriptor and matching the feature vectors. Subject to moderate in-plane rotation, BRIEF provides a high recognition rate.

4.6   Oriented FAST and Rotated BRIEF (ORB)

ORB [95] is a fast and robust feature point detector, proposed by Rublee et al. in 2011. Many tasks in computer vision like object identification, 3D image reconstruction, image similarity analysis etc. can be done using ORB. It is based on the FAST feature point detector and BRIEF (Binary Robust Independent Elementary Features) visual descriptor. It is invariant to rotation and noise resistant. ORB provides a fast and efficient alternative to SIFT and has been shown to be two orders of magnitude faster than SIFT. A method to detect moving objects during camera motion is presented in [96]. To compensate the camera motion, Oriented FAST and Rotated BRIEF (ORB) is used for the feature matching task. The mismatched features between two frames are rejected for obtaining accuracy in compensation. The work also evaluates SIFT and SURF against the presented method to estimate performance in terms of speed and precision.

5   Proposed Methodology

A flow diagram of the proposed method is given in Fig. 6. The various steps of the method are detailed in further sub-sections.

5.1   Extraction of Time Sequenced Image Frames from a Video

The first step towards the video summarization process is to disintegrate the video into a set of time-sequenced image frames to facilitate the process of extracting key-frames from it. This is done by using a standard codec, corresponding to the file type i.e. MP4, MPEG, AVI etc. The images thus obtained are stored as bitmaps for further processing.

5.2   Detection of Video Segments

The transition between two shots is usually classified into two categories i.e. abrupt and gradual. The abrupt transitions are also referred to as hard cuts, whereas, gradual

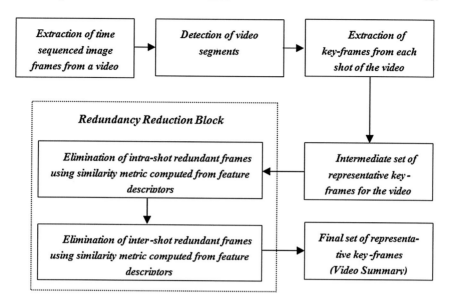

Fig. 6 Flow diagram of proposed method

transitions include dissolves, fade in and fade out. 95 % of these transitions are hard cuts. The decomposed video frames in the previous step are analyzed for detection of shot boundaries. An effective mechanism for video segmentation has been developed by the authors in [97], where a spatio-temporal fuzzy hostility index was used. The same mechanism is employed for detection of shot boundaries in this work.

5.3 Shot-Wise Extraction of Key-Frames and Formation of Representative Set

The key-frames in a video are the representative frames which aptly represents its contents. Given a video, $V = s_1 \otimes s_2 \otimes s_3 \ldots \otimes s_n$ where s_i is a composing shot of V, the task of static video summarization is to assign a Boolean value to the pair (f_{ij}, rs_j) where f_{ij} is the ith frame of the jth shot and rs_j is the representative set of the jth shot. Thus, the initial summary generated after the shot-wise extraction of key-frames is $RS = \{rs_1, rs_2, rs_3, \ldots, rs_n\}$. Initially, the frame having the highest average Fuzzy Hostility Index (FHI) [98] within a shot is chosen as the first key-frame. A search is conducted in both directions of the chosen key-frame such that a frame is reached which has dissimilarity more than $(\mu + 3\sigma)$ where μ is the mean of the average FHIs of the frames in the shot and σ is the standard deviation. The key-frame extraction method has been depicted in Fig. 7. To ensure proper content coverage,

Intermediate set of representative Key frame for the video

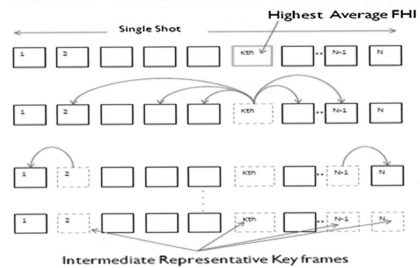

Fig. 7 Key-frame selection process

representative frames are chosen from each shot. The set of key-frames which are extracted from each shot of the video form the representative set.

5.4 Redundancy Reduction to Generate the Final Video Summary

Redundancy of content may occur at the intra-shot and inter-shot levels. Intra-shot content duplication takes place when multiple frames containing the same visual content are chosen as key-frames from within a particular shot. This occurs when there are enough discriminating features between the frames to render a conclusion that the frames are dissimilar in spite of same visual content. It may also occur in cases where the similarity metric or function chosen for the purpose, yields a value below a pre-determined threshold. Inter-shot redundancy occurs when shots with similar content are intermingled with other shots. This leads to similar frames being chosen from multiple shots. The process of intra-shot redundancy reduction on the set can be viewed as a task of eliminating a set of frames $F_i = \{f_1, f_2, f_3, \ldots, f_k\}$ from the representative set rs_i of the ith shot. The same operation is performed on all the shots and the set obtained may be referred to as reduced representative set (*RRS*). Thus, $RRS = \{rs_1 - F_1, rs_2 - F_2, \ldots, rs_n - F_n\}$. The inter-shot redundancy reduction is elimination of a key-frame set $F_R = \{f_1, f_2, f_3, \ldots, f_m\}$ such that the final representative set *FRS* or final summary generated is $FRS = RRS - F_R$. The result of such elimination process ensures that the similarity (δ) between two elements in *FRS* is less than a pre-determined threshold (γ). Thus, if we consider a set $T = \{y : \delta(x, y) > \gamma, x \in FRS, y \in FRS\}$, then $T = \phi$.

The pre-determined threshold may be computed in accordance with an empirical statute in statistics, called the three-sigma rule. According to this rule (refer Fig. 8), 68.2 % values in a normal distribution lie in the range $[M - \sigma, M + \sigma]$, 95.4 % values in $[M - 2\sigma, M + 2\sigma]$ and 99.6 % in the range $[M - 3\sigma, M + 3\sigma]$, where M denotes the arithmetic mean and σ denotes the standard deviation of the normally distributed values. This rule can be effectively utilized for computing the threshold (γ) used for redundancy elimination. A set of p feature point descriptors are extracted from an image frame I_1. The same set of descriptors are matched in another image frame I_2. Assuming that q out of p descriptors match, the similarity between the two image frames, $\delta(I_1, I_2) = \frac{p}{q}$. It can easily be seen that the extent of similarity between the two images is expressed as a real number in the range [0, 1]. Values closer to 1 denote a high similarity. It may further be noted that since δ is calculated on the basis of feature point descriptors, the metric used is closely related to the visual content of an image rather than other low level descriptors such as color model, histogram, statistical measures on pixel values etc. Therefore, for a shot $S_i = \{I_1, I_2, I_3, \ldots, I_n\}$ the mean and standard deviation of the similarity values is computed as:

$$\mu = \frac{\sum_{i,j=1}^{n} \delta(I_i, I_j)}{\binom{n}{2}} \tag{1}$$

$$\sigma = \sqrt{\frac{\sum_{i,j=1}^{n} (\delta(I_i, I_j) - \mu)^2}{\binom{n}{2}}}, i \neq j \tag{2}$$

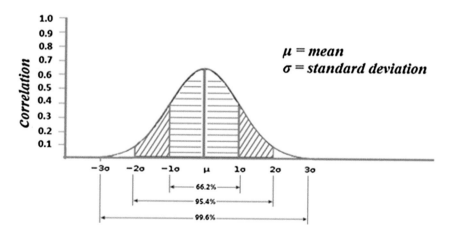

Fig. 8 Normal distribution with three standard deviations from mean

If the similarity value for a pair of frames is greater than $\mu + 3\sigma$, one of the two is eliminated. After intra-shot redundancy is eliminated from the composing shots of the video, the set *RRS* is obtained. Although intra-shot redundancy elimination ensures coverage, it compromises with conciseness of representation. The technique described above may be applied as a whole on the image frames of the set to generate the set after intra-shot redundancy reduction. The proposed method is able to tackle duplication of visual content not only at the intra-shot level but also on the video as a whole. In addition, the user can exercise control over the amount of redundancy by selecting a threshold above or below $\mu + 3\sigma$ which is based on statistical measure.

6 Metrics for Video Summary Evaluation

The evaluation of a video summary is not a simple task due to the unavailability of ground truth for the videos in the dataset under consideration. Moreover, the quality of a summary is based on human perception. It is sometimes difficult for humans to decide as to which summary is the better one. This has rendered difficulties for researchers in designing the different metrics necessary for both evaluation of the summaries and comparison of the different approaches. A brief explanation of the various approaches followed for video summary evaluation is presented in further sub-sections.

6.1 Key-Frame Evaluation

This method was proposed in [99] and focuses on an indirect evaluation of the system generated summary. The key-frames selected by the system are rated on a scale of 5 by independent evaluators [100, 101]. A score of 1 denotes least significance and 5 denote that the chosen key-frame is most significant and relevant for the summary. Appraisal of the video summary is also done by the evaluators to ensure the quality of the summary. The quality of a summary depends on two important factors:

1. Amount of information content (entropy)
2. Coverage of the video content through the key-frames

The mean score of the key-frames is computed to quantify the quality of summary. The formula used in [99] is:

$$score = \frac{sum\ of\ keyframe\ score}{number\ of\ keyframes} \qquad (3)$$

6.2 Shot Reconstruction Degree

The extent to which a set of key-frames is able to reconstruct a shot by means of interpolation is called shot reconstruction degree (SRD) [102]. SRD represents the ability of a key-frame set to reconstruct the original contents. Maximizing the SRD ensures that the motion dynamics of the shot content is preserved. The remaining frames of the video are generated from the key-frame set by employing an interpolation function. The summarization capability of the system is judged by the extent to which the original shot is reconstructed. A similarity function is used to compute the distance between the frames of the original video and those generated by interpolating the key-frames. Different schemes involving SRD have been proposed in [103, 104].

6.3 Coverage

The coverage of a set of key-frames extracted from the original video is defined as the number of frames which are represented by the key-frame set. In [42] a Minimal Spanning Tree (MST) is constructed from the frames of a shot. An adaptive threshold is calculated separately for each shot based on the mean and standard deviation of the edge weights of the MST. The density of a node is the number of frames lying within a disc, the radius of which is equal to the computed threshold. A greedy method is used to choose frames from the list with maximum density. Frames represented by the chosen key-frame are eliminated from the list. This ensures that the most appropriate representatives are chosen as key-frames. It can easily be seen that the chosen key-frames provide a full coverage of the shot. In [105] the coverage has been defined as the number of visually similar frames represented by a chosen key-frame. Hence, coverage may be computed by the following formula:

$$coverage = \frac{number\ of\ frames\ represented}{total\ number\ of\ frames} \tag{4}$$

In [67], the coverage is based on the number of feature points covered by a frame from the unique pool of feature points created from the composing frames of a shot. Initially all the feature points are part of the set $K_{uncovered}$. The coverage of a frame is computed using the formula:

$$C = \eta(K_{uncovered} \bigcap FP_i) \tag{5}$$

The redundancy of a frame is given by:

$$R = \eta(K_{covered} \bigcap FP_i) \tag{6}$$

where, $\eta(X)$ is the cardinality of set X and FP_i is the set of feature points in a frame. Coverage is thus another metric which reveals the quality of a video summary.

6.4 Recall, Precision and F_1 Score

The output of a video summarizer is referred to as the System Generated Summary (SGS). It is essential to evaluate the quality of this summary. The appropriate way for appraisal of the SGS is to compare it with a ground truth. Since the SGS is generated for users, it is natural to bring the ground truth as close as possible to human perception. The ground truth has been referred to as User Summary [30, 84] in the literature. The User Summary (US) is generated by a group of users. The videos under consideration are browsed by the users after disintegrating into constituent frames. The important frames according to user perception are chosen in order to form the US. The Final User Summary (FUS) is formed by an amalgamation of the user summaries. The amount of overlap between the FUS and SGS portrays the efficacy of the summary. The recall and precision are computed as follows:

$$recall = \frac{\eta(FUS \cap SGS)}{\eta(FUS)} \tag{7}$$

$$precision = \frac{\eta(FUS \cap SGS)}{\eta(SGS)} \tag{8}$$

FUS: Set of frames in user summary
SGS: Set of frames in system generated summary
$\eta(X)$: Cardinal no. of set X
The harmonic mean of precision and recall is taken for computing the F_1 score. It provides a consistent measure for determining the overall efficiency of an information retrieval system. The following expression is used to calculate the F_1 score:

$$F_1 = 2\frac{precision \times recall}{precision + recall} \tag{9}$$

The F_1 score varies in the range [0, 1] where a score of 1 indicates that the system is most efficient.

6.5 Significance, Overlap and Compression Factors

Mundur et al. [48] introduces three new factors for determining the quality of a summary. The *Significance Factor* denotes the importance of the content represented by a cluster of frames. The significance of the ith cluster is given as:

$$Significance\_Factor(i) = \frac{C_i}{\sum_{j=1}^{k} C_j} \tag{10}$$

where C_i is the total number of frames in the ith cluster and k is the total number of clusters.

The *Overlap Factor* determines the total significance of the overlapped clusters found in two summaries. In other words, we compute the cumulative significance of those clusters which have a common key-frame set with the ground-truth summary. This is an important metric for comparing two summaries. This factor is computed as:

$$Overlap\_Factor = \frac{\sum_{p \in Common \ keyframe \ clustes}^{C_p}}{\sum_{j=1}^{k} C_j} \tag{11}$$

A higher value of the *Overlap Factor* denotes a better representative summary with respect to the ground-truth.

The *Compression Factor* for a video denotes the size of the summary with respect to the original size of the video. It is defined as:

$$Compression\_Factor = \frac{No \ of \ keyframes \ in \ summary}{Total \ number \ of \ keyframes} \tag{12}$$

7 Experimental Results and Analysis

The proposed method for storyboard generation was tested on a dataset consisting of nine videos. The dataset is divided into two parts. The first part (Table 1) consists of short videos having average length of 3 min and 21 s. The second part (Table 2) consists of longer videos of average length 53 min and 34 s.

All the videos in the dataset have a resolution of 640×360 pixels at 25 fps (except video V7 which is at 30 fps). The videos are in MP4 file format (ISO/IEC 14496-14:2003), commonly named as MPEG-4 file format version 2.

The efficacy of the proposed method is evaluated by computing the recall, precision and F_1 score of the system generated summary (*SGS*) against the final user summary (*FUS*) as explained in Sect. 6.4. A frame to frame comparison is performed between the *SGS* and *FUS* by an evaluator program written for the purpose. A pair of frames is considered to be matched if the correlation is more than 0.7. It has been seen that the frames are visually similar when the correlation exceeds 0.7. This is significantly higher than the threshold used in [30], where the match threshold was considered as 0.5.

Table 1 Test video dataset-I

Video	V1	V2	V3	V4	V5
Duration (mm:ss)	02:58	02:42	04:10	03:27	03:31
No. of frames	4468	4057	6265	4965	5053
No. of hard cuts	43	70	172	77	138
Average no. of frames in each shot	101.54	57.14	36.21	63.65	36.35

Table 2 Test video dataset-II

Video	V6	V7	V8	V9
Duration (mm:ss)	44:14	52:29	58:06	59:29
No. of frames	66339	94226	87153	89226
No. of hard cuts	626	543	668	1235
Average no. of frames in each shot	105.80	173.21	130.27	72.18

7.1 The Video Dataset

The video dataset considered for testing comprised of videos of short and long duration. The first video (V1) is the Wimbledon semifinal match highlights between Djokovic and Del Potro. The video consists of small duration shots and rapid movement of objects. The second video (V2) is a Hindi film song "Dagabaaz" from the movie "Dabangg2". It consists of shots taken in the daylight and night time. The third video (V3) is another song "Chammak Challo" from the Hindi film "Ra.One". This video consists of shots taken indoors, as well as some digitally created frames intermingled with real life shots. A violin track by Lindsey Stirling forms the fourth video (V4) of the data set. Simultaneous camera and performer movements are observed in the video. Also there are quick zoom-in and zoom-out shots which are taken outdoors. The last of the small videos (V5) is the official song of the FIFA world cup called "Waka Waka". It consists of shots with varied illumination and background.

The videos in Table 2 are four documentaries (V6–V9) from different TV channels. The videos V6–V9 are four documentaries of longer duration from different TV channels. The videos are "Science and Technology developments in India", "Under the Antarctic Ice", "How to build a satellite" and "Taxi Driver". All the videos in the dataset are available on YouTube.

7.2 Experimental Results

The initial storyboard generated by the proposed method is called the representative set (RS). It is formed by extracting the key-frames as described in Sect. 5.3. The key-frames in RS are compared with user summary prior to redundancy removal and the results are presented in Table 3. The results show high precision, recall and F_1 scores indicating the efficacy of the proposed system. In the next step, intra-shot redundancy reduction is carried out using both the SURF and GIST feature descriptors on both *RS* and user summary. The amount of reduction achieved is summarized in Table 4. The recall and precision are again computed and the results are presented in Table 5. In the final step, redundancy is further removed from *RS* and user summary at the inter-shot level using SURF and GIST descriptors. The amount reduction achieved is enumerated in Table 6. The recall and precision values computed after the inter-shot redundancy phase are presented in Table 7. It can be easily seen from the results that elimination of duplicate frames does have effect on the precision and recall. In certain cases the post-redundancy metric values are better than the pre-redundancy phase.

Table 3 Comparison between user and system generated summary prior to redundancy removal

Video	Precision (%)	Recall (%)	F_1 Score (%)
V1	98.43	92.64	95.45
V2	90.85	97.54	94.08
V3	99.10	95.67	97.35
V4	94.69	93.98	94.33
V5	96.59	91.89	94.18
V6	99.65	99.55	99.59
V7	98.88	97.12	97.99
V8	98.39	97.18	97.78
V9	98.52	100	99.25

Table 4 Intra-shot redundancy reduction

Video	% reduction (SURF)	% reduction (GIST)
V1	23	28.12
V2	44.57	55.14
V3	21.645	28.76
V4	26.51	42.10
V5	17.83	23.24
V6	53.18	68.68
V7	57.83	74.19
V8	50.36	72.91
V9	40.38	55.24

Table 5 Comparison after intra-shot redundancy removal

Video	Precision (SURF) (%)	Recall (SURF) (%)	Precision (GIST) (%)	Recall (GIST) (%)
V1	98.03	98.03	100	100
V2	97.42	99.47	99.03	100
V3	100	100	100	100
V4	97.93	98.95	98.70	97.43
V5	100	100	100	100
V6	97.45	97.24	98.65	99.20
V7	98.30	98.85	96.45	97.22
V8	98.34	98.85	98.67	98.66
V9	96.56	95.55	98.36	98.42

Table 6 Inter-shot redundancy reduction

Video	% reduction (SURF)	% reduction (GIST)
V1	27	37.5
V2	70.28	78
V3	35.49	39.38
V4	41.66	54.13
V5	35.67	39.45
V6	63.07	73.60
V7	64.94	77.59
V8	48.17	73.11
V9	51.05	68.62

Table 7 Comparison after inter-shot redundancy removal

Video	Precision (SURF) (%)	Recall (SURF) (%)	Precision (GIST) (%)	Recall (GIST) (%)
V1	97.67	95.45	100	100
V2	100	100	100	100
V3	98.65	97.35	100	97.81
V4	96.10	96.10	95.23	96.77
V5	97.39	100	98.19	97.32
V6	98.35	97.63	98.75	100
V7	99.4	98.55	97.65	98.25
V8	97.68	98.44	99.32	99.74
V9	95.36	96.45	99.24	98.86

8   Discussions and Conclusion

Redundancy removal remains an important step in the task of video summarization. The proposed method is able to illustrate that the quality of the generated summary is not degraded by removing duplicate frames having nearly the same visual content. An additional contribution of this work is the determination of an automatic threshold for elimination of redundant frames based on the three-sigma rule. The results illustrate the efficacy of the threshold used. The experimental results leads us to conclude that the prominent features of a video may be represented in a succinct way, thereby saving the retrieval and browsing time of a user. This is particularly useful for low bandwidth scenarios.

Although the problem of removing visual redundancy has been tackled to a great extent by the use of feature descriptor, yet there is a long way to go in terms of semantic understanding of the video. For semantic understanding, development of semantic descriptors need to be designed which in turn require extraction of high level features. These high level features need to be presented in a manner which provides comparison and matching between the high level feature vectors. This would propel research in abstraction based representation of the video contents.

References

1. Zhang, H.J., Wu, J., Zhong, D., Smoliar, S.W.: An integrated system for content-based video retrieval and browsing. Pattern Recogn. **30**(4), 643–658 (1997)
2. Chang, S.F., Chen, W., Meng, H.J., Sundaram, H., Zhong, D.: A fully automated content-based video search engine supporting spatiotemporal queries. IEEE Trans. Circuits Syst. Video Technol. **8**(5), 602–615 (1998)
3. Lew, M.S., Sebe, N., Djeraba, C., Jain, R.: Content-based multimedia information retrieval: state of the art and challenges. ACM Trans. Multimedia Comput. Commun. Appl. (TOMM) **2**(1), 1–19 (2006)
4. Papadimitriou, C.H., Tamaki, H., Raghavan, P., Vempala, S.: Latent semantic indexing: a probabilistic analysis. In: Proceedings of the Seventeenth ACM SIGACT-SIGMOD-SIGART Symposium on Principles of Database Systems, pp. 159–168. ACM (1998)
5. Kim, H.S., Lee, J., Liu, H., Lee, D.: Video linkage: group based copied video detection. In: Proceedings of the 2008 International Conference on Content-Based Image and Video Retrieval, pp. 397–406. ACM (2008)
6. Kim, C., Hwang, J.N.: Object-based video abstraction for video surveillance systems. IEEE Trans. Circuits Syst. Video Technol. **12**(12), 1128–1138 (2002)
7. Ekin, A., Tekalp, A.M., Mehrotra, R.: Automatic soccer video analysis and summarization. IEEE Trans. Image Proc. **12**(7), 796–807 (2003)
8. Babaguchi, N., Kawai, Y., Ogura, T., Kitahashi, T..: Personalized abstraction of broadcasted American football video by highlight selection. IEEE Trans. Multimedia **6**(4), 575–586 (2004)
9. Pan, H., Van Beek, P., Sezan, M.I.: Detection of slow-motion replay segments in sports video for highlights generation. In: Proceedings of IEEE International Conference on Acoustics, Speech, and Signal Processing, vol. 3, pp. 1649–1652 (2001)
10. Tjondronegoro, D.W., Chen, Y.P.P., Pham, B.: Classification of self-consumable highlights for soccer video summaries. In: 2004 IEEE International Conference on Multimedia and Expo, 2004. ICME'04, vol. 1, pp. 579–582. IEEE (2004)

11. Nam, J., Tewfik, A.H.: Dynamic video summarization and visualization. In: Proceedings of the Seventh ACM International Conference on Multimedia (Part 2), pp. 53–56. ACM (1999)
12. Pfeiffer, S., Lienhart, R., Fischer, S., Effelsberg, W.: Abstracting digital movies automatically. J. Vis. Commun. Image Represent. 7(4), 345–353 (1996)
13. Yeung, M.M., Yeo, B.L.: Video visualization for compact presentation and fast browsing of pictorial content. IEEE Trans. Circuits Syst. Video Technol. 7(5), 771–785 (1997)
14. Moriyama, T., Sakauchi, M.: Video summarisation based on the psychological content in the track structure. In: Proceedings of the 2000 ACM Workshops on Multimedia, pp. 191–194. ACM (2000)
15. Yeung, M.M, Yeo, B.L.: Video content characterization and compaction for digital library applications. In: Electronic Imaging'97, pp. 45–58 (1997)
16. Lienhart, R., Pfeiffer, S., Effelsberg, W.: Scene determination based on video and audio features. In: IEEE International Conference on Multimedia Computing and Systems, 1999, vol. 1, pp. 685–690. IEEE (1999)
17. Thakore, V.H.: Video shot cut boundary detection using histogram. Int. J. Eng. Sci. Res. Technol. (IJESRT) 2, 872–875 (2013)
18. Baber, J., Afzulpurkar, N., Dailey, M.N., Bakhtyar, M.: Shot boundary detection from videos using entropy and local descriptor. In: 2011 17th International Conference on Digital Signal Processing (DSP), pp. 1–6. IEEE (2011)
19. Cernekova, Z., Nikou, C., Pitas, I.: Shot detection in video sequences using entropy based metrics. In: 2002 International Conference on Image Processing. 2002. Proceedings, vol. 3, p. III-421. IEEE (2002)
20. Hampapur, A., Jain, R., Weymouth, T.E.: Production model based digital video segmentation. Multimedia Tools Appl. 1(1), 9–46 (1995)
21. Zhang, H., Kankanhalli, A., Smoliar, S.W.: Automatic partitioning of full-motion video. Multimedia Syst. 1(1), 10–28 (1993)
22. Tonomura, Y.: Video handling based on structured information for hypermedia systems. In: International conference on Multimedia Information Systems' 91, pp. 333–344. McGraw-Hill Inc. (1991)
23. Barron, J.L., Fleet, D.J., Beauchemin, S.S.: Performance of optical flow techniques. Int. J. Comput. Vis. 12(1), 43–77 (1994)
24. Wang, T., Wu, Y., Chen, L.: An approach to video key-frame extraction based on rough set. In: International Conference on Multimedia and Ubiquitous Engineering, 2007. MUE'07, pp. 590–596. IEEE (2007)
25. Li, B., Sezan, M.I.: Event detection and summarization in sports video. In: IEEE Workshop on Content-Based Access of Image and Video Libraries, 2001. (CBAIVL 2001), pp. 132–138. IEEE (2001)
26. Potapov, D., Douze, M., Harchaoui, Z., Schmid, C.: Category-specific video summarization. In: Computer Vision-ECCV 2014, pp. 540–555. Springer (2014)
27. Wang, F., Ngo, C.W.: Rushes video summarization by object and event understanding. In: Proceedings of the International Workshop on TRECVID Video Summarization, pp. 25–29. ACM (2007)
28. Guan, G., Wang, Z., Lu, S., Deng, J.D., Feng, D.D.: Keypoint-based keyframe selection. IEEE Trans. Circuits Syst. Video Technol. 23(4), 729–734 (2013)
29. Panagiotakis, C., Pelekis, N., Kopanakis, I., Ramasso, E., Theodoridis, Y.: Segmentation and sampling of moving object trajectories based on representativeness. IEEE Trans. Knowl. Data Eng. 24(7), 1328–1343 (2012)
30. Cahuina, E.J., Chavez, C.G.: A new method for static video summarization using local descriptors and video temporal segmentation. In: 26th SIBGRAPI-Conference on Graphics, Patterns and Images (SIBGRAPI), 2013, pp. 226–233. IEEE (2013)
31. Patel, A., Kasat, D., Jain, S., Thakare, V.: Performance analysis of various feature detector and descriptor for real-time video based face tracking. Int. J. Comp. Appl. 93(1), 37–41 (2014)
32. Kapela, R., McGuinness, K., Swietlicka, A., Oconnor, N.E.: Real-time event detection in field sport videos. In: Computer Vision in Sports, pp. 293–316. Springer (2014)

33. Khvedchenia, I.: A battle of three descriptors: surf, freak and brisk. Computer Vision Talks
34. Uijlings, J.R., Smeulders, A.W., Scha, R.J.: Real-time visual concept classification. IEEE Trans. Multimedia **12**(7), 665–681 (2010)
35. Li, J.: Video shot segmentation and key frame extraction based on sift feature. In: 2012 International Conference on Image Analysis and Signal Processing (IASP), pp. 1–8. IEEE (2012)
36. Papadopoulos, D.P., Chatzichristofis, S.A., Papamarkos, N.: Video summarization using a self-growing and self-organized neural gas network. In: Computer Vision/Computer Graphics Collaboration Techniques, pp. 216–226. Springer (2011)
37. Lux, M., Schöffmann, K., Marques, O., Böszörmenyi, L.: A novel tool for quick video summarization using keyframe extraction techniques. In: Proceedings of the 9th Workshop on Multimedia Metadata (WMM 2009). CEUR Workshop Proceedings, vol. 441, pp. 19–20 (2009)
38. Chatzichristofis, S.A., Boutalis, Y.S.: Cedd: color and edge directivity descriptor: a compact descriptor for image indexing and retrieval. In: Computer Vision Systems, pp. 312–322. Springer (2008)
39. Chatzichristofis, S., Boutalis, Y.S., et al.: Fcth: Fuzzy color and texture histogram-a low level feature for accurate image retrieval. In: Ninth International Workshop on Image Analysis for Multimedia Interactive Services, 2008. WIAMIS'08, pp. 191–196. IEEE (2008)
40. Chatzichristofis, S.A., Boutalis, Y.S.: Content based radiology image retrieval using a fuzzy rule based scalable composite descriptor. Multimedia Tools Appl. **46**(2–3), 493–519 (2010)
41. Chatzichristofis, S.A., Boutalis, Y.S., Lux, M.: Spcd-spatial color distribution descriptor-a fuzzy rule based compact composite descriptor appropriate for hand drawn color sketches retrieval. In: ICAART (1), pp. 58–63 (2010)
42. Bhaumik, H., Bhattacharyya, S., Das, M., Chakraborty, S.: Enhancement of Perceptual Quality in Static Video Summarization Using Minimal Spanning Tree Approach. In: 2015 International Conference on Signal Processing, Informatics, Communication and Energy Systems (IEEE SPICES), pp. 1–7. IEEE (2015)
43. Liu, D., Shyu, M.L., Chen, C., Chen, S.C.: Integration of global and local information in videos for key frame extraction. In: 2010 IEEE International Conference on Information Reuse and Integration (IRI), pp. 171–176. IEEE (2010)
44. Qian, Y., Kyan, M.: Interactive user oriented visual attention based video summarization and exploration framework. In: 2014 IEEE 27th Canadian Conference on Electrical and Computer Engineering (CCECE), pp. 1–5. IEEE (2014)
45. Qian, Y., Kyan, M.: High definition visual attention based video summarization. In: VISAPP, vol. 1, pp. 634–640 (2014)
46. Zhuang, Y., Rui, Y., Huang, T.S., Mehrotra, S.: Adaptive key frame extraction using unsupervised clustering. In: 1998 International Conference on Image Processing, 1998. ICIP 98. Proceedings, vol. 1, pp. 866–870. IEEE (1998)
47. Gong, Y., Liu, X.: Video summarization and retrieval using singular value decomposition. Multimedia Syst. **9**(2), 157–168 (2003)
48. Mundur, P., Rao, Y., Yesha, Y.: Keyframe-based video summarization using delaunay clustering. Int. J. Digit. Libr. **6**(2), 219–232 (2006)
49. Wan, T., Qin, Z.: A new technique for summarizing video sequences through histogram evolution. In: 2010 International Conference on Signal Processing and Communications (SPCOM), pp. 1–5. IEEE (2010)
50. Cayllahua-Cahuina, E., Cámara-Chávez, G., Menotti, D.: A static video summarization approach with automatic shot detection using color histograms. In: Proceedings of the International Conference on Image Processing, Computer Vision, and Pattern Recognition (IPCV), p. 1. The Steering Committee of The World Congress in Computer Science, Computer Engineering and Applied Computing (WorldComp) (2012)
51. Furini, M., Geraci, F., Montangero, M., Pellegrini, M.: Stimo: still and moving video storyboard for the web scenario. Multimedia Tools Appl. **46**(1), 47–69 (2010)

52. de Avila, S.E.F., Lopes, A.P.B., da Luz, A., de Albuquerque Araújo, A.: Vsumm: a mechanism designed to produce static video summaries and a novel evaluation method. Pattern Recogn. Lett. **32**(1), 56–68 (2011)
53. Xie, X.N., Wu, F.: Automatic video summarization by affinity propagation clustering and semantic content mining. In: 2008 International Symposium on Electronic Commerce and Security, pp. 203–208. IEEE (2008)
54. Liu, Z., Zavesky, E., Shahraray, B., Gibbon, D., Basso, A.: Brief and high-interest video summary generation: evaluating the at&t labs rushes summarizations. In: Proceedings of the 2nd ACM TRECVid Video Summarization Workshop, pp. 21–25. ACM (2008)
55. Ren, J., Jiang, J., Eckes, C.: Hierarchical modeling and adaptive clustering for real-time summarization of rush videos in trecvid'08. In: Proceedings of the 2nd ACM TRECVid Video Summarization Workshop, pp. 26–30. ACM (2008)
56. Dumont, E., Merialdo, B.: Rushes video summarization and evaluation. Multimedia Tools Appl. **48**(1), 51–68 (2010)
57. Gao, Y., Dai, Q.H.: Clip based video summarization and ranking. In: Proceedings of the 2008 International Conference on Content-Based Image and Video Retrieval, pp. 135–140. ACM (2008)
58. Tian, Z., Xue, J., Lan, X., Li, C., Zheng, N.: Key object-based static video summarization. In: Proceedings of the 19th ACM International Conference on Multimedia, pp. 1301–1304. ACM (2011)
59. Liu, X., Song, M., Zhang, L., Wang, S., Bu, J., Chen, C., Tao, D.: Joint shot boundary detection and key frame extraction. In: 2012 21st International Conference on Pattern Recognition (ICPR), pp. 2565–2568. IEEE (2012)
60. Casella, G., George, E.I.: Explaining the gibbs sampler. Am. Stat. **46**(3), 167–174 (1992)
61. Yilmaz, A., Javed, O., Shah, M.: Object tracking: a survey. ACM Comput. Surv. (CSUR) **38**(4), 13 (2006)
62. Aggarwal, A., Biswas, S., Singh, S., Sural, S., Majumdar, A.K.: Object tracking using background subtraction and motion estimation in mpeg videos. In: Computer Vision-ACCV 2006, pp. 121–130. Springer (2006)
63. Pritch, Y., Ratovitch, S., Hende, A., Peleg, S.: Clustered synopsis of surveillance video. In: Sixth IEEE International Conference on Advanced Video and Signal Based Surveillance, 2009. AVSS'09, pp. 195–200. IEEE (2009)
64. Kokare, M., Chatterji, B., Biswas, P.: Comparison of similarity metrics for texture image retrieval. In: TENCON 2003. Conference on Convergent Technologies for the Asia-Pacific Region, vol. 2, pp. 571–575. IEEE (2003)
65. Liu, Y., Zhang, D., Lu, G., Ma, W.Y.: A survey of content-based image retrieval with high-level semantics. Pattern Recogn. **40**(1), 262–282 (2007)
66. Belongie, S., Carson, C., Greenspan, H., Malik, J.: Color-and texture-based image segmentation using em and its application to content-based image retrieval. In: Sixth International Conference on Computer Vision, 1998, pp. 675–682. IEEE (1998)
67. Szeliski, R.: Foundations and trends in computer graphics and vision. Found. Trends Comput. Graphics Vis. **2**(1), 1–104 (2007)
68. Marzotto, R., Fusiello, A., Murino, V.: High resolution video mosaicing with global alignment. In: Computer Vision and Pattern Recognition, 2004. CVPR 2004. Proceedings of the 2004 IEEE Computer
69. Matsushita, Y., Ofek, E., Ge, W., Tang, X., Shum, H.Y.: Full-frame video stabilization with motion inpainting. IEEE Trans. Pattern Anal. Mach. Intell. **28**(7), 1150–1163 (2006)
70. Zitova, B., Flusser, J.: Image registration methods: a survey. Image Vis. Comput. **21**(11), 977–1000 (2003)
71. Hu, W., Xie, N., Li, L., Zeng, X., Maybank, S.: A survey on visual content-based video indexing and retrieval. IEEE Trans. Syst. Man Cybern. Part C Appl. Rev. **41**(6), 797–819 (2011)
72. Lee, J.H., Kim, W.Y.: Video summarization and retrieval system using face recognition and mpeg-7 descriptors. In: Image and Video Retrieval, pp. 170–178. Springer (2004)

73. Fatemi, N., Khaled, O.A.: Indexing and retrieval of tv news programs based on mpeg-7. In: International Conference on Consumer Electronics, 2001. ICCE, pp. 360–361. IEEE (2001)

74. Lowe, D.G.: Object recognition from local scale-invariant features. In: The Proceedings of the Seventh IEEE International Conference on Computer Vision, 1999, vol. 2, pp. 1150–1157. IEEE (1999)

75. Ke, Y., Sukthankar, R.: Pca-sift: a more distinctive representation for local image descriptors. In: Proceedings of the 2004 IEEE Computer Society Conference on Computer Vision and Pattern Recognition, 2004. CVPR 2004, vol. 2, p. II-506. IEEE (2004)

76. Azad, P., Asfour, T., Dillmann, R.: Combining harris interest points and the sift descriptor for fast scale-invariant object recognition. In: IEEE/RSJ International Conference on Intelligent Robots and Systems, 2009. IROS 2009, pp. 4275–4280. IEEE (2009)

77. Khosla, A., Hamid, R., Lin, C.J., Sundaresan, N.: Large-scale video summarization using web-image priors. In: 2013 IEEE Conference on Computer Vision and Pattern Recognition (CVPR), pp. 2698–2705. IEEE (2013)

78. Liu, C., Yuen, J., Torralba, A.: Sift flow: dense correspondence across scenes and its applications. IEEE Trans. Pattern Anal. Mach. Intell. **33**(5), 978–994 (2011)

79. Pandey, R.C., Singh, S.K., Shukla, K., Agrawal, R.: Fast and robust passive copy-move forgery detection using surf and sift image features. In: 2014 9th International Conference on Industrial and Information Systems (ICIIS), pp. 1–6. IEEE (2014)

80. Yuan, Z., Lu, T., Wu, D., Huang, Y., Yu, H.: Video summarization with semantic concept preservation. In: Proceedings of the 10th International Conference on Mobile and Ubiquitous Multimedia, pp. 109–112. ACM (2011)

81. Bay, H., Ess, A., Tuytelaars, T., Van Gool, L.: Speeded-up robust features (surf). Comput. Vis. Image Underst. **110**(3), 346–359 (2008)

82. Struzik, Z.R., Siebes, A.: The haar wavelet transform in the time series similarity paradigm. In: Principles of Data Mining and Knowledge Discovery, pp. 12–22. Springer (1999)

83. Sathyadevan, S., Balakrishnan, A.K., Arya, S., Athira Raghunath, S.: Identifying moving bodies from cctv videos using machine learning techniques. In: 2014 First International Conference on Networks & Soft Computing (ICNSC), pp. 151–157. IEEE (2014)

84. Bhaumik, H., Bhattacharyya, S., Dutta, S., Chakraborty, S.: Towards redundancy reduction in storyboard representation for static video summarization. In: 2014 International Conference on Advances in Computing, Communications and Informatics (ICACCI), pp. 344–350. IEEE (2014)

85. Tola, E., Lepetit, V., Fua, P.: Daisy: an efficient dense descriptor applied to wide-baseline stereo. IEEE Trans. Pattern Anal. Mach. Intell. **32**(5), 815–830 (2010)

86. Mikolajczyk, K., Schmid, C.: A performance evaluation of local descriptors. IEEE Trans. Pattern Anal. Mach. Intell. **27**(10), 1615–1630 (2005)

87. Oliva, A., Torralba, A.: Modeling the shape of the scene: a holistic representation of the spatial envelope. Int. J. Comput. Vis. **42**(3), 145–175 (2001)

88. Pass, G., Zabih, R.: Histogram refinement for content-based image retrieval. In: Proceedings 3rd IEEE Workshop on Applications of Computer Vision, 1996. WACV'96., pp. 96–102. IEEE (1996)

89. Li, X., Wu, C., Zach, C., Lazebnik, S., Frahm, J.M.: Modeling and recognition of landmark image collections using iconic scene graphs. In: Computer Vision-ECCV 2008, pp. 427–440. Springer (2008)

90. Sikirić, I., Brkić, K., Šegvić, S.: Classifying traffic scenes using the gist image descriptor (2013). arXiv preprint arXiv:1310.0316

91. Hays, J., Efros, A.A.: Scene completion using millions of photographs. ACM Trans. Graphics (TOG) **26**(3), 4 (2007)

92. Torralba, A., Fergus, R., Weiss, Y.: Small codes and large image databases for recognition. In: IEEE Conference on Computer Vision and Pattern Recognition, 2008. CVPR 2008, pp. 1–8. IEEE (2008)

93. Weiss, Y., Torralba, A., Fergus, R.: Spectral hashing. In: Advances in neural information processing systems, pp. 1753–1760 (2009)

94. Calonder, M., Lepetit, V., Strecha, C., Fua, P.: Brief: binary robust independent elementary features. Comput. Vis.-ECCV **2010**, 778–792 (2010)
95. Rublee, E., Rabaud, V., Konolige, K., Bradski, G.: Orb: an efficient alternative to sift or surf. In: 2011 IEEE International Conference on Computer Vision (ICCV), pp. 2564–2571. IEEE (2011)
96. Xie, S., Zhang, W., Ying, W., Zakim, K.: Fast detecting moving objects in moving background using orb feature matching. In: 2013 Fourth International Conference on Intelligent Control and Information Processing (ICICIP), pp. 304–309. IEEE (2013)
97. Bhaumik, H., Bhattacharyya, S., Chakraborty, S.: Video shot segmentation using spatio-temporal fuzzy hostility index and automatic threshold. In: 2014 Fourth International Conference on Communication Systems and Network Technologies (CSNT), pp. 501–506. IEEE (2014)
98. Bhattacharyya, S., Maulik, U., Dutta, P.: High-speed target tracking by fuzzy hostility-induced segmentation of optical flow field. Appl. Soft Comput. **9**(1), 126–134 (2009)
99. De Avila, S.E., da Luz, A., de Araujo, A., Cord, M.: Vsumm: an approach for automatic video summarization and quantitative evaluation. In: XXI Brazilian Symposium on Computer Graphics and Image Processing, 2008. SIBGRAPI'08, pp. 103–110. IEEE (2008)
100. De Avila, S.E., da Luz Jr, A., De Araujo, A., et al.: Vsumm: A simple and efficient approach for automatic video summarization. In: 15th International Conference on Systems, Signals and Image Processing, 2008. IWSSIP 2008, pp. 449–452. IEEE (2008)
101. Liu, X., Mei, T., Hua, X.S., Yang, B., Zhou, H.Q.: Video collage. In: Proceedings of the 15th international conference on Multimedia, pp. 461–462. ACM (2007)
102. Liu, T., Zhang, X., Feng, J., Lo, K.T.: Shot reconstruction degree: a novel criterion for key frame selection. Pattern Recogn. Lett. **25**(12), 1451–1457 (2004)
103. Lee, H.C., Kim, S.D.: Iterative key frame selection in the rate-constraint environment. Sign. Process. Image Commun. **18**(1), 1–15 (2003)
104. Liu, R., Kender, J.R.: An efficient error-minimizing algorithm for variable-rate temporal video sampling. In: 2002 IEEE International Conference on Multimedia and Expo, 2002. ICME'02. Proceedings, vol. 1, pp. 413–416. IEEE (2002)
105. Chang, H.S., Sull, S., Lee, S.U.: Efficient video indexing scheme for content-based retrieval. IEEE Trans. Circuits Syst. Video Technol. **9**(8), 1269–1279 (1999)

A Real Time Dactylology Based Feature Extractrion for Selective Image Encryption and Artificial Neural Network

Sirshendu Hore, Tanmay Bhattacharya, Nilanjan Dey,
Aboul Ella Hassanien, Ayan Banerjee and S.R. Bhadra Chaudhuri

Abstract Dactylology or Finger Spelling is popularly known as sign speech, is a kind of gesture based language used by the deaf and dumb people to communicate with themselves or with other people in and around them. In many ways Finger-Spelling provides a connection between the sign and oral language. Dactylology can also be used for secret communication or can be used by the security personnel to communicate secretly with their counterpart. In the proposed work a two phase encryption technique has been proposed wherein the first phase a 'Gesture Key', generated from Indian Sign Language in real time has been used for encrypting the Region of Interests (ROIs) and in the second phase a session key has been used to encrypt the partially encrypted image further. The experimental results show that the scheme provides significant security improvement without compromising the image quality. The speed of encryption and decryption process is quite good. The Performance of the proposed scheme is compared with the few other popular encryption methods to establish the relevance of the work.

S. Hore (✉)
Department of CSE, Hooghly Engineering & Technology College, Hooghly, India
e-mail: shirshendu.hore@hetc.ac.in

T. Bhattacharya
Department of IT, Techno India, Saltlake, India
e-mail: tanmayb29@gmail.com

N. Dey
Department of Information Technology, Techno India College of Technology, Kolkata, India
e-mail: neelanjan.dey@gmail.com

A.E. Hassanien
Cairo University, Scientific Research Group, Giza, Egypt
e-mail: aboitcairo@gmail.com

A. Banerjee · S.R. Bhadra Chaudhuri
Department of ETC, Indian Institute of Engineering Science and Technology, Shibpur, India
e-mail: ayan@telecom.becs.ac.in

S.R. Bhadra Chaudhuri
e-mail: prof.srbc@gmail.com

© Springer International Publishing Switzerland 2016
A.I. Awad and M. Hassaballah (eds.), *Image Feature Detectors and Descriptors*,
Studies in Computational Intelligence 630, DOI 10.1007/978-3-319-28854-3_8

Keywords ANN · Dactylology · Indian sign language · ROI-security · Real time · SURF

1 Introduction

The Sign language plays a very important role for deaf and dumb people, i.e. who have spoken and hearing deficiency. It is the only mode of communication for such people to convey their messages to others. Thus, it is important to understand and apprehend the sign language [1] correctly. FingerSpelling is an accepted practice and widely used for the aforesaid communication. Finger signs are used in different sign languages for different purposes. It is popularly used to symbolize words of a verbal language, which has no specific sign convention. So a lot of explanations and examples should be given for teaching and learning of a sign language. Sign languages are categorized based on geographical regions [2]. Figure 1 illustrate different signs used in Finger Spelling American sign language (ASL).

Similarly, Indian Sign Language was also evolved for Indian deaf and dumb community. It is different in the phonetics, grammar, gestures and syntax from other country's sign languages. Designing a hand gesture recognition system for ISL [3–5] is more challenging than other sign languages due to the following reasons:

Fig. 1 American sign language

1. ISL is not standardized.
2. ISL uses both static and dynamic hand gestures.
3. ISL uses both hands to represent most of the alphabet.

2 Background

Dactylology based security is quite challenging since most of the activities are to be performed in real time [6–10]. A real time environment presents a three-dimensional (3D) environment. The environment allows person to interact directly with the intended objects naturally and interactively [11, 12]. In real time environment one can locate objects and its direction; can produced different kinds of gesture such as speech, sound, facial expression, hand, eye, and symbolic responses for better interaction between human and computer [13–15].

Sometimes communications are done through a digital image form (Figs. 2, 3 and 4). But digital images are prone to attack since it can be manipulated easily using basic image processing tools. Hence, protecting the user's identity and the computer's data both are becoming gradually more and more difficult. The greatest challenge is to ensure authenticity and integrity of digital image, which means the origin and the content of the digital file, should remain unaffected. As the problem stated, is potentially massive and is looming large on the horizon, therefore data security is the most focused subjects which increasingly gaining more and more importance and attention. Cryptography [16–21] is the heart of security as it is the most appropriate and practiced approach.

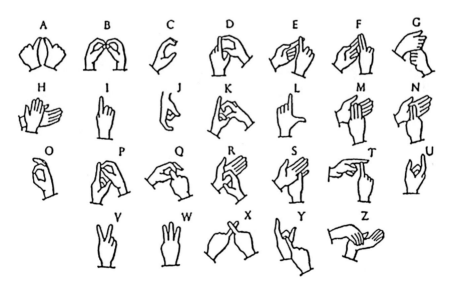

Fig. 2 Indian sign language

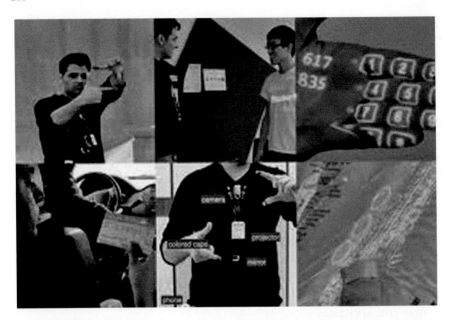

Fig. 3 Real-time enviornment

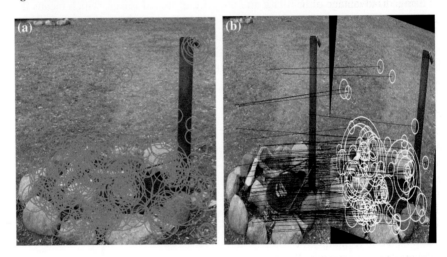

Fig. 4 **a** SURF based interested points dectation, **b** matching of interested points with targepoints

Selective encryption is a subset of traditional cryptography where encryption is performed only on the Region of Interest (ROI). Since technology is at our doorstep therefore novice user or physically challenged people will find it easier to select the region of his or her interest in the intended object in real time interactively. Selecting ROI in this way brings flexibility, reduces the use of complex algorithm to determine the ROIs within the intended object. ROI based Selective cryptography [22] at real time have lots of advantages over traditional cryptography:

1. Since we are no longer interested about the whole image rather we are focusing on some portion of the image which is our ROIs therefore we can reduce our work to some extent i.e. to determine our ROIs through some complex algorithm.
2. The size of the original image and encrypted image is all most same therefore we can avoid one extra step that is compression to reduce the consumption of bandwidth while sending data across proposed by many researchers.
3. Since performance of any encryption and decryption is measured in terms of speed i.e. time taken to encrypt or decrypt the data, which will be largely reduced in this process.
4. Selection of ROIs is user dependent, Real time and Session based.

Various application domains of ROI are as follows:

1. In 'Test Identity Parade', the identity of the victim is partially protected by putting up a black mask over the face.
2. The Identity of criminals and terrorist are also protected in the same way.
3. Sometime the interested part (face) of the person who is revealing some unwanted incidence or secret information before the media.
4. In medical science, by hiding or masking the faces of patients, who are suffering from critical diseases to reduce the social harassment.

Some disadvantage of ROIs over traditional cryptography system are:

1. For a cryptanalyst it will be easy to track and identify the important portion of the image.
2. Required expertise domain knowledge in some special cases.

Automatic pattern recognition, feature description, grouping and classification of patterns are significant problems in the various fields of scientific and engineering disciplines such as medicine, marketing, computer vision, psychology, biology, artificial intelligence, and remote sensing. A pattern could be a gait, fingerprint image, on or offline handwritten, a human face, retina, iris or a speech signal. Pattern recognition is the discipline where we observed how machines can study or analyze the environment, and then discriminates patterns of interest from their knowledge base, and makes logical decisions about the categories of the patterns. Among the various traditional approaches of pattern recognition through machine intelligence, the statistical approaches are mainly practiced and thoroughly studied [23–25]. Recently the ANN based techniques has receiving significant attention [26–34]. Sign language database/template is generally created using features, extracted from finger images. Some of the popular techniques used for feature extraction and feature matching are SIFT, SURF [3, 35], FAST etc.

Inspired by the success of SIFT descriptor, SURF i.e. Speeded-Up Robust Features was proposed by Bay et al. [35] in 2006, is a robust local feature detector, which can be used in computer vision for object recognition or reconstruction. The experimental results show that standard version of SURF is several times faster than SIFT. SURF is based on sums of 2D Haar wavelet responses and makes an efficient use of integral images. Integral images provided fast computation of box type convolution Filters.

The entry of an integral image I_(x) at a location x = (x; y) T represents the sum of all pixels in the input image I within a rectangular region formed by the origin and x.

$$T \sum (x) = \sum_{i=0}^{i<=x} \sum_{j=0}^{j<=y} I(i, j) \tag{1}$$

Interest points can be detected using a hessian based blob detector. The determinant of a Hessian matrix expresses the extent of the response and is an expression of the local change around the area.

$$H(x, \sigma) = \begin{bmatrix} L_{xx}(X, \sigma) & L_{xy}(X, \sigma) \\ L_{xy}(X, \sigma) & L_{yy}(X, \sigma) \end{bmatrix} \tag{2}$$

where

$$L_{xx}(X, \sigma) = I(x) * \frac{\delta^2}{\delta x^2} g(\sigma) \tag{3}$$

$$L_{xy}(X, \sigma) = I(x) * \frac{\delta^2}{\delta_{xy}} g(\sigma) \tag{4}$$

3   Related Work

Since research works in real time environment, encryption and machine learning are significantly gaining popularity, a large number of researches have been carried out in these areas. Some of the researchers' findings and methods adopted are listed below:

Gurjal and Kunnur [3] proposed an algorithm, in which they divided a captured video into various frames and then used the Gaussian difference feature to solve Real Time hand gesture recognition. Here the authors used SIFT algorithm to extract Scale space Feature from each and every frame. A complex and critical method was presented by Ghotkar et al. [4] where the authors investigated hand gesture recognition for Indian Sign Language using Camshift and HSV model and then Genetic Algorithm (GA) was used for recognizing gestures. Since Camshift and HSV model have compatibility issues with different versions of MATLAB and use of GA takes a huge amount of time for its development, making the application to run relatively slow compared to other applications. Rajan and Balakrishnan [5] in their work suggested that each gesture could be recognized through 7 bit orientation, and features have been generated through LEFT and RIGHT scan. Approximately six modules were used for recognition of signs and thus it was quite lengthy and tiresome. Tamer and Assaleh [36] used Arabic Sign Language gestures in a client independent mode. In their proposed work, the authors suggested, wearing of gloves by the signer would simplify the process of hand segmentation. Correa-Tome and Sanchez-Yanez

[10] proposed an alternative to the commonly used Hausdorff distance (HD) problem. Here the author used a visual similarity metric based on precision–recall graphs; a bipartite graph to represent the relationship between a reference shape and a test template; the Hopcroft–Karp algorithm to solve the matching problem which reduces computational complexity. Ishii et al. [11] improved the Viola–Jones face detector for fast face detection with boosting-based face tracking algorithm. Here the authors shown that for an 8 bit color image of 512×512 pixels the tracker rate is 500 fps in real time. Badrinath et al. [12] shown that indexing scheme can be very efficient for a palmprint-based identification system. In their work geometric hashing on SURF key-points are used to index the palmprint into hash map table and creates score level fusion of voting strategy based on geometric hashing and SURF score to identify the palmprint in real time.

Azzaz et al. [37] developed new design architecture for hardware based chaotic key generator for low-cost image encryption for embedded systems in real time. Their proposed design is robust, compact, highly secured and speed of encryption is quite good while preventing key analysis and statistical attacks. Dardas and Georganas [13] creates a grammar that generates gesture commands to control game application Here the authors generated feature (bag of features) using SIFT, histogram vector (bag of words) using K-means clustering and multiclass support vector machine for training. Faudzi et al. [14] shows that how effectively we can control a robot at real-time using hand gesture. Here the authors control the robot using bounding box and center of mass of the object of the gesture sing. Ohn-Bar and Trivedi [15] proposed to classify hand gestures that combined RGB and depth descriptor for a vision-based system. Here the author used two interconnected modules one that detects a hand in the region and performs user classification, other which performs gesture recognition using RBGD dataset

Min et al. [25] suggested that the visual recognition can be either static or dynamic gesture. In their work they recognized hand gestures without using any external devices. Gestures are picked up from the visual images on a 2D image plane. Gestures were spotted by a task specific state transition based on natural human articulation. Static gestures were recognized using image moments of hand posture, while dynamic gestures were recognized by analyzing their moving trajectories on the Hidden Markov Models (HMMs).

A lot of research was also carried out in the field of cryptography. Since no single algorithm is sufficient different researchers proposed different approaches. Yang and Wang [16] developed a new transforming LSB substitution called matching approach, to find a better solution and to reduce long running time. Rahouma [17] proposed a block cipher technique for security of data and computer networks. The technique was used for text, binary and hexadecimal information. Belmeguenai et al. [18] introduced a new stream cipher which is based on nonlinear filtering function. To get the best possible result, prevent possible attacks and to satisfy all the cryptographic criteria here the authors introduced a Boolean function in their algorithm that is a resilient function. Bhattacharya et al. [19] proposed session based bit level cipher technique using helical and columnar transpositions where 'Session Key' was used as a sequence of decimal digits. Mandal and Dutta [20] used a 256-bit recursive pair parity encoder for encryption. Bhattacharya et al. [21] shows that an enhancement of information security can be obtain using substitution of bits through prime detection in blocks.

The growing popularity of neural network models to resolve the pattern recognition related problems has drawn attention for researcher's to get involved in this domain. Ali Shah et al. [28] proposed Interactive Voice Response (IVR) with pattern recognition based on Neural Networks. Identity verification is resolved by combining the password and voice sample. Lamar et al. [29] proposed American and Japanese alphabet recognition using PCA for extracting features like position of the finger, shape of the finger and direction of the image described by mean values, Eigen values and Eigen vectors respectively. Yewale and Bharne [30] investigate different hand gesture recognition process and some algorithm related to skin and edge detection in this regard using Matlab as a development tools. Ibraheem and Khan [31] conducted a survey on numerous hand gesture recognition that make use of Neural Networks and drawn comparisons between these methods, advantages and drawbacks and implementation for each of the stated method. Zhang et al. [33] introduced a new framework for hand gesture recognition that is based on two sensors namely ACC and EMG. The intensity of the EMG signals is used as a base for automatic detection of meaningful gesture segments. Outputs are obtained after successful fusion of decision. Here the authors used multistream hidden Markov models and decision tree. A highly secured session based data encryption technique using robust fingerprint based authentication was proposed by Bhattachaya et al. [34] where Artificial Neural Network used for fingerprint template generation. Biometric key was generated from the template.

4 Proposed System

In the proposed method we are going to secure any digital data from unauthorized access using both ROIs and traditional approach. The block schematic diagram of our proposed scheme is shown in Fig. 5.

The proposed system is divided into five major steps.

4.1 Generation of ISL Database

The ISL Database generation process consists of following steps.

1. Different sample Images [A–Z] for ISL Database have been generated.
2. Unwanted noises are removed from the generated images through pre-processing.
3. Background part of ISL images are removed using skin detection.
4. Optimized features of captured images are extractbased on optimized points.
5. Trained, Tested and validated the sample acquired images using ANN.
6. ISL database has been generated and categorized based on extracted features.

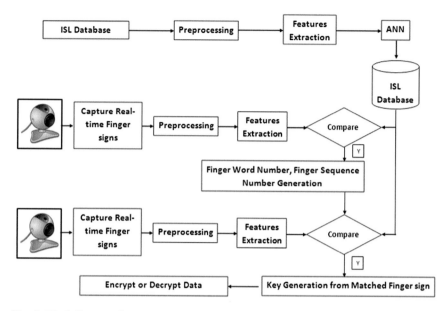

Fig. 5 Block diagram of our proposed system

4.2 Real-Time Finger-Word Generation

The Real-time Finger-word generation process consists of following steps

1. In real time user indicate the number of finger signs to be captured to form the desired Finger-Word.
2. Finger sign is captured using webcam until desired number is satisfied.
3. Captured image is pre-processed to remove unwanted noise and backgrounds.
4. Optimal features are extracted and matched with the pre-build ISL Database
5. On successful match a Finger sequence number has been generated from the corresponding ISL database sequence number and goto step 2 otherwise goto step 7.
6. Finally store the Finger sequence number and number of finger signs to be captured.
7. Finger-word generation process has been terminated.

4.3 Real-Time Finger-Word Recognition and Key Generation

The Real-time Finger-word recognition and key generation process consists of following steps:

1. All the necessary information generated at Finger-Word generation process has been retrieved.
2. In real-time user indicate the number of finger sign to be capture to form the desired Finger-Word.
3. On successful match with the indicated number of finger sign, goto step 4 otherwise goto step 11.
4. Finger sign is captured in real time using webcam.
5. Captured image is pre-processed to remove the unwanted noise.
6. Background part of ISL image are removed using skin detection.
7. Optimized features of captured image is extracted based on optimized points.
8. Extracted features are matched with pre-build ISL database.
9. On successful match a Finger sequence number has been generated from the corresponding ISL data base sequence number and goto step 2 otherwise goto step 11.
10. On successful match of finger a sequence number a 'Gesture Key' have been generated from the gesture images for the first phase encryption.
11. Finger-Word Recognition has been terminated.

4.4   Encryption of Data

Encryption of data process consists of two steps.

- 1st Phase Encryption of ROIs
- 2nd Phase Encryption of whole Image.

4.5   Decryption of Data

Step 1: It is the just opposite of the encryption process.

5   Explanation of Proposed Approach

5.1   Generation of ISL Database

The ISL Database generation process consists of following steps.

1. **Different sample Images [A-Z] for ISL Database have been generated**: At the outset different Finger images are captured through webcam which are used as signs in ISL. Since ISL signs are not standardized, flexibly these can be used for different purposes, which is ideal from security point of view. Here 24 signs are captured corresponding to 24 alphabets (English).

2. **Pre-processing and skin detection**: To keep the system simple, background color of the captured images is kept white. Median filter has been used twice for the purpose of noise removal. Since we are only interested in finger shapes, skin detection algorithm has been used to extract the sign portion from the image. Figure 11a–c illustrate finger signs, Finger sign after skin detection and extraction. Further sign part is cropped to remove unwanted portion.

3. **Optimized features have been generated**: Once Finger parts are separated from the image, the SURF feature detection algorithm is used for extracting the features from image.

   1. Given a image the key points $K = \{k1, k2, k3, \ldots, kn\}$ have been generated where n is the no of key points
   2. From the above detected keypoints, the corresponding descriptor $F = \{f1, f2, f3, \ldots, fn\}$ have been generated
   3. For each keypoint k_i and descriptor f_i, Stored the keypoints and descriptors in a $M \times N$ matrix.

Since number of SURF Points differ from the image to image, in the proposed approach Ten (10) strongest features have been extracted.

For i = 1 to M,
    For j = 1 to N.
        Calculate the Euclidian distances $ED = \{L1, L2, L3, \ldots, Ln - 1\}$ for (n − 1) other neighboring keypoints.
        Find the ten good matches $m = \{m1, m2, m3, \ldots, m10\}$ based on minimum Euclidian distance.

Figure 11d, e illustrate Finger sign with SURF points and finger Sign with 10 optimized points.

4. **ISL database has been generated and categorized based on extracted features**: ISL database have been generated with the optimal features. Figures 9 and 10 shows the ANN validation and classification assessment results.

5.2 Real Time Finger-Sequence Generation

The Real-time Finger-word generation process consists of following steps

1. **Real-time Finger-Word have been generated**: Initially the user indicated the number of finger signs to be captured to form his or her desired Finger-Word. Now if the user wants to communicate the word 'YES'. The user indicates three (3) and the system is going to captured 3 finger signs using a web cam. Some delay is used to ensure proper capture. Now the user shows the first finger sign. System captured the sign and does all pre-processing to remove the unwanted noise and background information.

2. **Optimized features extraction and feature matching with the pre-build ISL Database**: If the matching score of extracted features from Finger-Word is within the threshold value then a string type finger sequence has been generated. The finger sequence is the sequence number obtained from ISL database. In our example for 'y' it is '25'. If the matching score does not satisfy the threshold value, then Finger-Word generation process has been terminated. This process will continue until the desired Finger-Word is produced. Finally the finger sequence number and number of sign to be captured are stored. For our example the finger sequence is '250518' which is concatenated value of 'y', 'e', 's' and number of sign to be captured is '3'.

3. **Finally store the Finger sequence number and number of finger signs to be captured**.

5.3 Real-Time Finger-Word Recognition and Key Generation

In this part real time Finger-Word is recognized and 'Gesture Key' has been generated. The process consists of following steps

1. **Finger-Word Information has been retrieved**: In the very beginning system retrieved the necessary information stored in Finger-Word generation step. In our case it is '250518' and '3'.

2. **No of Finger-sign has been captured in realtime**: The user indicates the number of finger sign to be captured using webcam. If the numbers of signs to be captured are not matched with the retrieved sign number i.e. '3' then the recognition process has been terminated. On successful match with the indicated number of finger sign go to the next step.

3. **Desired Finger signs are captured in real time**: On a successful verification first sign is captured using a web cam. As stated in the earlier section, median filter has been used twice for noise removal after that skin detection algorithm has been used to extract the finger parts from the captured image and then cropped, to avail the portion intended for feature extraction.

4. **Finger sequence has been generated**: Ten (10) strongest features point are being extracted using SURF algorithm and simulated the extracted feature value with the pre-build ISL database. If the matching score has within the threshold value then produce a string type finger sequence as discussed in Finger Word generation step. This process will continue until the desired Finger-Word is produced. Finally finger sequence number is matched with the retrieved information. In our example it is '250518'. On successful verification password from a sequence of captured finger image have been generated.

5. **Gesture Key has been generated**: On a successful match of finger sequence number a 128 bit 'Gesture Key' has been generated using MD5 hash algorithm for the first phase encryption of ROIs. Finally encrypted the selected portions of the image using ROIs as well as the whole image traditional approach using a session based key.

5.4 Encryption of Data

Encryption of data process consists of two steps.

1st Phase Encryption of ROIs

1. The secret image is selected for the purpose of encryption.
2. Multiple regions of Interest have been selected interactively at real time.
3. Binary masks, based on ROIs is generated and placed over the ROIs.
4. The selected regions of the image are encrypted using 'Gesture Key' which is generated from the gesture images.

2nd Phase Encryption of whole Image

1. To ensure better security partially encrypted image is further perturbed using Session based pass word. If the size of key generated is small then necessary padding being done. Bit level XOR operation is performed using
2. 'Session Key' to encrypt the image. Finally the encrypted image is saved and sends to the intended recipient.

5.5 Decryption of Data

Decryption of data process consists of following steps

1. Session based, password should be given.
2. The same dactylology gestures are reproduced to generate the 'Gesture Key'. The image is fully recovered by applying the reverse process of masking multiple ROIs.

6 Results and Discussion

In this section, experimental results of the proposed approach have been shown. The proposed system has been implemented using MATLAB (R2013) on an Intel Dual core 2.0 GHz Processor, with 2 GB RAM, Windows XP as a platform and with an 12 Mega pixel, USB webcam capture (with different resolutions such as 640×480, 320×240 and 160×120, at 15 frames-per second) for image, which is adequate for real time image application. In our proposed approach, we have used a set of popular standard image such as Lena, Cameraman, Baboon, Gold Hill, Barbara etc. with size varies from 64×64 to 512×512 as the original images (plaintext image) and apply the encryption or decryption process. The experiments were performed several times to assure that the results are consistent and are valid. The Fig. 6a shows the proposed system; Fig. 6b shows the real time environment. Figure 6c shows the generated Finger-Word using the different Finger sign for our proposed work (Figs. 7 and 8).

(a)

Fig. 6 **a** UI of the proposed system, **b** real-time environment of proposed system, **c** generated ISL Dataset for our proposed work

Fig. 7 Finger-Word 'yes' has been generated using sequence of ISL based finger sign at real time using the proposed scheme

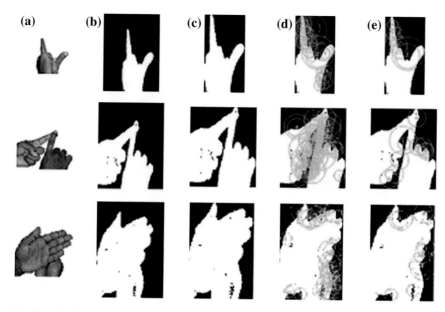

Fig. 8 **a** Sign image captured at real time for our proposed work, **b** removal of background information from the captured image with skin detection, **c** cropped sign image, **d** extraction of features points from the sign image using SURF, **e** optimal feature have been generated from extracted features

Fig. 9 **a** Performance assessment of ANN process using validation curves, **b** ratio of correct and incorrect classification

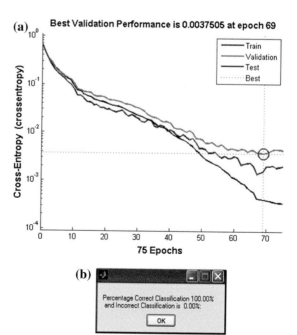

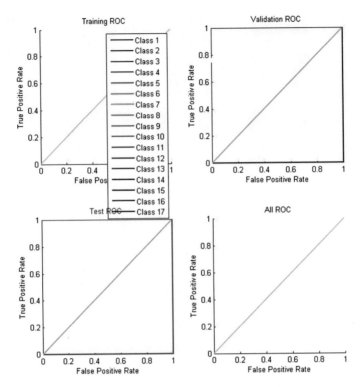

Fig. 10 Performance assessment of ANN training process using ROC curve

Figure 9 showing the performance of ANN training using validation curves, classi-fication performance ratio correct versus incorrect classification in percentage while Fig. 10 showing the classification assessment using ROC Curve (Figs. 11, 12 and 13).

Based on the results shown in Fig. 14 there is no visual information observed in the encrypted image, and the encrypted images are visually indistinguishable even with a big difference with respect to the original images.

6.1 Comparitative Anlysis of Proposed Work

In this section, a number of parameters have been discussed. Using these parameters the effectiveness and integrity of an image encryption scheme can be evaluated. In order to evaluate the performance and the strength/security of the computed system,

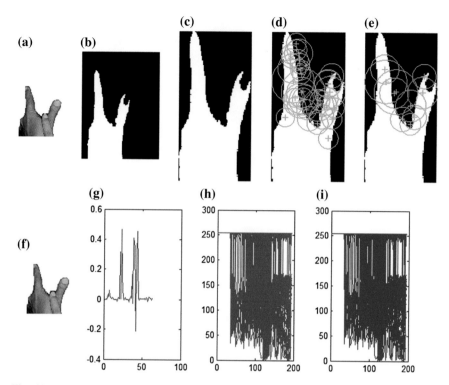

Fig. 11 **a** Finger sign captured in real-time, **b** background removal through skin detection, **c** showing cropped sign image, **d** extraction of features points from the sign image using SURF, **e** optimal feature have been generated from extracted features, **f** closest matched sign image have been retrieved from ISL database, **g** comparison of features between (**a**) and (**f**), **h** pixel distribution of (**a**), **i** pixel distribution of (**f**)

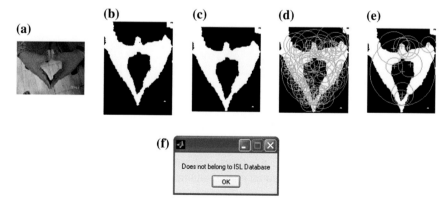

Fig. 12 **a** Finger sign captured in real-time, **b** background removal through skin detection, **c** cropped sign image, **d** extraction of features points from the sign image using SURF, **e** optimal feature have been generated from extracted features, **f** message showing the captured image does not belong to ISL database

Fig. 13 **a** Original image, **b** *Blue color rectangle* and *square* is showing ROIs in original image, **c** ROI's are encrypted with noise, **d** encrypted image

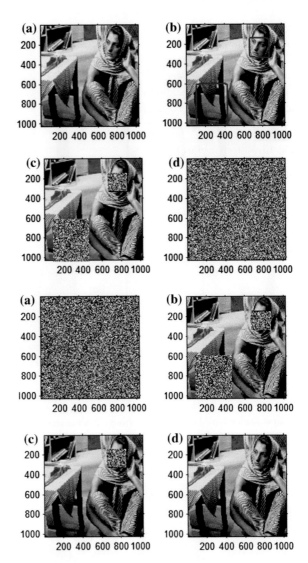

Fig. 14 **a** Encrypted image, **b** decrypted image with noise 1 and noise 2 **c** image after noise 1 is removed **d** fully recovered image [both noise 1 and 2 are removed]

the parameters that are used for algorithm testing are Histogram, Correlation coefficient, Entropy, MSE, and PSNR. Figure 15 shows Histogram of Original image and Encrypted Image. Tables 1 and 2 measures, encryption quality through correlation coefficient (plain image vs. encrypted image) and image entropy analysis. In Table 3 MSE and PSNR results shows the quality of the encryption process. Another chosen factor to determine the performance or response of the proposed system is the speed or time taken to encrypt /decrypt data of various sizes. Tables 4, 5 and 6 display the Throughput, CPU utilization time and Memory utilization, encryption and Decryption time. The proposed system is quite secure since the key stream Used for encryption was generated randomly, thus it will not be very hard to predict future key stream nor the attacker should not be able to recover the cipher's key. In the Tables 7 and 8, we compare performance of proposed work with Salsa20 [38].

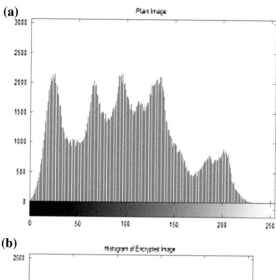
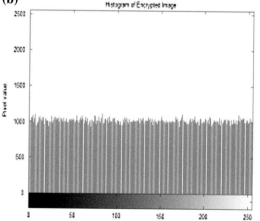

Fig. 15 Histogram of **a** original image, **b** encrypted image

Table 1 Correlation and coefficient value showing the encryption quality between plain Image versus encrypted image of the proposed work

Correlation and coefficient analysis (plain image vs. encrypted image)

Image	Direction of adjacent pixel	Plain image	Encrypted image
Barbara	Horizontal	0.9859	0.0317
	Vertical	0.9686	0.0247
	Diagonal	1	0.0022
Lena	Horizontal	0.9992	0.0042
	Vertical	0.8917	0.0471
	Diagonal	0.9568	0.0018
Gold Hill	Horizontal	0.9762	0.078
	Vertical	0.9782	0.0433
	Diagonal	0.9532	0.0014
Baboon	Horizontal	0.9665	0.0187
	Vertical	0.9281	0.2
	Diagonal	1	0.0002

Table 2 Entropy value estimates the encryption quality of proposed work

Name of the image	m source of value						
	2	4	8	16	32	64	256
	H(m) ideal value of entropy						
	1	2	3	4	5	6	8
Barbara	0.9971	1.9978	2.915	3.9999	4.9999	5.9998	7.9971
Lena	0.9962	1.9965	2.9956	3.9998	4.9998	5.9996	7.9992
Baboon	0.9989	1.9878	2.998	3.9999	4.9999	5.9998	7.9991
Gold Hill	0.9974	1.9998	2.9938	3.9989	4.9999	5.9999	7.9992

Table 3 PSNR and MSE values measuring the quality of encryption process of proposwork

Image	PSNR	MSE
Barbara	7.88	10670
Lena	8.59	9075
Baboon	9.76	7248
Gold Hill	9.09	8085

Table 4 Throughput, CPU time (Hz), memory sized (KB), encryption and decryption speed in ms measuring the performance of the work

Image	Throughput	CPU (Hz)	Memory (Bytes)	Encryption time (ms)	Decryption time (ms)
Baboon 64	853333.33	0	0	0.000075	0.000076
Baboon 128	514590.16	0	20480	0.00244	0.000248
Baboon 256	245508.11	0	69632	0.001047	0.000937
Baboon 512	324255.85	0	532480	0.001579	0.002015
Baboon 1024	121025.88	0.0313	2113536	0.008461	0.008891
Average	413542.67	0.00626	547225.6	0.0022812	0.0024334

Table 5 Throughput, CPU time (Hz), memory sized (KB), encryption and decryption speed in ms reflecting the performance of the work

Image	Throughput	CPU (Hz)	Memory (Bytes)	Encryption time (ms)	Decryption time (ms)
Barbara 64	955223.88	0	0	0.000067	0.000076
Barbara 128	556521.73	0	20480	0.00023	0.000245
Barbara 256	274383.70	0	69632	0.000933	0.000971
Barbara 512	325699.74	0	532480	0.001572	0.003705
Barbara 1024	122502.69	0.0313	2113536	0.008359	0.008551
Average	446866.35	0.00626	547225.6	0.0022322	0.0027096

Table 6 Throughput, CPU time (Hz), memory sized (KB), encryption and decryption speed in ms showing the performance of the work

Image	Throughput	CPU (Hz)	Memory (Byte)	Encryption time (ms)	Decryption time (ms)
Lena 64	984615.38	0	0	0.000065	0.000076
Lena 128	522448.97	0	20480	0.000245	0.000246
Lena 256	281009.87	0	69632	0.000911	0.000941
Lena 512	232515.89	0	532480	0.002202	0.002028
Lena 1024	122634.73	0.0156	2113536	0.008351	0.008407
Average	428644.93	0.00312	547225.6	0.0023546	0.0023396

Table 7 Correlation and coefficient values showing performance of proposed versus existing Salsa method [38]

Correlation and coefficient comparison between proposed work versus existing Salsa method			
Image	Adjacent pixel direction		
	Horizontal	Vertical	Diagonal
Plain-image	1.0000	0.9986	0.9988
Salsa 20/8 cipher image	0.0430	0.0383	0.0117
Salsa 20/12 cipher image	0.0348	0.0021	0.0195
Salsa 20/20 cipher image	0.0030	0.0204	0.0653
Proposed method cipher image	0.0042	0.0271	0.0018

Table 8 Encryption and decryption time showing performance of proposed versus existing Salsa method [38]

Comparison of encryption and decryption time between proposed method versus existing Salsa method		
Image	Average encryption time (ms)	Average decryption time (ms)
Salsa 20/8 cipher image	1.3	1.3
Salsa 20/12 cipher image	1.7	1.7
Salsa 20/20 cipher image	2.6	2.6
Proposed method cipher image	0.000951	0.00095

7   Conclusion

Dactylology based selective image encryption using Speeded-Up robust features extraction technique and artificial neural network at real time is proposed in this work. In this scheme ISL images were used as a secret symbol or sign to performed encryption or decryption operations. Through experimental results we show that the work can be used to maintain the integrity of the ROIs in digital data, thus it has good prospects in the security domain. To judge the integrity and effectiveness of the proposed scheme we used Histogram, Correlation coefficient, Entropy, MSE, and PSNR values between the Original image and the Encrypted image. The performance or response of our work is measured using different parameter such as time taken to encrypt or decrypt data of varying size, CPU occupancy time, Memory it has consumed and Throughput. A comparative study between our scheme and existing Salsa method also suggest that work has sufficient prospect in the security domain. The work can be further extended by observing a sequence of video image instead of captured still image.

References

1. Padden, C.: How the alphabet came to be used in a sign language. Sign Language Studies. 4.1. Gallaudet University Press (2003)
2. Zeshan, U.: Indo-Pakistani sign language grammar: a typological outline. Sign Lang. Stud. **3**(3), 157–212 (2003)
3. Gurjal, P., Kunnur, K.: Real time hand gesture recognition using SIFT. Int. J. Electr. Eng. **2**(3), 19–33 (2012)
4. Ghotkar, S.A., Khatal, R., Khupase, S., Asati, S., Hadap, M.: Hand gesture recognition for indian sign language. In: Proceedings of the IEEE International Conference on Computer Communication and Informatics, pp. 1–4 (2012)
5. Rajam, P.S., Balakrishnan, G.: Real time Indian sign language recognition system to aid deaf and dumb people. In: Proceedings of the 13th International Conference on Communication Technology (ICCT), pp. 737–742 (2011)
6. Nandy, A., Prasad, J.S., Chakraborty, P., Nandi, G.C., Mondal, S.: Classification of Indian Sign Language in real time. Int. J. Comput. Eng. Inf. Technol. (IJCEIT) **10**(15), 52–57 (2010)
7. Nandy, A., Prasad, J.S., Mondal, S., Chakraborty, P. & Nandi, G.C.: Recognition of isolated Indian sign language gesture in real time. In: Proceedings of BAIP, Springer LNCS-CCIS, vol. 70, pp. 102–107 (2010)
8. Rekha, J., Bhattacharya, J., Majumder, S.: Shape, texture and local movement hand gesture features for Indian sign language recognition. In: Proceedings of the 3rd International Conference on Trendz in Information Sciences and Computing (TISC) IEEE, pp. 30–35 (2011)
9. Geetha, M., Manjusha, U.C.: A vision based recognition of Indian sign language alphabets and numerals using B-spline approximation. Int. J. Comput. Sci. Eng. **4**(3), 406–415 (2012)
10. Correa-Tome, F.E., Sanchez-Yanez, R.E.: Fast similarity metric for real-time template-matching applications. J. Real-Time Image Process. 1–9 (2013)
11. Ishii, I., Chida, T., Gu, Q., Takaki, T.: 500-fps face tracking system. J. Real-Time Image Process. **8**(4), 379–388 (2013)
12. Badrinath, G.S., Gupta, P., Mehrotra, H.: Score level fusion of voting strategy of geometric hashing and SURF for an efficient palmprint-based identification. J. Real-Time Image Process. **8**(3), 265–284 (2013)
13. Dardas, N.H., Georganas, N.D.: Real-time hand gesture detection and recognition using bag-of-features and support vector machine techniques. IEEE Trans. Instr. Meas. **60**(11), 3592–3607 (2011)
14. Faudzi, A.A.M., Ali, M.H.K., Azman, M.A., Ismail, Z.H.: Real-time hand gestures system for mobile robots control. In: International Symposium on Robotics and Intelligent Sensors, vol. 41, pp. 798–804 (2012)
15. Ohn-Bar, E., Trivedi, M.M.: Hand gesture recognition in real time for automotive interfaces: a multimodal vision-based approach and evaluations. IEEE Trans. Intell. Transp. Syst. **99**, 1–10 (2014)
16. Yang, C., Wang, S.: Transforming LSB substitution for image-based steganography in matching algorithm. J. Inf. Sci. Eng. **26**, 1199–1212 (2010)
17. Rahouma, K.H.: A block cipher technique for security of data and computer networks. In: Proceedings of the IEEE Internet Workshop (IWS'99), pp. 25–31 (1999)
18. Belmeguenaï, A., Derouiche, N., Khaled, M.: Image encryption using stream cipher algorithm with nonlinear filtering function. Int. J. Adv. Comput. Sci. Appl. **3**(9), 150–156 (2012)
19. Bhattacharya, T., Bhattacharya, T.K., Bhadra Chaudhuri, S.R.: A session based bit level cipher technique using helical and columnar transpositions. In: Proceedings of the National Conference on Cryptography and Network Security, pp. 87–91 (2009)
20. Mandal, J.K., Dutta, S.: A 256-bit recursive pair parity encoder for encryption. In: Proceedings of the Association for the Advancement of Modelling and Simulation Techniques in Enterprises (AMSE, France), vol. 1, pp. 1–14 (2004)

21. Bhattacharya, T., Bhattacharya, T.K., Bhadra Chaudhuri, S.R.: A general bit level data encryption technique using helical and session based columnar transpositions. In: Proceedings of the IEEE International Advance Computing Conference, pp. 364–368 (2009)

22. Kirishima, T., Sato, K., Chihara, K : Real-time gesture recognition by learning and selective control of visual interest points. IEEE Trans. Pattern Anal. Machine Intell. **27**(3), 351–364 (2005)

23. Lee, L.K., An, S.Y., Oh, S.Y.: Robust fingertip extraction with improved skin color segmentation for finger gesture recognition in human-robot interaction. In: Proceedings of the Evolutionary Computation (CEC) IEEE Congress on Computational Intelligence, pp 1–7 (2012)

24. Wu, Y., Huang, T.: Hand modeling analysis and recognition for vision-based human computer interaction. IEEE Signal Process. Mag. Spec. Issue Immers. Interact. Technol. **18**(3), 51–60 (2001)

25. Min, B.-W., Yoon, H.-S., Soh, J., Ohashi, T., Jima, T.: Visual recognition of static/dynamic gesture: gesture-driven editing system. J. Vis. Lang. Comput. **10**(3), 291–309 (1999)

26. Jain, A.K., Mao, J., Mohiuddin, K.M.: Artificial neural networks: a tutorial. IEEE J. Comput. Special Issue: Neural Comput. **29**(3), 31–44 (1996)

27. Kohonen, T.: Self-organizing maps. Springer Series in Information Sciences, vol. 30. Springer, Berlin (1995)

28. Ali Shah, S.A., Asar, A.Ul., Shaukat, S.F.: Neural network solution for secure interactive voice response. World Appl. Sci. J. **6**(9), 1264–1269 (2009)

29. Lamar, M.V., Bhuiyan, S., Iwata, A.: Hand alphabet recognition using morphological PCA and neural networks. In: Proceedings of the International Joint Conference on Neural Network IEEE, vol. 4, pp. 2839–2844 (1999)

30. Yewale, S.K., Bharne, P.K.: Hand gesture recognition using different algorithms based on artificial neural network. In: Proceedings of the Emerging Trends in Networks and Computer Communications (ETNCC) IEEE. pp. 287–292 (2011)

31. Ibraheem, N.A., Khan, R.Z.: Vision based gesture recognition using neural networks approaches: a review. Int. J. Hum. Comput. Interact. (IJHCI) **3**(1), 1–14 (2012)

32. Yu, C., Wang, X., Huang, H., Shen, J., Wu, K.: Vision-based hand gesture recognition using combinational features. In: Proceedings of the IEEE Sixth International Conference on Intelligent Information Hiding and Multimedia Signal Processing, pp. 543–546 (2010)

33. Zhang, X., Chen, X., Li, Y., Lantz, V., Wang, K., Yang, J.: A framework for hand gesture recognition based on accelerometer and EMG sensors. IEEE Trans. Systems, Man Cybern-Part A: Syst. Hum. **41**(6), 1064–1076 (2012)

34. Bhattacharya, T., Hore, S., Mukherjee, A., Bhadra-Chaudhuri, S.R.: A novel highly secured session based data encryption technique using robust fingerprint based authentication. In: Proceedings of the (LNCS, Springer), pp. 422–431 (2010)

35. Bay, H., Ess, A., Tuytelaars, T., Gool, L.V.: SURF: speeded-up robust features. Comput. Vis. Image Underst. (CVIU) **110**(3), 346–359 (2008)

36. Tamer, S., Assaleh, K.: Arabic sign language recognition in user independent mode. In: Proceedings of the IEEE International Conference on Intelligent and Advanced Systems, pp. 597–600 (2007)

37. Azzaz, M.S., Tanougast, C., Sadoudi, S., Dandache, A.: Robust chaotic key stream generator for real- time images encryption. J. Real-Time Image Process. **8**(3), 297–306 (2013)

38. Jolfaei, A., Mirghadri, A.: Survey: image encryption using Salsa20. Int. J. Comput. Sci. Issues **7**(5), 213–220 (2010)

Spectral Reflectance Images and Applications

**Abdelhameed Ibrahim, Takahiko Horiuchi, Shoji Tominaga
and Aboul Ella Hassanien**

Abstract Spectral imaging has received a great deal of attention recently. Spectral reflectance observed from object surfaces provides crucial information in computer vision and image analysis which include the essential problems of feature detection, image segmentation, and material classification. The estimation of spectral reflectance is affected by several illumination factors such as shading, gloss, and specular highlight. The spectral invariant representations for dielectric materials only, for these factors, are inadequate for other characteristic materials like metal. In this chapter, a spectral invariant representation is introduced for obtaining reliable spectral reflectance images. The invariant formulas for spectral images of natural objects preserve spectral information and are invariant to highlights, shading, surface geometry, and illumination intensity. As an application, a material classification method is presented based on the invariant representation, which results in reliable segmentations for natural scenes and raw circuit boards spectral images.

1 Introduction

During the last few years, the importance of spectral imagery has sharply increased following the development of new optical devices and the introduction of new applications. The spectral imaging system is a system which captures and describes color information by a greater number of sensors than an RGB device resulting in a spectral representation that uses more than three parameters. Figure 1 shows an illustration of a spectral scene and the corresponding 3-color RGB image. The problem with conventional color imaging systems is that they have some limitations, namely, dependence on the illuminant and characteristics of the imaging system. On the

A. Ibrahim (✉)
Faculty of Engineering, Mansoura University, Mansoura, Egypt
e-mail: afai79@mans.edu.eg

T. Horiuchi · S. Tominaga
Graduate School of Advanced Integration Science, Chiba University, Chiba, Japan

A. Ella Hassanien
Faculty of Computers and Information, Cairo University, Cairo, Egypt

© Springer International Publishing Switzerland 2016
A.I. Awad and M. Hassaballah (eds.), *Image Feature Detectors and Descriptors*,
Studies in Computational Intelligence 630, DOI 10.1007/978-3-319-28854-3_9

Fig. 1 An illustration of the spectral data with ten bands and the corresponding image as it would be acquired with a color imager [3]

other hand, spectral imaging systems can provide spectral reflectance information and therefore the systems are illuminant independent [1, 2]. Color imaging naturally becomes spectrophotometric; therefore, spectral imaging must be the technique of the immediate future.

Spectral imaging is used, for example, in remote sensing, computer vision, and industrial applications. Spectral information has become an important quality factor in many industrial processes because of its high accuracy [4]. Spectral images can be obtained, for example, by a CCD-camera with narrow-band interference filters [5]. Tominaga [6, 7] described two generations of a multi-channel vision system based on the use of a CCD-camera and six color filters to reconstruct the surface spectral reflectance and illuminant spectral power distribution. Manabe et al. [8] proposed a measurement system of spectral distribution and shape of an object with the use of an imaging spectrograph installed in CCD-camera. Haneishi et al. [9] developed a six-band camera consisting of three high sensitive bands and three low sensitive bands. Several multispectral acquisition systems have been proposed for imaging artworks [10–21]. Pelagotti et al. [22] used a multispectral imaging system for the noninvasive analysis of works of art. Recently, the spectral image resolution and sensitivity are much improved [23–29]. Du et al. [30] presented a prism-based system for capturing multispectral videos. That system is consists of a triangular prism, a monochromatic camera, and an occlusion mask.

The main advantage of spectral images, compared with color images, is the large amount of information involved, which dramatically improves the ability to detect individual materials or separate areas with visually different spectral bands. The disadvantage of spectral images is that, since we have to process additional data, the required computation time and memory increase significantly. However, since the speed of the hardware will increase and the costs for memory will decrease in the future, it can be expected that spectral images will become more important in many fields of image analysis and computer vision. This chapter introduces a spectral imaging system that captures high-dynamic range (HDR) image with spatially and spectrally high resolutions in the region of visible wavelength.

Spectral images are useful for a variety of applications such as material identification, natural scene rendering, colorimetric analysis, and machine vision tasks [31]. The observed spectral images do not only depend on surface-spectral reflectance and illuminant spectrum, but also include various reflection effects such as shading, gloss, and specularity, which mainly depend on illumination geometries and surface materials. Therefore, image representations invariant to shading, shadow, lighting, and specularity have been proposed for spectral images [26, 32–35] and for color images [36–50] so far in several ways. These invariant representations play an important role in many applications such as image segmentation [26, 32, 33, 44, 45, 51, 52], feature detection, such as edge and corner detection, [37, 41, 49, 50, 53–57], object recognition [39, 58–60], image retrieval [40, 61–63], cast shadow segmentation [64, 65], optical flow calculation [42, 66, 67], and robots [68, 69].

Geusebroek et al. [37] investigated the differential photometric invariance. The authors provided a set of photometric invariant derivative filters which were used for invariant edge detection. van de Weijer et al. [49] introduced photometric quasi-invariants which were a set of photometric invariant derivatives with better noise, stability characteristics, and then introduced less edge displacement than full photometric invariants. Combining the photometric quasi-invariants with derivative based feature detectors led to features which could identify various physical causes [50]. This combination allows for detection of photometric invariant edges. Stokman and Gevers [35] proposed a method for edge classification from spectral images. Their method aimed at detecting edges and assigning one of the types of shadow, highlight, and material edge. Montoliu et al. [34] proposed a spectral invariant representation for dielectric materials. This method used for edge detection of spectral images.

However, most of those methods were constructed based on the dichromatic reflection model by Shafer [70]. This model assumes that an object surface is composed of inhomogeneous dielectric material, and the reflected light from the surface is decomposed into two additive components of body (diffuse) reflection and interface (specular) reflection. This decomposition results in the classification of physics events, such as shadows and highlights. However, the model-based method is valid for such limited materials as plastics and paints [71–73]. It should be noted that there are metallic objects in real-world scenes which cannot be described by the standard dichromatic reflection model. Tominaga [73] shows that the surface reflection of metal can be approximated by extended dichromatic reflection model.

This chapter introduces an invariant representation for spectral reflectance images. The invariant representation for a variety of objects in a real world is derived from the standard dichromatic reflection model for dielectric and the extended dichromatic reflection model for metal [25] in more details. We show that the invariant formulas for spectral images of both artificial object and natural ones preserve surface-spectral reflectance information and are invariant to highlight, shadow, surface geometry, and illumination intensity. Here the illumination spectrum is measured by using a Spectro-radiometer and a standard white reference and measured the spectral sensitivity functions of the imaging system by using a Spectro-radiometer and a Monochromator.

As an application of the presented invariant representation, the transformed spectral image with the invariant properties is used as an invariant operator for the image segmentation problem. Several algorithms were recently proposed for segmenting the spectral images into distinct surface areas (e.g., [24, 74–78]). However, those algorithms were not always robust for the highlight and shading effects occurred for different surface materials, and required a careful adjustment of the lighting position. As a typical example, it was difficult to segment materials on a raw circuit board with various tiny elements [79–102]. This chapter presents a segmentation algorithm based on the invariant representation for effectively segmenting spectral images of natural scenes and bare circuit boards [25, 26] in details. The presented invariant representation is also applied to other alternative segmentation algorithms. Experiments using real-world objects including metals and dielectrics show that the representation is invariant to highlight, shadow, and object surface geometry, and effective for image segmentation.

The reminder of this chapter is organized as follow: 'Reflection models' section describes different types of reflection models. 'Spectral invariant representation' section develops the spectral invariant representation in details. 'Spectral image segmentation' section discusses spectral image segmentation based on the transformed spectral images to the spectral invariant. 'Experimental setting' section introduces the imaging system. 'Experiments' section shows the test results of applying the invariant representation for different spectral images and presents the segmentation results for the spectral invariant using different methods. This chapter ends with 'Summary' section.

2 Reflection Models

Shafer's dichromatic reflection model was the first physics-based model of reflection to separate different types of reflection on a single surface [70]. Shafer's model proposed that inhomogeneous dieletrics such as paints, ceramics and plastics exhibit two types of reflection, surface reflection and body reflection, and that the different types of reflection caused specific types of changes in appearance. Klinker et al. used the model to develop an algorithm for separating surface and body reflection in controlled conditions with a single illuminant [103].

Gershon et al. proposed a similar model, but divided the reflection into three components: specular, ambient, and diffuse [104]. These reflection modes interacted with direct and ambient illuminants, but not in an orthogonal manner. Where Shafer's material reflection model was based on a physical analysis of inhomogeneous dielectrics, Gershon's model arbitrarily divides the reflection components, linking the type of reflection with the type of illumination, despite the fact that the mechanisms of reflection work similarly for all visible light energy, regardless of its source.

Healey proposed a unichromatic reflection model for metals. This method was able to extract useful information about the illumination, reflection and material type under controlled conditions and a single illuminant [105].

2.1 Dichromatic Reflection Models

Standard Dichromatic Reflection Model Light reflected from the surface of an inhomogeneous dielectric object is composed of two additive components, the interface reflection and the body reflection according to the standard dichromatic reflection model [70]. Figure 2 shows the dichromatic reflection model of surface reflectance in inhomogeneous materials.

The radiance of the reflected light $Y(\theta, \lambda)$ is a function of the wavelength λ and the geometric parameters θ, including the direction angles of the viewing angle and

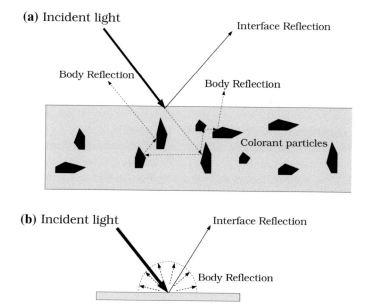

Fig. 2 The dichromatic reflection model of surface reflectance in inhomogeneous materials. **a** Light is scattered from a surface by two different mechanisms. Some incident light is reflected at the interface (interface reflection). Other light enters the material, interacts with the embedded particles, and then emerges as reflected light (body reflection). **b** Rays of light reflected by interface reflections is likely to be concentrated in one direction. Rays of light reflected by body reflection are reflected with nearly equal likelihood in many different directions [106]

the phase angle. The standard dichromatic reflection model describes the reflected light in the form

$$Y(\theta, \lambda) = m_I(\theta)L_I(\lambda) + m_B(\theta)L_B(\lambda) \tag{1}$$

where $L_I(\lambda)$ and $L_B(\lambda)$ are the spectral power distributions of the interface and body reflection components, respectively. The weights $m_I(\theta)$ and $m_B(\theta)$ are the geometric scale factors.

The reflection model is also described in terms of spectral reflectance. Let $E(\lambda)$ be the spectral-power distribution of a uniform illumination. The spectral reflectance function defined as $S(\theta, \lambda) = Y(\theta, \lambda)/E(\lambda)$, independent of illuminant, can be expressed as

$$S(\theta, \lambda) = m_I(\theta)S_I(\lambda) + m_B(\theta)S_B(\lambda) \tag{2}$$

where $S_I(\lambda)$ and $S_B(\lambda)$ are surface-spectral reflectances for the interface and body components, respectively. The standard model incorporates the neutral interface reflection (NIR) assumption which states that interface reflection component $S_I(\lambda)$ is constant over the range of visible wavelength as $S_I(\lambda) = S_I^c$. This allows Eq. (2) to be written as

$$S(\theta, \lambda) = m_I'(\theta) + m_B(\theta)S_B(\lambda) \tag{3}$$

where $m_I'(\theta) = m_I(\theta)S_I^c$. It is shown that this reflectance model is valid for a variety of natural and artificial dielectric objects including plastic and paint (e.g., see [71–73]).

Extended Dichromatic Reflection Model for Metals Metal is a homogeneous material that indicates essentially different reflection properties from the inhomogeneous dielectric materials. It consists of only interface reflection with the Fresnel reflectance. Thus, if the surface is shiny and stainless, the body reflection component in the reflected light is negligibly small. A sharp specular highlight is observed only at the viewing angle of the mirrored direction. Thus the surface reflection depends on the incident angle of illumination.

Tominaga [73] shows that the surface reflection of metal can be approximated by a linear combination of only two interface reflection components. This type of surface reflection is called the extended dichromatic reflection model. This model can be expressed as

$$S(\theta, \lambda) = m_{I1}(\theta)S_{I1}(\lambda) + m_{I2}(\theta)S_{I2}(\lambda) \tag{4}$$

where the first term in the right hand side corresponds to the specular reflection at the normal incident and the second corresponds to the grazing reflection at the horizontal incident. It is noted that surface-spectral reflectance is constant over the visible wavelength range at the grazing angle as $S_{I2}(\lambda) = S_{I2}^c$. Therefore, Eq. (4) can to be written as

$$S(\theta, \lambda) = m_{I1}(\theta)S_{I1}(\lambda) + m'_{I2}(\theta) \qquad (5)$$

where $m'_{I2}(\theta) = m_{I2}(\theta)S_{I2}^c$. It is important to note that the observed spectral reflectance can be expressed in a linear combination of the reflectance function at the normal incidence and a constant reflectance.

3 Spectral Invariant Representation

This section briefly explains a spectral invariant representation for spectral images [25, 26, 34, 35], which was limited to the standard reflection model for inhomogeneous dielectric objects, and then extend the invariant representation for all materials including inhomogeneous dielectric and homogeneous metal.

Let us suppose that a spectral image is captured at ρ points in the visible range [400, 700 nm]. Let i and j be two different wavelengths (spectral bands) in the range. From Eq. (3) of the standard reflection model, subtraction of one band from another provides a reflectance representation, independent of specular highlight as

$$
\begin{aligned}
S(\theta, \lambda_i) - S(\theta, \lambda_j) &= (m'_I(\theta) + m_B(\theta)S_B(\lambda_i)) - (m'_I(\theta) + m_B(\theta)S_B(\lambda_j)) \\
&= m_B(\theta)S_B(\lambda_i) - m_B(\theta)S_B(\lambda_j) \\
&= m_B(\theta)(S_B(\lambda_i) - S_B(\lambda_j))
\end{aligned}
\qquad (6)
$$

where the interface reflection component $m'_I(\theta)$ is eliminated. Moreover let l and n be the two other wavelengths. Then the following ratio of two subtractions between wavelengths can be illumination invariant, that is, invariant to highlight, shading, and surface geometry by eliminating the remaining weighting coefficient $m_B(\theta)$ [34, 35].

$$
\begin{aligned}
\frac{S(\theta, \lambda_i) - S(\theta, \lambda_j)}{S(\theta, \lambda_l) - S(\theta, \lambda_n)} &= \frac{m_B(\theta)(S_B(\lambda_i) - S_B(\lambda_j))}{m_B(\theta)(S_B(\lambda_l) - S_B(\lambda_n))} \\
&= \frac{S_B(\lambda_i) - S_B(\lambda_j)}{S_B(\lambda_l) - S_B(\lambda_n)}
\end{aligned}
\qquad (7)
$$

Next, the reflection model of metal object is considered. The extended reflection model in Eq. (5) is the same fashion mathematically as the standard model in Eq. (3) for dielectric, although the two reflection models are physically different. Therefore a unified invariant representation is derived for all materials including inhomogeneous dielectric and homogeneous metal. In fact, the following invariant equation can be derived for metal from Eq. (5),

$$
\begin{aligned}
\frac{S(\theta, \lambda_i) - S(\theta, \lambda_j)}{S(\theta, \lambda_l) - S(\theta, \lambda_n)} &= \frac{(m_{I1}(\theta)S_{I1}(\lambda_i) + m'_{I2}(\theta)) - (m_{I1}(\theta)S_{I1}(\lambda_j) + m'_{I2}(\theta))}{(m_{I1}(\theta)S_{I1}(\lambda_l) + m'_{I2}(\theta)) - (m_{I1}(\theta)S_{I1}(\lambda_n) + m'_{I2}(\theta))} \\
&= \frac{m_{I1}(\theta)S_{I1}(\lambda_i) - m_{I1}(\theta)S_{I1}(\lambda_j)}{m_{I1}(\theta)S_{I1}(\lambda_l) - m_{I1}(\theta)S_{I1}(\lambda_n)} \\
&= \frac{m_{I1}(\theta)(S_{I1}(\lambda_i) - S_{I1}(\lambda_j))}{m_{I1}(\theta)(S_{I1}(\lambda_l) - S_{I1}(\lambda_n))} \\
&= \frac{S_{I1}(\lambda_i) - S_{I1}(\lambda_j)}{S_{I1}(\lambda_l) - S_{I1}(\lambda_n)}
\end{aligned}
\tag{8}
$$

where the geometric weighting coefficients $m_{I1}(\theta)$ and $m'_{I2}(\theta)$ in Eq. (5) are eliminated. It is noted that the operation of Eq. (8) results in an equivalent operation to Eq. (7), which depends on only surface-spectral reflectance and is invariant to highlights, shading, and geometries.

The above spectral operations based on subtraction and division provides mathematically simple and robust spectral invariant representations. However the above computation depends on wavelengths such as λ_l and λ_n, and therefore, for real spectral data it is unstable because the denominator includes the subtraction. The above operation can be generalized to a stable form. Note that use of the minimum value of spectral reflectance preserves the original spectral characteristics. Then an invariant representation for spectral reflectance is defined as

$$
S'(\theta, \lambda) = \frac{S(\lambda) - min S(\lambda)}{\sqrt{\int_{\lambda_{min}}^{\lambda_{max}} (S(\lambda) - min S(\lambda))^2 d\lambda}}
\tag{9}
$$

For practical computation, let us sample spectral reflectance at ρ wavelengths in the visible range $[\lambda_{min}, \lambda_{max}]$. Then the discrete version of the invariant representation is written as

$$
S'(\theta, \lambda_i) = \frac{S(\theta, \lambda_i) - min\{S(\theta, \lambda_1), \ldots, S(\theta, \lambda_\rho)\}}{\sqrt{\sum_{j=1}^{\rho}(S(\theta, \lambda_j) - min\{S(\theta, \lambda_1), \ldots, S(\theta, \lambda_\rho)\})^2}},
$$
$$
(i = 1, 2, \ldots, \rho)
\tag{10}
$$

This representation is spectral invariant for all materials, including inhomogeneous dielectric and homogeneous metal. The observed spectral reflectance $S(\theta, \lambda)$ can be reduced into a normalized surface-spectral reflectance $S'(\theta, \lambda) = S'(\lambda)$ that is independent of the geometric parameter θ. This representation is useful as an invariant operator for a variety of spectral image analysis, including material classification and image segmentation.

Lemma 1 *Assuming the standard dichromatic reflection model for inhomogeneous dielectric materials, the normalized surface-spectral reflectance $S'(\theta, \lambda)$ is independent of highlight, shading, surface geometry, and illumination intensity.*

Proof By substituting Eq. (3) in Eq. (10) we have Eq. (11), factoring out dependencies on highlight, shading, surface geometry, and illumination intensity where the geometric weighting coefficients $m'_I(\theta)$ and $m_B(\theta)$ in Eq. (3) are eliminated.

$$
\begin{aligned}
S'(\theta, \lambda_i) &= \frac{m_B(\theta)S_B(\lambda_i) - min\{m_B(\theta)S_B(\lambda_1), \ldots, m_B(\theta)S_B(\lambda_\rho)\}}{\sqrt{\sum_{j=1}^{\rho}(m_B(\theta)S_B(\lambda_j) - min\{m_B(\theta)S_B(\lambda_1), \ldots, m_B(\theta)S_B(\lambda_\rho)\})^2}} \\
&= \frac{m_B(\theta)(S_B(\lambda_i) - min\{S_B(\lambda_1), \ldots, S_B(\lambda_\rho)\})}{m_B(\theta)\sqrt{\sum_{j=1}^{\rho}(S_B(\lambda_j) - min\{S_B(\lambda_1), \ldots, S_B(\lambda_\rho)\})^2}} \\
&= \frac{S_B(\lambda_i) - minS_B(\lambda)}{\sqrt{\sum_{j=1}^{\rho}(S_B(\lambda_j) - minS_B(\lambda))^2}}, \quad (i = 1, 2, \ldots, \rho)
\end{aligned}
\tag{11}
$$

Lemma 2 *Assuming the extended dichromatic reflection model for homogeneous metal, the normalized surface-spectral reflectance $S'(\theta, \lambda)$ is independent of highlight, shading, surface geometry, and illumination intensity.*

Proof By substituting Eq. (5) in Eq. (10) we have Eq. (12), factoring out dependencies on highlight, shading, surface geometry, and illumination intensity where the geometric weighting coefficients $m_{I1}(\theta)$ and $m'_{I2}(\theta)$ in Eq. (5) are eliminated.

$$
\begin{aligned}
S'(\theta, \lambda_i) &= \frac{m_{I1}(\theta)S_{I1}(\lambda_i) - min\{m_{I1}(\theta)S_{I1}(\lambda_1), \ldots, m_{I1}(\theta)S_{I1}(\lambda_\rho)\}}{\sqrt{\sum_{j=1}^{\rho}(m_{I1}(\theta)S_{I1}(\lambda_j) - min\{m_{I1}(\theta)S_{I1}(\lambda_1), \ldots, m_{I1}(\theta)S_{I1}(\lambda_\rho)\})^2}} \\
&= \frac{m_{I1}(\theta)(S_{I1}(\lambda_i) - min\{S_{I1}(\lambda_1), \ldots, S_{I1}(\lambda_\rho)\})}{m_{I1}(\theta)\sqrt{\sum_{j=1}^{\rho}(S_{I1}(\lambda_j) - min\{S_{I1}(\lambda_1), \ldots, S_{I1}(\lambda_\rho)\})^2}} \\
&= \frac{S_{I1}(\lambda_i) - minS_{I1}(\lambda)}{\sqrt{\sum_{j=1}^{\rho}(S_{I1}(\lambda_j) - minS_{I1}(\lambda))^2}}, \quad (i = 1, 2, \ldots, \rho)
\end{aligned}
\tag{12}
$$

4 Spectral Image Segmentation

The segmentation problem of spectral images based on the illumination-invariant representation is considered. Several algorithms have been proposed for segmenting the spectral images into distinct surface areas [24, 74–76, 107, 108]. Xing et al. [108] proposed a fast spectral image segmentation approach based on mean-shift filtering and kernel-based clustering. An effective method was introduced by Martínez-Usó et al. [76] for spectral image segmentation in fruit inspection applications. Haneishi et al. [107] proposed a multispectral image segmentation method of paintings drawn with natural mineral pigments using the kernel based nonlinear subspace based on 16-bands camera. However, those algorithms were not always robust for the highlight and shading effects occurred for different surface materials, and required a careful adjustment of the lighting position.

In this chapter, we use a segmentation algorithm "PCA+Ncut method" for material classification using dimension-reduced spectral information which was implemented in [24]. This method was used to execute the normalized cut [109] in a low-dimensional space obtained with the principal components analysis. We use another algorithm which was fundamentally based on the normalized cut approximated by the Nyström method [110] to incorporate both spectral information and spatial information in Ref. [74]. Those methods were useful for material classification of printed circuit boards. However the material classification was not illumination-invariant, but was performed under the ideal lighting condition of eliminating specular reflection and shadow.

Here we present an image segmentation algorithm combined with the illumination-invariant representation, which results in reliable segmentation results under arbitrary lighting conditions. Our image segmentation process is composed of the following four steps,

1. The surface-spectral reflectances are estimated from the camera outputs, and then transformed into the invariant representation to highlight, shadow, and surface geometry.
2. The similarity matrix between pixels is constructed in the normalized cut scheme.
3. The computation burden is reduced by the Nyström approximation.
4. The K-means algorithm is applied to cluster the leading normalized eigenvectors to get the final segmentation.

4.1 Illumination Estimation

Here suppose that the illuminant spectrum $E(\lambda)$ is estimated by such a method as described in Sect. 4.2. Then the surface-spectral reflectance $S(\theta, \lambda)$ of an object is obtained in a straightforward way using Eq. (15). Moreover the normalized spectral reflectance $S'(\lambda)$ for invariant representation is obtained from the transformation of Eq. (10). The following spectral segmentation method is adapted to the invariant representation for classifying each pixel to a specific object.

4.2 Similarity Matrix Construction

In the normalized cut scheme, a similarity matrix $\mathbf{W}$ is constructed by all combination of two pixels based on the input ρ-dimensional feature vector, which corresponds to the illumination-invariant spectral reflectances. The similarity $w(x, y) \in \mathbf{W}$ between two pixels x and y is defined as product of two similarity measures of the normalized spectral reflectance and the spatial location. Using the Gaussian kernel function to represent the similarity measure,

$$w(x, y) = \exp\left\{-\frac{\|\mathbf{F}(x) - \mathbf{F}(y)\|_2^2}{\sigma_S^2}\right\} \cdot \exp\left\{-\frac{\|\mathbf{Z}(x) - \mathbf{Z}(y)\|_2^2}{\sigma_Z^2}\right\} \quad (13)$$

where $\mathbf{F}(x) \in [0, 1]^\rho$ is the ρ-dimensional feature vector at pixel x, which corresponds to the illumination-invariant spectral reflectances $S'(\lambda)$. The function $\mathbf{Z}(x)$ represents the spatial location that is effectively used to connect different regions to a similar segment, and ranges from 0 to the tested image size. The vector norm operator $\|.\|_2$ finds the Euclidean distance. The standard deviation σ represents the sensitivity of the Gaussian distribution. The spectral sensitivity σ_S depends on the materials appearance while the location sensitivity σ_Z depends on the size of tested image.

4.3 Nyström Approximation

Exact normalized cut classification requires calculations of eigenvalues and eigenvectors of the huge similarity matrix $\mathbf{W}$. The Nyström method is a technique for finding a numerical approximation to eigendecomposition, and it was widely applied to areas involving large dense matrices. In this algorithm, the Nyström method is applied here to approximate similarity matrix for reducing the computation burden; the method estimates the eigenvalues and eigenvectors from a smaller matrix that is obtained by sampling the original similarity matrix $\mathbf{W}$. In this study, I select the m samples by K-means algorithm. From the $m \times m$ approximated similarity matrix, one can derive the approximated eigenvectors matrix $\mathbf{V}$ of the similarity matrix $\mathbf{W}$. The largest k eigenvectors are selected from matrix $\mathbf{V}$ according to the required number of k materials.

4.4 Eigenvectors Clustering

I use the k eigenvectors stacked in columns simultaneously to get the final clusters for k objects. From the matrix $\mathbf{V} \in \mathfrak{R}^{L \times k}$ form matrix $\bar{\mathbf{V}}$ by renormalizing each of $\mathbf{V}$'s rows to have unit length [111] as

$$\bar{\mathbf{V}}_{ij} = \frac{\mathbf{V}_{ij}}{\sqrt{\sum_j \mathbf{V}_{ij}^2}}, i = 1, \ldots, L, j = 1, \ldots, k. \quad (14)$$

where L is the number of pixels. By treating each row of the leading normalized eigenvectors $\bar{\mathbf{V}}$ as a point in $\mathfrak{R}^k$, we cluster them into k different objects via K-means that attempts to minimized the distortion.

Finally, we assign the pixel S_i' of the normalized spectral reflectance to cluster j if and only if row i of the matrix $\bar{\mathbf{V}}$ was assigned to cluster j. Thus the input normalized spectral reflectance data $S_i'(\lambda)$ of spectral images is assigned to different segments to get the final segmentation.

5 Experimental Setting

5.1 Spectral Camera System

Figure 3 shows an experimental setup of the spectral acquisition system. The main components are a cooling monochromatic CCD camera (Retiga 1300) with 12-bit dynamic range, a macro lens of C-mount connected directly to the camera, Liquid Crystal Tunable Filter (LCTF), IR-cut filter, and personal computer. The VariSpec^{TM} LCTF [112] is convenient for spectral imaging because the wavelength band can be changed easily and electronically [23]. The LCTF used in this study has the spectral properties of narrow band filtration of 10 nm and wavelength range [400–700 nm]. The XY stage helps to easily control the camera system distance and position. The rotating stage controls rotation and position of the measured object.

The imaging system automatically captures and saves spectral images with arbitrary number of bands and shutter speeds. The actual measurement time required for capturing one spectral image with size 1280×1024 pixels for the area $35\,\text{mm} \times 30\,\text{mm}$ and 31-bands of a printed circuit board (PCB) is 4.75 s. Figure 4a depicts a set of relative functions for representing the whole spectral sensitivities

Fig. 3 The experimental setup of the spectral imaging system [74]

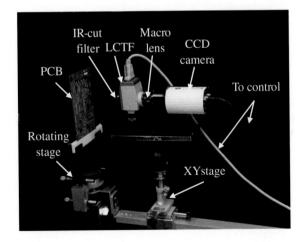

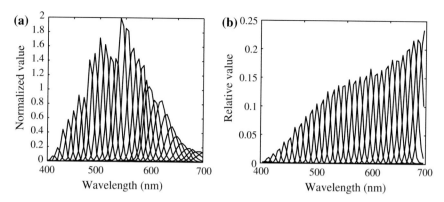

Fig. 4 Spectral characteristics of the imaging system: **a** whole spectral sensitivity functions; **b** transmittances of the VariSpec^{TM} LCTF [74]

that were obtained by combining the LCTF transmittances in Fig. 4b, IR-cut filter transmittances, and the spectral sensitivity function of the monochromatic CCD camera.

5.2 Illumination Estimation

The estimation of scene illumination from image data is one of interesting topics in image analysis and computer vision [31]. Although many algorithms were proposed for scene illuminant estimation, most algorithms assumed uniform illumination from a single light source [6, 113]. Concerning spectral imaging, illuminant estimation methods from the outputs of a multi-channel vision system were presented [6, 7], where the dichromatic reflection model was assumed to object surfaces. Therefore the illuminant spectrum could be estimated from the interface reflection component. Recently, a method for estimating the illuminant spectral-power distributions from omnidirectional observations by a multi-channel omnidirectional imaging system was proposed [114, 115].

An easy way to estimate the illuminant spectrum is to place a standard white reference in a scene and capture the corresponding sensor outputs from the spectral image of the scene. The sensor outputs are described as $\int_{400}^{700} E(\lambda)R_i(\lambda)d\lambda$ ($i = 1, 2, \ldots, 31$), where $E(\lambda)$ is the illuminant spectral-power distribution (SPD), and $R_i(\lambda)$ is the ith sensor spectral sensitivity function in Fig. 4a. When the spectral sensitivities $R(\lambda)$ are narrow, the illuminant is estimated directly from the sensor outputs. This method is utilized in inspection systems for industrial application.

Here suppose that the illuminant spectrum $E(\lambda)$ is estimated by such a method as described in the above. Then the surface-spectral reflectance $S(\theta, \lambda)$ of an object is obtained in a straightforward way from the sensor outputs $Y(\theta, \lambda_1), Y(\theta, \lambda_2), \ldots, Y(\theta, \lambda_{31})$ as

$$S(\theta, \lambda) = \frac{Y(\theta, \lambda_i)}{\int_{400}^{700} E(\lambda) R_i(\lambda) d\lambda}, \quad (i = 1, 2, \ldots, 31). \tag{15}$$

The sensor output at each wavelength in Eq. (15) is normalized with a factor on illuminant and sensing sensitivity. Thus, the spectral reflectance is recovered by eliminating the lighting and sensing effects from the sensor outputs.

6 Experiments

6.1 Experiments for Spectral Invariant

A test scene including a metal object of copper and two dielectric objects of ceramic (cup) and plastic (frog) is used. Figure 5a shows the color image of the original surface-spectral reflectances, which was obtained using the CIE-color matching functions to the observed reflectance $S(\theta, \lambda)$. It is noted that the scene has the illumination effects of shadows and highlights observed over different parts on the object surfaces. From this scene I obtained a set of spectral reflectance images with the size of $439 \times 297 \times 31$. Figure 5b shows the spectral invariant representation of the normalized spectral reflectance. Shadows, highlights, and surface geometry are much reduced in the invariant representation of both dielectric objects and metal object. For detailed inspection, Fig. 5c depicts a 3D view of the component image of spectral reflectance at 550 nm for a small rectangular area including metal and dielectrics. The observed reflectance $S(\theta, \lambda)$ is strongly influenced by shading and illumination effects. Figure 5d shows the 3D view of the transformed invariant spectral reflectances $S'(\theta, \lambda) = S'(\lambda)$ for the same part at 550 nm. Shadows and highlights disappear from the test part, so that the transformed spectral image depends on only the inherent spectral reflectance to each object surface. Therefore, the image is clearly segmented into two different material regions. Thus, the presented spectral invariant representation is valid for both dielectric and metal objects, and is much more robust under a variety of illumination effects than the observed reflectance data.

In order to confirm the effectiveness of the spectral invariant, the invariant representation is compared with a representative technique proposed by Tominaga [116]. This technique normalizes the deviation vectors using a constant average value to cancel out the geometric factors. The method was used for dielectric materials. Figure 6 shows the representation by this normalization technique. It is noted that using a constant average value affects the performance of the normalization result specially the background and this will affect the spectral image segmentation. The strong shadows still affect the results as shown in the small rectangular area. The edge between two materials is not so sharp, compared with Fig. 5d. However, I note that the highlights from the metal object surface are removed. Therefore, the conventional technique can

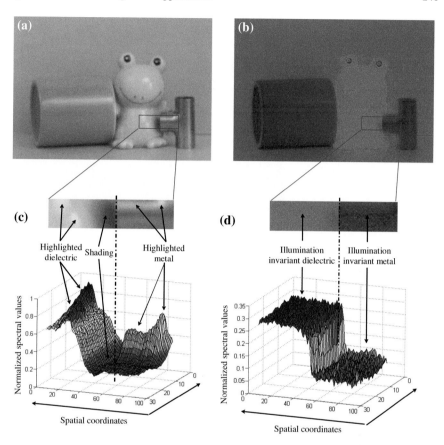

Fig. 5 Invariant evaluation for spectral reflectance image including dielectric and metal objects: **a** color image of the original spectral reflectances; **b** spectral invariant representation; **c** 3D view of the component image of spectral reflectance at 550 nm for a small area including metal and dielectrics; **d** 3D view of the transformed illumination-invariant spectral reflectances for the same part at 550 nm [25]

be extended to remove the illumination effects from the observed spectral reflectance image for metal objects as well as dielectric objects.

To show the stability of the illumination-invariant representation, Fig. 7 shows the invariant representation produced by subtraction of the spectral reflectances as in Eq. (8) using two different sets of wavelengths. The results are invariant to highlights, shading, and geometries for both metals and dielectric objects. However, the boundary between two materials is not clear, and the segmentation fails. Thus the invariant is unstable and depends on wavelengths. This shows the important result that, using the minimum value of spectral reflectance by the invariant representation in Eq. (9) preserves the original spectral characteristics and thus improves material segmentation.

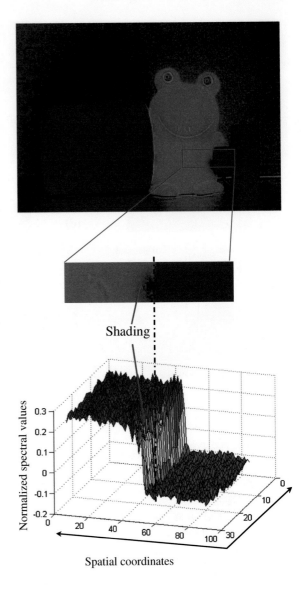

Fig. 6 Invariant representation by the normalization technique [116] and the relevant 3D view for a small area at 550 nm

Another spectral image including a metal object and two dielectric objects of plastic is tested to confirm the effectiveness of the presented spectral invariant. Figure 8a shows the color image of the original surface-spectral reflectances. The image has the illumination effects of shadows and highlights. Figure 8b shows the color image of the presented invariant representation. We note that shadows, highlights, and surface geometry are much reduced in the invariant representation of both the dielectric objects and the metal object.

(a)   **(b)**

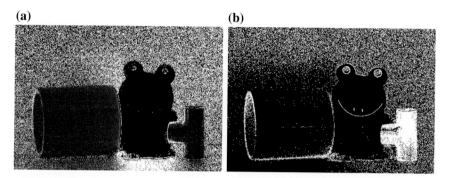

Fig. 7 Invariant representation produced by Eq. (8): **a** using the wavelengths of 450, 500, 550, and 600 nm; **b** using the wavelengths of 550, 600, 650, and 700 nm [26]

(a)   **(b)**

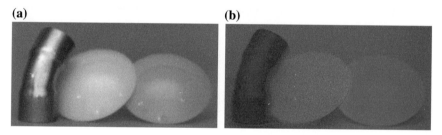

Fig. 8 Invariant evaluation for spectral reflectance image including dielectric and metal objects: **a** color image of the original spectral reflectances; **b** color image of the presented invariant representation [25]

6.2 Experiments for Spectral Image Segmentation

The performance of the illumination-invariant based segmentation algorithm is examined for the spectral images of natural scenes containing metal and dielectric objects. The spectral imaging system shown in Fig. 3a is used to perform these experiments. Nature scenes spectral images are captured under arbitrary lighting conditions. First we used the same spectral reflectance image as in Fig. 5. Figure 9a is the original image of a copper object and two dielectric objects. Figure 9b depicts the ground truth of material classification by segmenting manually the image. Figure 9c shows the segmentation result by the previous algorithm using the original spectral reflectance data [74]. The result contains many shadows and highlights effects. Figure 9d shows the segmentation result by the presented algorithm based on the normalized spectral reflectance data. The image is clearly segmented into different material regions and background. Thus the presented segmentation algorithm is independent of the illumination effects of highlight and shadow, and the geometry of object shape.

In order to confirm the effectiveness of the illumination-invariant representation, two alternative methods were also examined. Although I tried several different

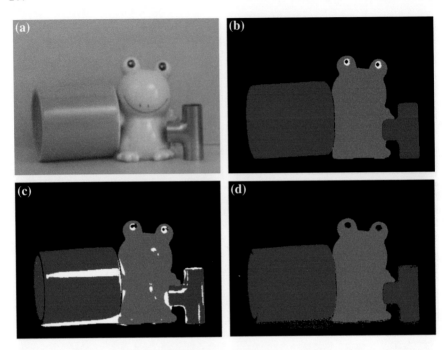

Fig. 9 Evaluation of image segmentation for natural objects: **a** original image; **b** ground truth manually segmented; **c** segmentation result based on the original spectral reflectance data; **d** segmentation result based on the normalized spectral reflectance data

parameter settings in each method, only the best results are shown for the comparison in this chapter. First, the K-means classification method [117] was used to segments the spectral images in Fig. 9a by clustering the original spectral reflectance data in the spectral space using random seeds, wherein the number of clusters was set to 4. The result is shown in Fig. 10a. Second, the PCA+Ncut method [24] was used to execute the normalized cut in a low-dimensional space obtained with the principal components analysis. The segmentation result of clustering the original spectral reflectance data is shown in Fig. 10b. We should note that the both results are strongly affected by different illumination events on the objects surfaces. On the other hand, Fig. 10c, d show the segmentation results by the K-means and the PCA+Ncut methods, respectively, based on the illumination-invariant representation. The image is well segmented into different material regions and background, compared with the results in Fig. 10a, b.

Next, the presented segmentation algorithm is applied to material classification of a printed circuit board with various tiny elements. A raw circuit board surface layer is composed of various elements, which are a mixture of different materials including dielectrics (i.e. photo-resist, silk-screen print, and substrate) and metals

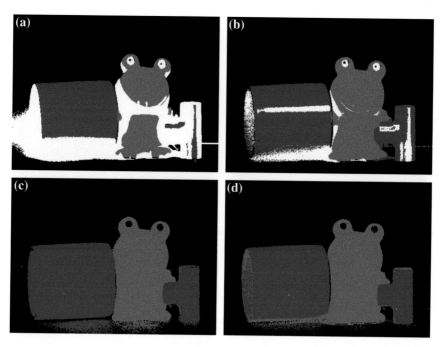

Fig. 10 Segmentation results by using two alternative segmentation methods: **a** K-means [117] of the original spectral reflectance data; **b** PCA+Ncut [24] of the original spectral reflectance data; **c** K-means of the normalized spectral reflectance data; **d** PCA+Ncut of the normalized spectral reflectance data

(i.e. copper), as shown in Fig. 11. I used a part of a real circuit board. Figure 11a shows the color image of the original spectral reflectance data. The board is illuminated by one light source from the left direction. The size of the captured spectral image by our imaging system is $433 \times 363 \times 31$. Figure 11b depicts the ground truth of material classification by segmenting manually the image. The ground truth image consists of four segments such as white—silk-screen print, yellow—metal, green—resist-coated metal, and black—background (substrate). Figure 11c shows the material classification result based on the original spectral reflectance data with the illumination effects. I found that miss classification occurs around metal and resist-coated metal lines at the left side of the segmented image. This can be caused by specular reflection of these materials surfaces. Figure 11d shows the classification result based on the normalized spectral reflectance data without the illumination effects. The original scene is well segmented into four material regions.

Then, this chapter compare the material classification method based on the illumination-invariant representation, with the material classification method based on the estimated spectral reflectance at every pixel point. Figure 12a shows the color image of the original spectral reflectance data for a board illuminated by one light

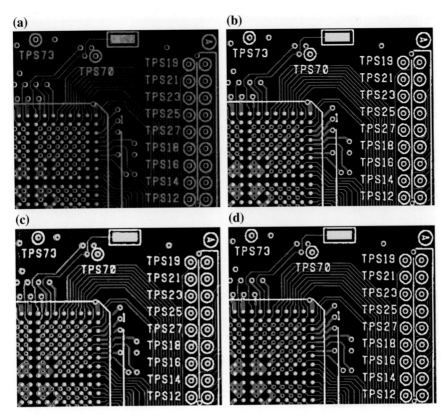

Fig. 11 Material classification of a raw printed-circuit board: **a** color image of the original spectral reflectances; **b** ground truth image; **c** classification result based on the original spectral reflectance data; **d** classification result based on the normalized spectral reflectance data

source from the left direction. Figure 12b shows the color image of the original spectral reflectance data for the same board illuminated by two light sources from the left and right directions. The classification results based on the illumination-invariant representation and the estimated spectral reflectance of the spectral images in Fig. 12a, b are shown in Fig. 12c, d, respectively. It is noted that the original scene is well segmented into four material regions based on the invariant representation.

The accuracy of the segmentation results between the segmented image and the ground truth image is numerically demonstrated by two different measures. First, the similarity measure with the window size 16 × 16 is used for labeled images shown in [118]. The similarity measure is calculated based on binary relations of arbitrary pixels in the labeled images. Thus this measure can evaluate both area-based labeled images and pixel-based labeled images for the segmented color images. Second, a pixel by pixel comparison is used to calculate the segmentation quality numerically for the whole regions using Eq. (16).

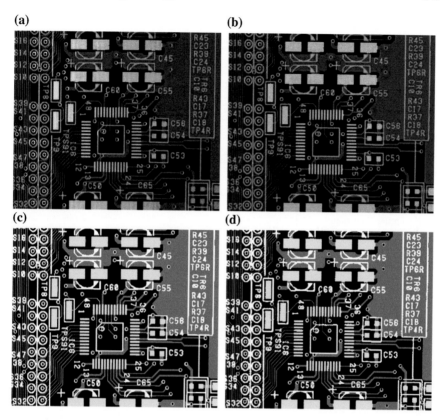

Fig. 12 Material classification of a raw printed-circuit board: **a** spectral image of a board illuminated by one light source from the *left* direction; **b** spectral image of the same board illuminated by two light sources from the *left* and *right* directions; **c** classification result based on the illumination-invariant representation; **d** classification result based on the estimated spectral reflectance

$$Quality\ rate = \frac{Correct\ classified\ pixels}{Total\ number\ of\ pixels} \tag{16}$$

Tables 1 and 2 list the numerical accuracy of segmentation results for the natural scene and the circuit board images. The accuracy using the similarity measure and the quality rate show that the segmentation results based on the presented illumination-invariant representation achieve high accuracy for both natural scene and circuit board scenes, compared with the segmentation results using the original spectral reflectance data and the estimated spectral reflectances. Table 3 shows that the material classification based on the illumination-invariant representation is more accurate than the material classification based on the estimated spectral reflectance.

Table 1 Accuracy of segmentation results for the natural scene

Method and representation	Similarity measure (%)	Quality rate (%)
Algorithm using the original spectral reflectance data	95.76	95.26
Algorithm using the normalized spectral reflectance data	98.03	97.90
K-means using the original spectral reflectance data	91.86	79.80
K-means using the normalized spectral reflectance data	93.85	92.64
PCA+Ncut using the original spectral reflectance data	92.11	90.26
PCA+Ncut using the normalized spectral reflectance data	94.11	93.87

Table 2 Accuracy of classification results for the circuit board in Fig. 11

Method and representation	Similarity measure (%)	Quality rate (%)
Algorithm using the original spectral reflectance data	86.02	91.51
Algorithm using the normalized spectral reflectance data	98.77	99.18

Table 3 Accuracy of classification results for the circuit board in Fig. 12

Method and representation	Similarity measure (%)	Quality rate (%)
Algorithm based on the illumination-invariant representation	98.81	99.20
Algorithm based on the estimated spectral reflectance	96.63	98.03

7 Summary

The present chapter has presented an illumination-invariant representation of spectral images which can be applied to dielectric and metal objects. The invariant representation was derived from the standard dichromatic reflection model for dielectric and the extended dichromatic reflection model for metal. It is shown that normalized surface-spectral reflectance by the minimum reflectance is illumination-invariant, that is independent of geometric parameters. The presented invariant representation is applied to image segmentation problems based on spectral information. An illumination-invariant segmentation algorithm was presented for effectively segmenting spectral images of a natural scene and a raw circuit board. Experimental results showed the feasibility of the presented illumination-invariant representation method in image segmentation with high accuracy. The invariant representation was also applied to other alternative segmentation algorithms to verify the performance. Thus, we can conclude that the illumination invariant representation and the segmentation method are useful for segmenting various kinds of materials.

References

1. Berns, R.S.: Art Spectral Imaging. Research Program (2005). http://www.art-si.org/
2. Bochko, V., Tsumura, N., Miyake, Y.: Spectral color imaging system for estimating spectral reflectance of paint. J. Imaging Sci. Technol. **51**(1), 70–78 (2007)
3. Parmar, M., Lansel, S., Wandell, B.A.: Spatio-spectral reconstruction of the multispectral datacube using sparse recovery. In: Proceedings of IEEE International Conference on Image Processing, pp. 473–476 (2008)
4. Hauta-Kasari, M., Miyazawa, K., Toyooka, S., Parkkinen, J.: Spectral vision system for measuring color images. J. Opt. Soc. Am. A **16**(10), 2352–2362 (1999)
5. Kawata, S., Sasaki, K., Minami, S.: Component analysis of spatial and spectral patterns in multispectral images. J. Opt. Soc. Am. A **4**(11), 2101–2106 (1987)
6. Tominaga, S.: Multi-channel vision system for estimating surface and illumination functions. J. Opt. Soc. Am. A **13**(11), 2163–2173 (1996)
7. Tominaga, S.: Spectral imaging by a multi-channel camera. J. Electr. imaging **8**(4), 332–341 (1999)
8. Manabe, Y., Kurosaka, S., Chihara, K.: Simultaneous measurement of spectral distribution and shape. In: Proceedings of IEEE International Conference on Pattern Recognition, vol. 3, pp. 803–806 (2000)
9. Haneishi, H., Miyahara, S., Yoshida, A.: Image acquisition technique for high dynamic range scenes using a multiband camera. Color Res. Appl. **31**(4), 294–302 (2006)
10. Antonioli, G., Fermi, F., Oleari, C., Reverberi, R.: Spectrophotometric scanner for imaging of paintings and other work of art. In: Proceedings of European Conference on Color in Graphics, Imaging and Vision, pp. 219–224 (2004)
11. Barni, M., Pelagotti, A., Piva, A.: Image processing for the analysis and conservation of paintings: opportunities and challenges. IEEE Signal Process. Mag. **22**(5), 141–144 (2005)
12. Baronti, S., Casini, A., Lotti, F., Parcinai, S.: Multispectral imaging system for the mapping of pigments in works of art by use of principal-component analysis. Appl. Optics **37**(8), 1299–1309 (1998)
13. Bonifazzi, C., Carcagní, P., Della Patria, A., Ferriani, S., Fontana, R., Greco, M., Mastroianni, M., Materazzi, M., Pampaloni, E., Romano, A.: A scanning device for multispectral imaging of paintings. In: Proceedings of SPIE, Spectral Imaging: Eighth International Symposium on Multispectral Color Science, vol. 6062, pp. 1–10 (2006)
14. Carcagní, P., Della Patria, A., Fontana, R., Greco, M., Mastroianni, M., Pampaloni, E., Pezzati, L.: Multispectral imaging of paintings by optical scanning. Optics Lasers Eng. **45**(3), 360–367 (2007)
15. Colantoni, P., Pillayn, R., Lahanier, C., Pitzalis, D.: Analysis of multispectral images of paintings. In: Proceedings of 14th European Signal Processing Conference, pp. 4–8 (2006)

16. Cornelis, B., Dooms, A., Leen, F., Munteanu, A., Schelkens, P.: Multispectral imaging for digital painting analysis: a Gauguin case study. In: Proceedings of SPIE, Applications of Digital Image Processing XXXIII, vol. 7798, pp. 77980I–77980I-13 (2010)
17. Haneishi, H., Hasegawa, T., Tsumura, N., Miyake, Y.: Design of color filters for recording artworks. In: Proceedings of the IS&T's 50th Annual Conference, pp. 369–372 (1997)
18. Liang, H., Saunders, D., Cupitt, J.: A new multispectral imaging system for examining paintings. J. Imaging Sci. Technol. 49(6), 551–562 (2005)
19. Martinez, K., Cupitt, J., Saunders, D., Pillay, R.: Ten years of art imaging research. Proc. IEEE 90(1), 28–41 (2002)
20. Novati, G., Pellegri, P., Schettini, R.: An affordable multispectral imaging system for the digital museum. Int. J. Digital Libr. 5(3), 167–178 (2005)
21. Ribés, A., Schmitt, F., Pillay, R., Lahanier, C.: Calibration and spectral reconstruction for CRISATEL: an art painting multispectral acquisition system. J. Imaging Sci. Technol. 49(6), 563–573 (2005)
22. Pelagotti, A., Mastio, A.D., Rosa, A.D., Piva, A.: Multispectral imaging of paintings. IEEE Signal Process. Mag. 25(4), 27–36 (2008)
23. Ibrahim, A., Tominaga, S., Horiuchi, T.: Material classification for printed circuit boards by spectral imaging system. In: Proceedings of IAPR Computational Color Imaging Workshop, vol. 5646, pp. 216–225 (2009)
24. Ibrahim, A., Tominaga, S., Horiuchi, T.: Unsupervised material classification of printed circuit boards using dimension-reduced spectral information. In: Proceedings of IAPR Conference on Machine Vision Applications, pp. 435–438 (2009)
25. Ibrahim, A., Tominaga, S., Horiuchi, T.: Invariant representation for spectral reflectance images and its application. EURASIP J. Image Video Process. 2011, 2 (2011)
26. Ibrahim, A., Tominaga, S., Horiuchi, T.: A spectral invariant representation of spectral reflectance. Opt. Rev. 18 (2011)
27. Manfron, G., Alessandro, P., Mirco, B., Mail, B., Nutini, F., Nelson, A.: Comparative analysis of normalised difference spectral indices derived from modis for detecting surface water in flooded rice cropping systems. PLOS One 9(2), e88741 (2014)
28. Morimoto, T., Ikeuchi, K.: Multispectral image segmentation using normalized cut. In: Proceedings of IEICE Meeting on Image Recognition and Understanding, pp. 760–766 (2008)
29. Morimoto, T., Ikeuchi, K.: Multispectral imaging for material analysis in an outdoor environment using normalized cuts. In: Proceedings of IEEE Color and Reflectance in Imaging and Computer Vision Workshop, in conjunction with ICCV'09, pp. 1909–1916 (2009)
30. Du, H., Tong, X., Cao, X., Lin, S.: A prism-based system for multispectral video acquisition. In: Proceedings of IEEE International Conference on Computer Vision, pp. 175–182 (2009)
31. Trëmeau, A., Tominaga, S., Plataniotis, K.N.: Color in image and video processing: most recent trends and future research directions. EURASIP J. Image Video Process. 2008, 26 (2008)
32. Ibrahim, A., Tominaga, S., Horiuchi, T.: Illumination-invariant spectral representation for image segmentation. In: Proceedings of IEICE Meeting on Image Recognition and Understanding, pp. 1784–1791 (2010)
33. Ibrahim, A., Tominaga, S., Horiuchi, T.: Spectral invariant representation for spectral reflectance image. In: Proceedings of IEEE 20th International Conference on Pattern Recognition, pp. 2776–2779 (2010)
34. Montoliu, R., Pla, F., Klaren, A.K.: Multispectral invariants. In: Technical Report, DLSI, Universitat Jaume I, Castellon, Spain (2004)
35. Stokman, H.M.G., Gevers, T.: Detection and classification of hyper-spectral edge. In: Proceedings of 10th British Machine Vision Conference, pp. 643–651 (1999)
36. Finlayson, G.D.: Color in perspective. IEEE Trans. Pattern Anal. Mach. Intell. 18(10), 1034–1038 (1996)
37. Geusebroek, J.-M., Boomgard, R., Smeulders, A.W.M., Geerts, H.: Color invariance. IEEE Trans. Pattern Anal. Mach. Intell. 23(12), 1338–1350 (2001)

38. Geusebroek, J.-M., Smeulders, A.W.M., van den Boomgaard, R.: Measurement of color invariants. In: Proceedings of IEEE Computer Society Conference on Computer Vision and Pattern Recognition, vol. 1, pp. 50–57 (2000)
39. Gevers, T., Smeulders, A.W.M.: Color based object recognition. Pattern Recogn. **32**(3), 453–464 (1999)
40. Gevers, T., Smeulders, A.W.M.: PicToSeek: combining color and shape invariant features for image retrieval. IEEE Trans. Image Process. **9**(1), 102–119 (2000)
41. Gevers, T., Stokman, H.M.G.: Classification of color edges in video into shadow, geometry, highlight, or material transitions. IEEE Trans. Multimed. **5**(2), 237–243 (2003)
42. Mallick, S.P., Zickler, T.E., Kriegman, D.J., Belhumeur, P.N.: Beyond Lambert: reconstructing specular surfaces using color. In: Proceedings of IEEE Computer Society Conference on Computer Vision and Pattern Recognition, vol. 2, pp. 619–626 (2005)
43. Marchant, J.A., Onyango, C.M.: Shadow-invariant classification for scenes illuminated by daylight. J. Opt. Soc. Am. A **17**(11), 1952–1961 (2000)
44. Narasimhan, S.G., Ramesh, V., Nayar, S.K.: A class of photometric invariants: separating material from shape and illumination. In: Proceedings of IEEE International Conference of Computer Vision, vol. 2, pp. 1387–1394 (2003)
45. Park, J.B.: Efficient color representation for image segmentation under nonwhite illumination. In: Proceedings of SPIE, Intelligent Robots and Computer Vision XXI: Algorithms, Techniques, and Active Vision, vol. 5267, pp. 163–174 (2003)
46. Slater, D., Healey, G.: The illumination-invariant recognition of 3D objects using local color invariants. IEEE Trans. Pattern Anal. Mach. Intell. **18**(2), 206–210 (1996)
47. Smeulders, A.W.M., Geusebroek, J.-M., Gevers, T.: Invariant representation in image processing. In: Proceedings of IEEE International Conference on Image Processing, vol. 3, pp. 18–21 (2001)
48. Tan, R.T., Ikeuchi, K.: Separating reflection components of textured surface using a single image. IEEE Trans. Pattern Anal. Mach. Intell. **27**(2), 178–193 (2005)
49. van de Weijer, J., Gevers, T., Geusebroek, J.-M.: Edge and corner detection by photometric quasi-invariants. IEEE Trans. Pattern Anal. Mach. Intell. **27**(4), 625–630 (2005)
50. van de Weijer, J., Gevers, T., Smeulders, A.W.M.: Robust photometric invariant features from the color tensor. IEEE Trans. Image Process. **15**(1), 118–127 (2006)
51. Gevers, T.: Adaptive image segmentation by combining photometric invariant region and edge information. IEEE Trans. Pattern Anal. Mach. Intell. **24**(6), 848–852 (2002)
52. Wesolkowski, S., Tominaga, S., Dony, R.D.: Shading- and highlight-invariant color image segmentation using the MPC algorithm. In: Proceedings of SPIE, Color Imaging: Device-Independent Color, Color Hardcopy, and Graphic Arts, vol. 4300, pp. 229–240 (2000)
53. Gevers, T., Stokman, H.M.G.: Classifying color transitions into shadow-geometry, illumination highlight or material edges. In: Proceedings of IEEE International Conference on Image Processing, vol. 1, pp. 521–525 (2000)
54. Gevers, T., Voortman, S., Aldershoff, F.: Color feature detection and classification by learning. In: Proceedings of IEEE International Conference on Image Processing, vol. 2, pp. 714–717 (2005)
55. Koschan, A., Abidi, M.: Detection and classification of edges in color images. IEEE Signal Process. Mag. **22**(1), 64–73 (2005)
56. Stokman, H.M.G., Gevers, T.: Selection and fusion of color models for image feature detection. IEEE Trans. Pattern Anal. Mach. Intell. **29**(3), 371–381 (2007)
57. van de Weijer, J., Schmid, C.: Coloring local feature extraction. In: Proceedings of the European Conference on Computer Vision, vol. 3952, pp. 334–348 (2006)
58. Gevers, T., Stokman, H.M.G.: Robust histogram construction from color invariants for object recognition. IEEE Trans. Pattern Anal. Mach. Intell. **26**(1), 113–118 (2004)
59. van de Sande, K.E.A., Gevers, T., Snoek, C.G.M.: Evaluating color descriptors for object and scene recognition. IEEE Trans. Pattern Anal. Mach. Intell. **32**(9), 1582–1596 (2010)
60. van Gemert, J.C., Burghouts, G.J., Seinstra, F.J., Geusebroek, J.-M.: Color invariant object recognition using entropic graphs. Int. J. Imaging Syst. Technol. **16**(5), 146–153 (2006)

61. Jin, C.: A statistical image retrieval method using color invariant. In: Proceedings of Sixth International Conference on Computer Graphics, Imaging and Visualization, pp. 355–360 (2009)
62. Vacha. P., Haindl, M.: Demonstration of image retrieval based on illumination invariant textural MRF features. In: Proceedings of the 6th ACM international conference on Image and video retrieval, pp. 135–137 (2007)
63. Vacha. P., Haindl, M.: Image retrieval measures based on illumination invariant textural MRF features. In: Proceedings of the 6th ACM International Conference on Image and Video Retrieval, pp. 448–454 (2007)
64. Salvador, E., Cavallaro, A., Ebrahimi, T.: Shadow identification and classification using invariant color models. In: Proceedings of IEEE International Conference on the Acoustics, Speech, and Signal Processing, vol. 3, pp. 1545–1548 (2001)
65. Salvador, E., Cavallaro, A., Ebrahimi, T.: Cast shadow segmentation using invariant color features. Comput. Vis. Image Understand. 95(2), 238–259 (2004)
66. van de Weijer, J., Gevers, T.: Robust optical flow from photometric invariants. In: Proceedings of IEEE International Conference on Image Processing, vol. 3, pp. 1835–1838 (2004)
67. Zickler, T., Mallick, S.P., Kriegman, D.J., Belhumeur, P.N.: Color subspaces as photometric invariants. Int. J. Comput. Vis. 79(1), 13–30 (2008)
68. Maier, W., Bao, F., Mair, E., Steinbach, E., Burschka, D.: Illumination-invariant image-based novelty detection in a cognitive mobile robot's environment. In: Proceedings of IEEE International Conference on Robotics and Automation, pp. 5029–5034 (2010)
69. Maier, W., Bao, F., Steinbach, E., Mair, E., Burschka, D.: Illumination-invariant image-based environment representations for cognitive mobile robots using intrinsic images. In: Proceedings of Vision, Modeling, and Visualization Workshop, pp. 379–380 (2009)
70. Shafer, S.A.: Using color to separate reflection components. Color Res. Appl. 10(4), 210–218 (1985)
71. Lee, H.C., Breneman, E.J., Schulte, C.: Modeling light reflection for computer color vision. IEEE Trans. Pattern Anal. Mach. Intell. 12(4), 402–409 (1990)
72. Tominaga, S.: Dichromatic reflection models for a variety of materials. Color Res. Appl. 19(4), 277–285 (1994)
73. Tominaga, S.: Dichromatic reflection models for rendering object surfaces. J. Imaging Sci. Technol. 40(6), 549–555 (1996)
74. Ibrahim, A., Tominaga, S., Horiuchi, T.: Spectral imaging method for material classification and inspection of printed circuit boards. Opt. Eng. 49(5), 057201-(10) (2010)
75. Li, H., Bochko, V., Jaaskelainen, T., Parkkinen, J., Shen, I.F.: Kernel-based spectral color image segmentation. J. Opt. Soc. Am. A 25(11), 2805–2816 (2008)
76. Martínez-Usó, A., Pla, F., García-Sevilla, P.: Multispectral image segmentation by energy minimization for fruit quality estimation. In: Proceedings of 2nd Iberian Conference on Pattern Recognition and Image Analysis, vol. 3523, pp. 689–696 (2005)
77. Mohammad-Djafari, A., Bali, N., Mohammadpour, A.: Hierarchical Markovian models for hyperspectral image segmentation. In: Proceedings of International Workshop on Intelligent Computing in Pattern Analysis/Systems, pp. 416–424 (2006)
78. Paclík, P., Duin, R.P.W., van Kempen, G.M.P., Kohlus, R.: Segmentation of multispectral images using the combined classifier approach. Image Vis. Comput. 21(6), 473–482 (2003)
79. Chang, P.C., Chen, L.Y., Fan, C.Y.: A case-based evolutionary model for defect classification of printed circuit board images. J. Intell. Manuf. 19(2), 203–214 (2008)
80. Chomsuwan, K., Yamada, S., Iwahara, M., Wakiwaka, H., Shoji, S.: Application of Eddy-current testing technique for high-density double-layer printed circuit board inspection. IEEE Trans. Magnetics 41(10), 3619–3621 (2005)
81. de Almeida Barreto, C., Zuffo, J.A., Kofuji, S.T.: Automated optical inspection system for professional double face printed circuit boards. In: Proceedings of the IEEE International Symposium on Industrial Electronics, vol. 1, pp. 65–71 (1997)
82. Emary Eid, E., Taha, M., Moustafa, K.: Automatic optical inspection for pcb manufacturing: a survey. Int. J. Sci. Eng. Res. 5(7) (2014)

83. Huang, S.Y., Mao, C.W., Cheng, K.S.: Contour-based window extraction algorithm for bare printed circuit board inspection. IEICE Trans. Inf. Syst. **E88-D**(12), 2802–2810 (2005)
84. Ibrahim, Z., Al-Attas, S.A.R.: Wavelet-based printed circuit board inspection algorithm. Integr. Comput. Aided Eng. **12**(2), 201–213 (2005)
85. Leta, F.R., Feliciano, F.F., Martins, F.P.R.: Computer vision system for printed circuit board inspection. In: ABCM Symposium Series in Mechatronics, vol. 1, pp. 623–632 (2008)
86. Leta, F.R., Feliciano, F.F.: Computational system to detect defects in mounted and bare PCB based on connectivity and image correlation. In: Proceedings of the IEEE 15th International Conference on Systems, Signals and Image Processing, pp. 331–334 (2008)
87. Lin, S.-C., Chou, C.-H., Su, C.-H.: A development of visual inspection system for surface mounted devices on printed circuit board. In: Proceedings of the IEEE 33rd Annual Conference on Industrial Electronics Society, pp. 2440–2445 (2007)
88. Lin. S.-C., Su, C.-H.: A visual inspection system for surface mounted devices on printed circuit board. In: Proceedings of the IEEE Conference on Cybernetics and Intelligent Systems, pp. 1–4 (2006)
89. Lin, S.-C., Su, C.-H., Chou, C.-H.. Chen, H.-C.: A development of inspection techniques for printed circuit board: from 2-D to 3-D. In: Proceedings of the IEEE SICE Annual Conference, pp. 1110–1115 (2008)
90. Liu, C., Gu, J.: Discriminative illumination: per-pixel classification of raw materials based on optimal projections of spectral brdf. IEEE Trans. Pattern Anal. Mach. Intell. **36**(1), 86–98 (2014)
91. Loh, H.-H., Lu, M.-S.: Printed circuit board inspection using image analysis. IEEE Trans. Ind. Appl. **35**(2), 426–432 (1999)
92. Malge, P.S., Nadaf, R.S.: A survey: automated visual pcb inspection algorithm. Int. J. Eng. Res. Technol. (IJERT) **3**(1) (2014)
93. Mashohor, S., Evans, J.R., Arslan, T.: Elitist selection schemes for genetic algorithm based printed circuit board inspection system. In: Proceedings of the IEEE Congress on Evolutionary Computation, vol. 2, pp. 974–978 (2005)
94. S. Mashohor, J. R. Evans, and A. T. Erdogan. Automatic hybrid genetic algorithm based printed circuit board inspection. In Proceedings of the IEEE First NASA/ESA Conference on Adaptive Hardware and Systems, pages 390–400, 2006
95. Putera, S.H.I., Ibrahim, Z.: Printed circuit board defect detection using mathematical morphology and MATLAB image processing tools. In: Proceedings of the IEEE 2nd International Conference on Education Technology and Computer, vol. 5, pp. 359–363 (2010)
96. Li, D., Wang, Q., Cao, D., Zhang, W., Chen, H.: Unsupervised defect detection of flexible printed circuit board gold surfaces based on wavelet packet frame. In: Proceedings of the IEEE 2nd International Conference on Industrial and Information Systems, vol. 2, pp. 324–327 (2010)
97. Slee, D., Stepan, J., Swart, J., Wei, W.: Introduction to printed circuit board failures. In: Proceedings of the IEEE Symposium on Product Compliance Engineering, pp. 1–8 (2009)
98. Iwahori, Y., Nakagawa, T., Bhuyan, M.K.: Reduction of defect misclassification of electronic board using multiple svm classifiers. Int. J. Softw. Innov. **2**(1), 25–36 (2014)
99. Tsai, D.M., Yang, R.H.: An eigenvalue-based similarity measure and its application in defect detection. Image Vis. Comput. **23**(12), 1094–1101 (2005)
100. Wada, H., Nakajima, A., Sawaragi, T., Horiguchi, Y.: A teaching system fostering expertise for the tuning of printed circuit board inspection systems. In: Proceedings of the IEEE 32nd Annual Conference on Industrial Electronics, pp. 3739–3744 (2006)
101. Wu, H., Li, H., Feng, G., Zeng, X.: Automated visual inspection of surface mounted chip components. In: Proceedings of the IEEE International Conference on Mechatronics and Automation, pp. 1789–1794 (2010)
102. Lee, W.Y., Park, T.-H.: Correction method for geometric image distortion with application to printed circuit board inspection systems. In: Proceedings of the IEEE ICROS-SICE International Joint Conference, pp. 4001–4006 (2009)

103. Shafer, S.A., Klinker, G.J., Kanade, T.: A physical approach to image understanding. Int. J. Comput. Vis. **4**(1), 7–38 (1990)
104. Jepson, A.D., Gershon, R., Tsotsos, J.K.: Ambient illumination and the determination of material changes. J. Opt. Soc. Am. A **3**(10), 1700–1707 (1986)
105. Healey, G.E.: Using color for geometry-insensitive segmentation. J. Opt. Soc. Am. A **6**(6), 920–937 (1989)
106. Wandell, B.A.: Foundations of Vision. Sinauer Associates Inc., Sunderland (1995)
107. Haneishi, H., Ohtani, R., Kouno, H.: Multispectral image segmentation of paintings drawn with natural mineral pigments using the kernel based nonlinear subspace method. In: Proceedings of Fifteenth Color Imaging Conference: Color Science and Engineering Systems, Technologies, and Applications, pp. 95–99 (2007)
108. Xing, M., Li, H., Jia, J., Parkkinen, J.: Fast spectral color image segmentation based on filtering and clustering. In: Proceedings of SPIE Multispectral Image Processing, vol. 7494, pp. 74942Q–74942Q-8 (2009)
109. Shi, J., Malik, J.: Normalized cuts and image segmentation. IEEE Trans. Pattern Anal. Mach. Intell. **22**(8), 888–905 (2000)
110. Fowlkes, C., Belongie, S., Chung, F.R.K., Malik, J.: Spectral grouping using the Nyström method. IEEE Trans. Pattern Anal. Mach. Intell. **26**(2), 214–225 (2004)
111. Ng, A.Y., Jordan, M.I., Weiss, Y.: On spectral clustering: analysis and an algorithm. In: Proceedings of Advances in Neural Information Processing Systems, MIT Press, Cambridge, vol. 14, pp. 849–856 (2002)
112. VariSpec liquid crystal tunable filters. http://www.spectralcameras.com/varispec. Accessed 04 June 2015
113. Parkkinen, J.P.S., Hallikaine, J., Jaaskelainen, T.: Characteristic spectra of Munsell colors. J. Opt. Soc. Am. A **6**(2), 318–322 (1989)
114. Tominaga, S., Fukuda, T., Kimachi. A.: A high-resolution imaging system for omnidirectional illuminant estimation. J. Imaging Sci. Technol. **52**(4), 040907-(1)-040907-(9) (2008)
115. Tominaga, S., Matsuura, A., Horiuchi, T.: Spectral analysis of omnidirectional illumination in a natural scene. J. Imaging Sci. Technol. **54**(4), 040502-(9) (2010)
116. Tominaga, S.: Surface identification using the dichromatic reflection model. IEEE Trans. Pattern Anal. Mach. Intell. **13**(7), 658–670 (1991)
117. Duda, R.O., Hart, P.E., Stork, D.G.: Pattern Classification, 2nd edn. Wiley, New York (2001)
118. Horiuchi, T.: Similarity measure of labelled images. In: Proceedings of IEEE International Conference on Pattern Recognition, vol. 3, pp. 602–605 (2004)

Image Segmentation Using an Evolutionary Method Based on Allostatic Mechanisms

**Valentín Osuna-Enciso, Virgilio Zúñiga, Diego Oliva,
Erik Cuevas and Humberto Sossa**

Abstract In image analysis, segmentation is considered one of the most important steps. Segmentation by searching threshold values assumes that objects in a digital image can be modeled through distinct gray level distributions. In this chapter it is proposed the use of a bio-inspired algorithm, called Allostatic Optimisation (AO), to solve the multi threshold segmentation problem. Our approach considers that an histogram can be approximated by a mixture of Cauchy functions, whose parameters are evolved by AO. The contributions of this chapter are on three fronts, by using: a Cauchy mixture to model the original histogram of digital images, the Hellinger distance as an objective function, and AO algorithm. In order to illustrate the proficiency and robustness of the proposed approach, it has been compared to the well-known Otsu method, over several standard benchmark images.

V. Osuna-Enciso (✉) · V. Zúñiga
Departamento de Ingenierías, CUTonalá, Universidad de Guadalajara,
Tonalá, Jalisco, Mexico
e-mail: Valentin.Osuna@cutonala.udg.mx

V. Zúñiga
e-mail: Virgilio.Zuniga@cutonala.udg.mx

D. Oliva
Departamento de Ciencias Computacionales, Tecnológico de Monterrey,
Campus Guadalajara, Guadalajara, Jalisco, Mexico
e-mail: Diego.Oliva@itesm.mx

E. Cuevas
Departamento de Ciencias Computacionales, CUCEI, Universidad de Guadalajara,
Guadalajara, Jalisco, Mexico
e-mail: Erik.Cuevas@cucei.udg.mx

H. Sossa
Instituto Politécnico Nacional-CIC, México D.F., Mexico
e-mail: HSossa@cic.ipn.mx

© Springer International Publishing Switzerland 2016
A.I. Awad and M. Hassaballah (eds.), *Image Feature Detectors and Descriptors*,
Studies in Computational Intelligence 630, DOI 10.1007/978-3-319-28854-3_10

255

1 Introduction

Image segmentation is considered as an important operation for meaningful analysis and interpretation of images acquired. In particular, image segmentation aims to group pixels within meaningful regions. Commonly, gray levels belonging to an object, are substantially different from those featuring by other objects or by the background. Segmentation is typically conducted considering two main criteria: similarity of image regions and discontinuity between adjacent disjoint regions [1, 2]. Among the segmentation approaches based on similarity, thresholding is considered the simplest technique [3, 4]. It involves the basic assumption that the objects and the background in the digital image have distinct gray level distributions. Under such assumption, the gray level histogram contains two or more distinct peaks and threshold values separating them that can be obtained. Therefore, segmentation is performed by assigning regions having gray levels below the threshold to the background, and assigning those regions having gray levels above the threshold to the objects, or vice versa. Segmentation by thresholding has been used in several areas where a correct separation of the objects in images is a vital step to perform fully automatic vision systems for detection and classification such as medical imaging [5–11], aviation [12], spacecraft imagery [13] and nondestructive tests [14], among many other applications. Several thresholding segmentation approaches have been reported in the literature [15–20], being the most popular the Otsu [21] method.

In statistics, the Gaussian distribution [22] is a standard modeling tool which satisfies the central limit theorem. A Gaussian distribution assumes that the probability of any occurring value falls off rapidly as it is moved further away from the central value. However, several problems, such as those that involve the presence of several outliers in the population, cannot be appropriately modeled under such assumption. Similar to the Gaussian distribution, the Cauchy distribution [23] is a symmetric bell-shaped density function but with a greater probability mass in the tails. This fact allows that the probabilities of points with large deviations from the central value, such as outliers, do not drop off as precipitously as it is obtained by the Gaussian distribution [24]. Although the Cauchy distribution possesses better modeling capabilities (in presence of outlier data) than other distributions, it presents serious difficulties in estimating its behavior parameters [25]. The capacity of the Cauchy distributions to model complex process has been demonstrated in several engineering applications such as impulsive noise cancellation [26], image denoising algorithms [27, 28], among others.

In this work, the segmentation algorithm is based on a parametric model which groups a mixture of several Cauchy density functions (Cauchy mixture, CM). CM involves the model selection, i.e., to determine the number of components in the mixture, and the estimation of the parameters of each component in the mixture that better adjust the statistical model. In general, computing the parameters of each Cauchy function is considered a difficult task, sensible to the initialization [29–31] and full of possible singularities [32]. In order to calculate such parameters, several methods have been proposed in the literature [33–35], presenting most of them flaws such as high computational overhead and sub-optimal values as a result of getting trapped in a local minimum. In the proposed approach, the parameter estimation of the CM has been faced as an optimization problem.

On the other hand, an impressive growth in the field of biologically inspired evolutionary algorithms for search and optimization has emerged during the last decade. Several bio-inspired algorithms have been proposed in the literature. Some examples include methods such as the Evolutionary Algorithm (EA) proposed by Fogel et al. [36], De Jong [37], the Genetic Algorithms (GA) proposed by Holland [38], the Artificial Immune System proposed by De Castro and Von Zuben [39] the Particle Swarm Optimization (PSO) method proposed by Kennedy and Eberhart [40] and the Artificial Bee Colony (ABC) proposed by Karaboga [41].

The interesting and complex behavior of biological organs from the human body have fascinated and attracted the interest of researchers for many years. Biologists have studied these phenomena to model organ operations, and engineers applied these models as a framework for solving complex real-world problems. An important biological phenomenon is Allostasis which explains how the modifications of specialized organ conditions inside the body allow achieving stability when an unbalance health condition is presented. Therefore, if a body decompensation happens, according to the allostatic mechanisms, several set points compound by blood pressure, oxygen tension and others indexes are proved in order to get a stability state. Such set points are generated by using different specialized mechanisms.

In this chapter, a multi-thresholding segmentation algorithm based on a new evolutionary algorithm called Allostatic Optimization (AO) is presented. In the approach, the segmentation process is considered as an optimization problem by approximating the 1-D histogram of a given image in terms of a Cauchy mixture model, whose parameters are calculated through the AO algorithm. In AO, the searcher agents emulate different body conditions which interact to each other by using operators based on the biological principles of the allostasis mechanism. The proposed approach encodes the parameters of the CM as an individual. An objective function by using the Hellinger distance evaluates the matching quality between the CM candidate and the original histogram. Guided by the values of this objective function, the set of encoded candidate mixtures are evolved using the operators defined by AO so that they can fit into the original histogram. In order to illustrate the proficiency and robustness of the proposed approach, it has been compared to the well-known Otsu method. The comparison examines several standard benchmark images that are commonly considered within the literature. Experimental results show a high performance of the proposed method for searching appropriate threshold values in terms of accuracy and robustness.

2 Allostatic Optimization

2.1 Allostasis

Organ systems (OS) into the body are composed of organ groups working in coordinated ways in order to maintain vital functions [42]; the human being has eleven of such systems, and each one is responsible to perform several specialized tasks as can bee seen in Table 1.

Even though there are other forms to describe the body organization, in the explanation given next it is only considered the organization based on OS. Communication among cells belonging to different OS with the brain it is achieved by means of two systems: the nervous and the endocrine, who are responsible of the coordination among OS for regulation of each essential function inside the body. Once the brain detects some external or internal change (stress, pollution, social status, disease, etc.), it determines if the stability of the body is compromised, in whose case it uses those channels to communicate with the adequate OS, trying to cope with such perturbation in order to get again the stability of the body. Chemical messages from the endocrine system are sent through hormonal substances, which are in charge of triggering or inhibiting responses from several tissues through target cells (who usually belong to several OS), whereas the nervous systems mainly uses electrical messages, activated

Table 1 Organ systems in the body

No.	Organ system	Task(s)	Some elements
1	Integumentary	External protection, sensory receptors	Skin, hair
2	Skeletal	Internal protection of tissues and organs	Bones
3	Muscular	External and internal movements	Muscles
4	Circulatory	Carrying vital substances	Heart, blood vessels
5	Respiratory	Control and regulation of the breathing process	Lung, nose
6	Digestive	Turn substances in energy to cells	Stomach, intestines
7	Excretory	Disposal of wastes	Kidney
8	Lymphatic	Internal protection against toxins and substances	T-cells, B-cells
9	Reproductive	Sexual reproduction	Testicles, ovaries
10	Nervous	First communication center	Neurons, Nerves
11	Endocrine	Second communication center	Pituitary, hypothalamus

through neurotransmitters, also hormones, or a combination of them. Both chemical and electrical messengers are referred simply as mediators. One of the main theories that explains how the body achieves the stability of the body, or health, as well as the coordination with the different OS, it is called 'homeostasis', which means 'to maintain stability (of a system, organ, OS, condition, health, etc.) through constancy (of a determined mechanism's set point (SP))' [43–51]. Among the several types of homeostasis are counted: of glucose, intestinal, of immunologic resources, of lipids, of cholesterol, of zinc, of energy, pulmonary, of epidermis, of blood pressure, etc [42].

Let's consider the blood pressure of a person being sit: at such a moment, according to the homeostatic theory, one mechanism maintains the blood pressure into stability, by keeping up a SP inside of a narrow range (around of 10 beats/min in healthy people). As the hypothetical person is standing up, there is a difference between the actual SP and the required SP of blood pressure of a standing person, due simply to the gravity force; in other words, there is no stability of blood pressure. As the brain detects this instability, also starts sending signals (Mediators) to the related mechanism through the communication channels (Nervous and Endocrine systems), in order to activate the adequate response from the tissues involved (to increase the heart bumping, in this example changed to a new SP).

The problem being solved has only one liberty degree. In other words, it is required to find only one SP, and this must be inside a narrow margin. The homeostatic model has demonstrated its utility in medicine [43, 44]; however, in some cases that model is not enough to explain neither complex behaviors of OS in the body, nor even some disease patterns, and therefore medical prescriptions homeostasis-based tend to fail. By considering such a problem it was proposed an alternative model who is called allostasis, that means 'to maintain stability (of a system, organ, OS, condition, health, etc.) through change (of several mechanisms)' [51]; in this case, and different to the homeostasis model, are taken in account several SPs of mechanisms and the non-linear relationships among mediators, OS and the brain. A simple example of allostasis is shown in Fig. 1.

The fundamental difference between this model and the homeostasis is the existence of several SPs in several mechanisms involved in returning to stability. As can be seen from Fig. 1, some SPs are increased whereas other are decreased; moreover, there are relationships among mechanisms, mediators, other mechanisms, as well as OS, which are not fully understood by scientists working at medical areas [51]. Prior to the explanation of the proposed computational algorithm, it is important to consider both a standard vocabulary and considerations. For instance, we will consider that the communication groups of OS are Group A. In the computational algorithm we consider only three groups of mediators, even though that could exist more in natural allostasis; those groups are called Groups B1, B2 and B3, or simply, Group B. Both groups, A and B, are responsible of coordination between brain and mechanisms that directly changes the appropriated SPs that could cope with the perturbation found; also, we argue that those groups contain different versions of SPs

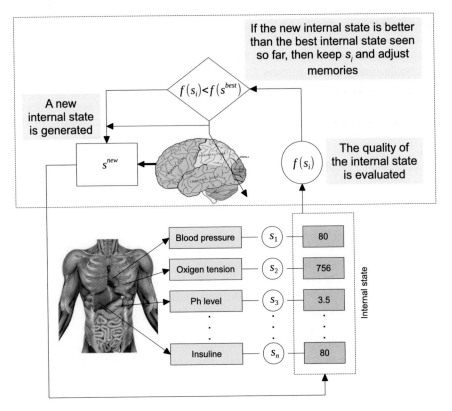

Fig. 1 Allostasis, a simple illustration scheme

(in hormonal as well as other kind of signaling means) historically used. In other words, we say that Groups A, B, and mechanisms, are different forms of SPs used in the past. By considering the aforementioned, in allostasis, the generation of new SPs is done by using the SPs historically used [49]. Whereas SPs of Group A do not suffer collective changes, those contained in Group B are constantly modified by collective operations. In order to generate new SPs, allostasis considers several procedures, being the main the so called 'combination' [52], which combines information of Groups A, B, and random variations. Once a new SP is generated, it is evaluated its capacity to reach a stable state, and whether the new SP improves the stability provided by the actual SP, a collective change is carried out over all elements contained in Group B [47, 48, 50].

2.2 Allostatic Optimization Algorithm

The computational approach of allostasis is called Allostatic Optimization (AO), which implements operations that resembles the interaction rules modeled by the mechanism of allostasis. In the algorithm, each *candidate solution* within the search space represents a *SP* vector, whereas the *fitness* value equals to a *degree of stability (or allostasis)* of each SP. The *population of candidate solutions* correspond to *stored SPs* in natural allostasis. The *optimum candidate solution* corresponds to the *whole stability*, or simply, *stability*. Finally, we consider a perturbation as a function of the difference between the allostasis and the stability; such a function it is called *err* in this chapter.

Also, the AO defines several operators, as the combination operation, which is considered the main operator and is applied over all individuals of the population. Other operators called collective (group operators: B1, B2 and B3) are also implemented in AO, and they affect only a group of elements. Following the biological model of allostasis, the AO approach divides the entire population in four different groups: Group A, group B1, group B2 and group B3. The elements of group A are only modified by the combination operator whereas the elements of groups B1, B2 and B3 are affected by the combination operator and other collective operations. Even though in the natural process each group could have different sizes, in the computational approach we consider the size of each group as one fourth of the entire population. Thus, the population size must be selected in such a way that it can be entirely divided by four (20, 40, etc.).

2.3 Description of the AO Algorithm

The AO algorithm starts by initializing the population randomly (candidate random solutions or SPs) and later, the evolutionary process acts as follows: The combination operator is applied to the first individual (SP) of the population, obtaining in such a way a new individual. Whether the new individual is better than the original one according to their allostasis (fitness value), the original individual is replaced by the new one whereas the groups B1, B2 and B3 are modified used collective operators. On the other hand, if the original individual is better than the new individual, no changes are executed to the population. An iteration is completed when the combination operator has been applied to the last individual. This procedure is applied until a termination criterion is met (i.e. the iteration number NI). Following the evolution process of AO, the following operators are employed:

1. Initialization.
2. Combination.
3. Collective B1.
4. Collective B2.
5. Collective B3.
6. Update the best element.

and the pseudo code of the proposal is

```
AO algorithm
1   Initialize S and determine its allostasis,
    divide S in groups A and B
2   while (criterion){
    for i=1 to Ns {
        Generate new individual snew
        by using the combination operator
        if f(snew)<f(si){
            if f(snew)<f(sbest){
                Calculate e and m
                Modify B1 by using collective operator B1
                Modify B2 by using collective operator B2
                Modify B3 by using collective operator B3
                sbest=snew; f(sbest)=f(snew);
            }
            si=snew; f(si)=f(snew);
        }
    }
}
```

Next, each operator is defined.

Initialization

In the first part, the algorithm initializes a population S of Ns set point vectors $(S = \{s_1, s_2, \ldots, s_{Ns}\})$, where each set point (SP) s_i is a D-dimensional vector containing parameter values to be optimized. Such values are randomly and uniformly distributed between the pre-specified lower initial parameter bound s_j^{low} and the upper initial parameter bound s_j^{high}:

$$s_{ij} = s_j^{low} + rand\,() \cdot \left(s_j^{high} - s_j^{low}\right); i = 1, \ldots, Ns; j = 1, \ldots, D \qquad (1)$$

with i and j being the individual and parameter indexes, respectively. Hence, s_{ij} is the jth parameter of the ith individual. After initialization of SPs, it is found the best individual from the population, e.g.

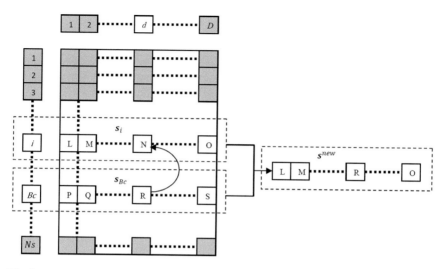

Fig. 2 The combination operator

$$s^{best} \in \left\{ S \mid f\left(s^{best}\right) = min(f\left(s_1\right), f\left(s_2\right), \ldots, f(s_{Ns})) \right\} \tag{2}$$

where $f(.)$ represents the cost function.

Combination

In allostasis, this operation combines each SP in the population, with information provided by other SPs. In this work, such effect is simulated by using a single operation of mutation, by replacing information of an original individual s_i with information extracted from other s_{Bc}, obtaining in such a way a new individual s^{new}, who combines information from both individuals. In order to implement this operator, two different integers are randomly generated, Bc inside the number of SPs $(1, \ldots, Ns)$ and d inside the dimension number $(1, \ldots, D)$. The combination takes place substituting the element s_{di} from the original s_i with the element s_{dBc} from the element s_{Bc}. Therefore, the only difference between s_i and s_{Bc} is the element in the position d. Figure 2 shows graphically the combination operation. Once the new individual s^{new} is generated by using the combination operator, it is compared whether such individual is better than the original one s_i and also the best found to far s^{best}, according to their fitness values. If s^{new} is better, the elements s_i as well as s^{best} are replaced by s^{new}, whereas the groups B1, B2 and B3 are modified used the collective operators. However, if s_i is better than s^{new}, no changes are executed to the population.

Collective B1

This operator modifies only the elements of group B1, namely SPs from $(Ns/4) + 1$ to $2 \cdot (Ns/4)$. In the allostasis mechanism, SPs from Group B1 are substituted by similar versions of the average answer produced by the entire set of SPs. In the AO approach, the average answer $a = \{a_1, \ldots, a_D\}$ is computed as:

$$a_j = \left(\frac{1}{Ns}\right) \cdot \sum_{i=1}^{Ns} s_{ij}; \ j = 1, \ldots, D \tag{3}$$

The modification, applied to each element, depends on the existent difference between s^{new} and s^{best}. Such relationship, defined as m, is calculated by using:

$$m = \psi \left[1.1 - \frac{1}{e^{\psi \cdot err}} \right] \tag{4}$$

where $\psi \in [0.01, 1.5]$ and $err = ((f(s^{new}) - f(s^{best})) / (f(s^{new}) + f(s^{best})))$
Therefore, the SPs of group B1 are updated according to:

$$s_{g1,j} = a_j - m + 2 \cdot m \cdot rand() \tag{5}$$

where $j \in \{1, \ldots, D\}, g1 \in \{\left(\frac{Ns}{4}\right) + 1, \left(\frac{Ns}{4}\right) + 2, \ldots, 2 \cdot \left(\frac{Ns}{4}\right)\}$ and $rand()$ is a number randomly generated between 0 and 1.

Collective B2

According to the allostasis mechanism, elements of group B2 are replaced by SPs randomly generated inside the average answer. Such effect is simulated modifying the elements of group B2 according to the following model:

$$s_{g2,j} = a_j \cdot rand() \tag{6}$$

where $g2 \in \{2 \cdot \left(\frac{Ns}{4}\right) + 1, 2 \cdot \left(\frac{Ns}{4}\right) + 2, \ldots, 3 \cdot \left(\frac{Ns}{4}\right)\}$.

Collective B3

Following the allostasis model, SPs of Group B3 are replaced by those who have demonstrated to be successful when a similar decompensation has happened. Such a behavior is emulated producing perturbed versions of the best SP $s^{best} = \{s_1^{best}, s_2^{best}, \ldots, s_D^{best}\}$ found so-far. Thus, the elements of group B3 are modified by using:

$$s_{g3,j} = s_j^{best} - m + (2 \cdot m \cdot rand()) \tag{7}$$

where $g3 \in \{3 \cdot \left(\frac{Ns}{4}\right) + 1, 3 \cdot \left(\frac{Ns}{4}\right) + 2, \ldots, Ns\}$.

Update the Best Element

In order to update de best SP s^{best} seen so-far, the best found individual from the current population $s^{best,k}$ is compared with the best individual $s^{best,k-1}$ of the last generation. If $s^{best,k}$ is better than $s^{best,k-1}$ according to their fitness values, s^{best} is updated with $s^{best,k}$, otherwise s^{best} remains without changes. Therefore, s^{best} stores the best historical SP found so-far.

3 Parametrical Model

3.1 Histogram Approximation by Using a CM

Let consider an image holding L gray levels $[0, \ldots, L-1]$ whose distribution is displayed within the histogram $h(g)$. In order to simplify the description, the histogram is normalized just as a probability distribution function, yielding:

$$h(g) = \frac{n_g}{N}, h(g) \geq 0, N = \sum_{g=0}^{L-1} n_g, \text{ and } \sum_{g=0}^{L-1} h(g) = 1, \tag{8}$$

where n_g specifies the number of pixels with gray level g, whereas N represents the total number of pixels contained in the image. The image histogram can thus be approximated by a CM of the form:

$$p(x) = \sum_{i=1}^{K} P_i \cdot p_i(x) = \sum_{i=1}^{K} P_i \left[\frac{\gamma_i^2}{(x - \rho_i)^2 + \gamma_i^2} \right] \tag{9}$$

where P_i is the a priori probability of class i, $p_i(x)$ is the probability distribution of gray-level random variable x in class i, ρ_i and γ_i are the location and the scale parameter of the ith Cauchy distribution and K is the number of classes contained in the image. In addition, the constraint $\sum_{i=1}^{K} P_i = 1$ must be satisfied.

In the proposed approach, the parameters $(P_i, \rho_i, \gamma_i, i = 1, \ldots, K)$ of the CM are encoded in an individual (a possible candidate solution). In order to correctly evaluate the matching quality between a candidate CM and the original histogram, the Hellinger distance E [53] is used. Such distance is defined as follows:

$$E = \sqrt{\sum_{j=0}^{L} \left[\sqrt{p(x_j)} - \sqrt{h(x_j)} \right]^2} \tag{10}$$

where $p(x_j)$ is the probability defined by the candidate CM, in the gray level point x_j whereas $h(x_j)$ represents its respective histogram value. Therefore, Eq. 10 is the objective function used by the AO algorithm to assess the quality of each individual.

Once obtained the best histogram approximation by a CM, the next step is to determine the optimal threshold values. At first, the location parameters are organized such as $\rho_1 < \rho_2 < \cdots < \rho_K$; then, the threshold values are calculated by estimating the overall probability error for two adjacent Cauchy functions:

$$E(T_i) = P_{i+1} \cdot E_1(T_i) + P_i \cdot E_2(T_i), \quad i = 1, 2, \ldots, K - 1 \tag{11}$$

considering

$$E_1(T_i) = \int_{-\infty}^{T_i} p_{i+1}(x)dx, \quad \text{and} \quad E_2(T_i) = \int_{T_i}^{\infty} p_i(x)dx \tag{12}$$

$E_1(T_i)$ is the probability of mistakenly classifying the pixels in the $(i + 1)$th class belonging to the ith class, while $E_2(T_h)$ is the probability of erroneously classifying the pixels in the ith class belonging to the $(i + 1)$th class. P_i's are the a-priori probabilities within the combined probability density function, and T_i is the threshold value between the ith and the $(i + 1)$th classes. The T_i value is chosen as to minimize the error $E(T_i)$. By differentiating $E(T_i)$ with respect to T_i and equating the result to zero, it is possible to use the following equation to define the optimum threshold value T_i:

$$AT_i^2 + BT_i + C = 0 \tag{13}$$

where

$$A = \gamma_i^2 - \gamma_{i+1}^2 \tag{14}$$

$$B = 2 \cdot (\rho_i \gamma_{i+1}^2 - \rho_{i+1} \gamma_i^2) \tag{15}$$

$$C = (\gamma_i \rho_{i+1})^2 - (\gamma_{i+1}\rho_i)^2 + 2 \cdot (\gamma_i \gamma_{i+1})^2 \cdot \ln\left(\frac{\gamma_{i+1} P_i}{\gamma_i P_{i+1}}\right) \tag{16}$$

From Eq. 13, it is only considered the solution whose value is positive and falls inside the valid interval. Figure 3 shows the determination process of threshold points, considering only two consecutive Cauchy functions.

Fig. 3 Determination of the threshold points

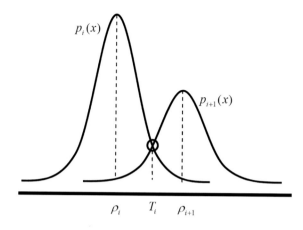

3.2 Otsu's Method

This method is a nonparametric technique for thresholding segmentation proposed by Otsu [21] that employs the maximum variance value of the different classes as a criterion to segment the image. In this approach the image histogram $h(g)$ is divided in m threshold values $T = \{T_1, \ldots, T_{m-1}\}$, considering $T_0 = 0$ and $T_m = L$. Each i-partition of m is defined as:

$$C_i = \{g | g \in (1, \ldots L - 1), \ T_{i-1} < g < T_i\}, \quad i = 1, \ldots, m \tag{17}$$

Such values are calculated as follows:

$$q_1 = \sum_{i=0}^{T_1} h(g_i), \quad \mu_1 = \sum_{i=0}^{T_1} \frac{h(g_i) \cdot i}{q_1}, \quad \sigma_1^2 = \sum_{i=0}^{T_1} \frac{(i - \mu_1)^2 \cdot h(g_i)}{q_1} \tag{18}$$

$$q_i = \sum_{T_i+1}^{T_{i+1}} h(g_i), \quad \mu_i = \sum_{T_i+1}^{T_{i+1}} \frac{h(g_i) \cdot i}{q_i}, \quad \sigma_i^2 = \sum_{T_i+1}^{T_{i+1}} \frac{(i - \mu_i)^2 \cdot h(g_i)}{q_i} \tag{19}$$

where $i = 1, \ldots, m - 1$. Therefore, the variance for the K-class is computed following the model:

$$\sigma_{WC}^2 = \sum_{j=1}^{K} q_j \cdot \sigma_j^2 \tag{20}$$

In order to find the threshold values, Eq. 21 must be minimized:

$$\sigma_{WC}^2(T_1^*, \ldots, T_K^*) = \min_{0 \leq T_1 \leq \cdots \leq T_K \leq L-1} \sigma_{WC}^2(T_1^*, \ldots, T_K^*) \qquad (21)$$

The Otsu method is considered the most popular segmentation algorithm with respectable results. Nevertheless, if the number of threshold values increases, the number of function evaluations increases. Such fact is considered its main drawback. Due to its wide popularity, the Otsu algorithm is used for comparing the performance of the approach proposed in this chapter.

4 Experimental Results

In the proposed approach, the parameters of the CM are encoded as an individual. An objective function by using the Hellinger distance evaluates the matching quality between the CM candidate (individual) and the original histogram. Guided by the values of this objective function, the set of encoded candidate mixtures are evolved using the operators defined by AO so that they can fit into the original histogram.

In this section, several experiments have been conducted considering different classes. Since each Cauchy function is defined by three parameters (P, ρ, γ), each individual l will have $3 \times K$ dimensions, if K different classes would be considered $[P_1^l, \rho_1^l, \gamma_1^l, \ldots, P_K^l, \rho_K^l, \gamma_K^l]$. Table 2 shows the general parameters utilized by AO. All the experiments are performed on a desktop computer with Intel® Core i7-2600 3.4 GHz, 8 GB of RAM and programmed in Matlab® 7.13.0.

Table 2 Parameters used in AO

Parameter	Value	Observations
L	256	Number of gray levels
N_p	90	Population size
N_{max}	200	Maximum number of iterations
x_l^{high}	$L - 1$	Higher limits of candidate l
x_l^{low}	0	Lower limits of candidate l
K	[2, 7]	Number of classes to find
T	$K - 1$	Number of thresholds to find
ψ	0.03	Tuning parameter of AO

In order to demonstrate the performance of the proposed algorithm, several images extracted from the Berkeley and the All-IDB databases [54, 55] have been used. Figures 4, 5, 6, 7, 8 and 9 present the experimental results after applying the

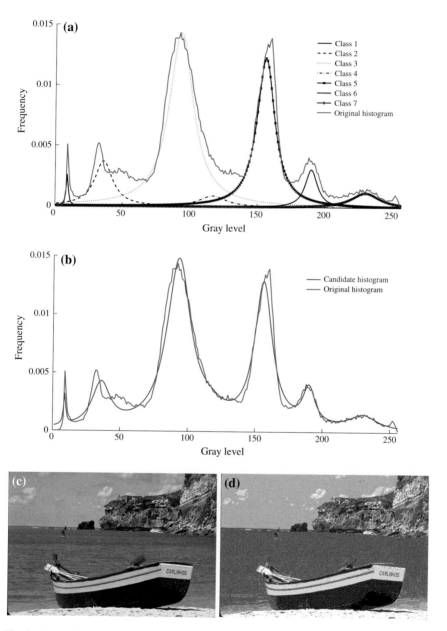

Fig. 4 Image 233, **a** class distribution with seven classes ($K = 7$), **b** approximation considering seven classes, **c** original image, **d** segmented image

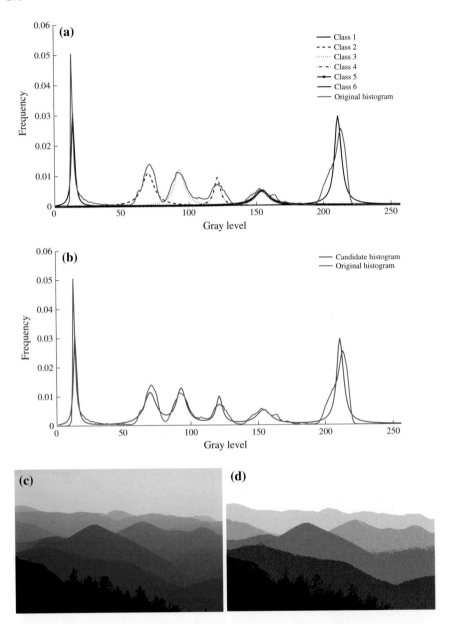

Fig. 5 Image Q24a, **a** class distribution with six classes ($K = 6$), **b** approximation considering six classes, **c** original image, **d** segmented image

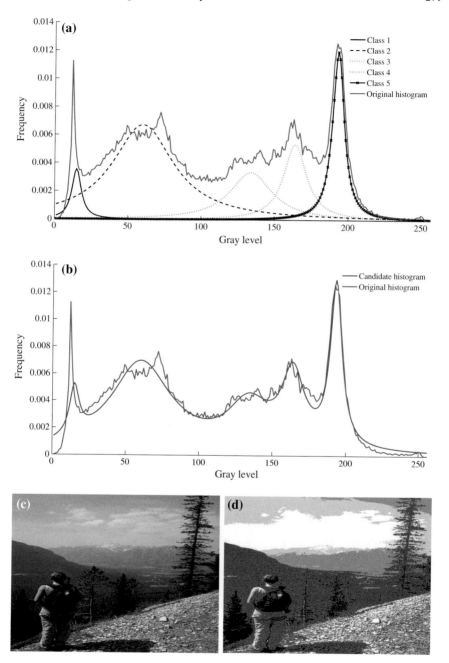

Fig. 6 Image Im001_1, **a** class distribution with five classes ($K = 5$), **b** approximation considering five classes, **c** original image, and **d** segmented image

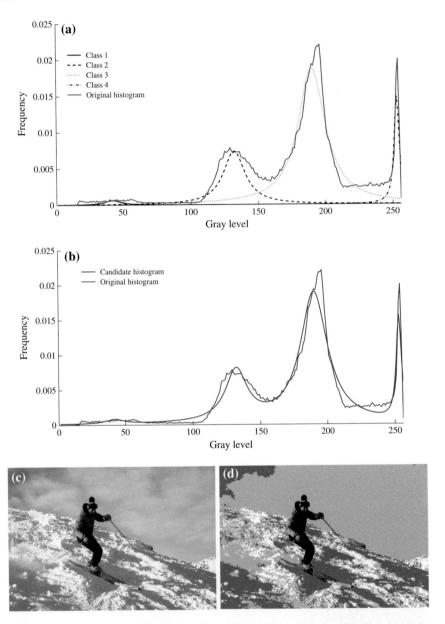

Fig. 7 Image Im002_1, **a** class distribution with five classes ($K = 5$), **b** approximation considering five classes, **c** original image, and **d** segmented image

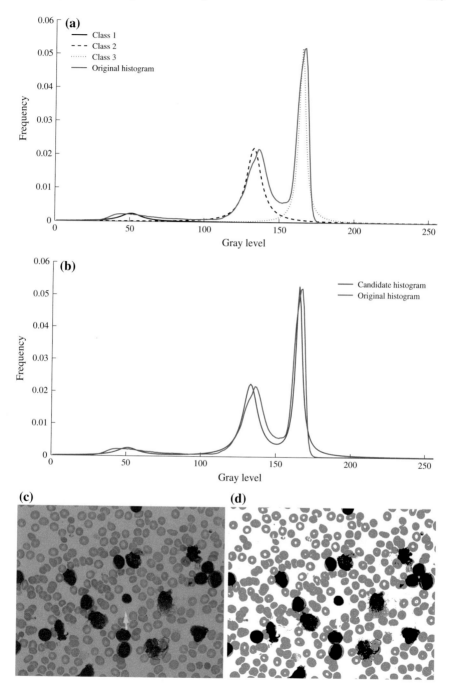

Fig. 8 Image 61060, **a** class distribution with three classes ($K = 3$), **b** approximation considering three classes, **c** original image, and **d** segmented image

Table 3 Experimental results obtained from the comparison between the Otsu and the AO methods

Image	Clases	AO $\mu(\sigma)$			Otsu $\mu(\sigma)$		
		T_1	T_2	T_3	T_1	T_2	T_3
233	2	98(1.09)	NA	NA	90(0)	NA	NA
Q24a	2	117(1.94)	NA	NA	125(0)	NA	NA
Im001_1	3	105(1.36)	154(0.54)	NA	97(0)	148(0)	NA
Im002_1	3	97(2.81)	154(0.63)	NA	96(0)	148(0)	NA
61060	4	84(1.18)	157(0.96)	241(1.89)	91(0)	162(0)	215(0)
253036	4	142(1.77)	189(3.84)	232(3.61)	138(0)	191(0)	222(0)

AO-based algorithm. In all figures, the approximation results over the original histogram are also shown.

In order to enhance the performance analysis, the proposed approach has been compared with the Otsu method [21]. Table 3 shows some results obtained by both methods, considering the mean and standard deviation of threshold values obtained by each algorithm when they have been executed 50 times for each image. The results have been presented in the format mean value μ (standard deviation, σ) whereas the elements that not correspond for a specific experiment are marked by NA (Figs. 4, 5, 6, 7).

Figure 10 shows two images proposed in [54] as segmentation benchmarks. Such problems consist in segmenting different cells, considering that their optimal results have been already obtained by a human expert (ground-truth). Under these conditions, it is possible to compare the segmentation results obtained by our approach and the Otsu method in terms of the optimal results. Figure 11 presents the results obtained by the Otsu method and the AO-based algorithm considering the benchmark images from [54]. A visual inspection of Fig. 11 demonstrates that the Otsu method presents more undesirable artifacts as a consequence of a poor segmentation.

In order to appropriately compare the results from Fig. 11, the Hausdorff distance in terms of the ground-truth has been used. Table 3 shows the averaged Hausdorff distances considering both images from Fig. 11. Considering the mean value μ of the Hausdorff distance, it is clear that the proposed method produces better results, than Otsu's method, as can be seen from Table 4.

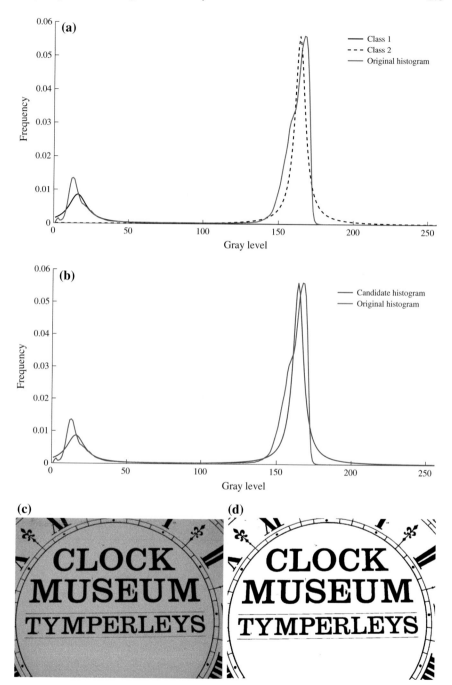

Fig. 9 Image 253036, **a** class distribution with two classes ($K = 2$), **b** approximation considering two classes, **c** original image, and **d** segmented image

(a)   **(b)**   **(c)**   **(d)**

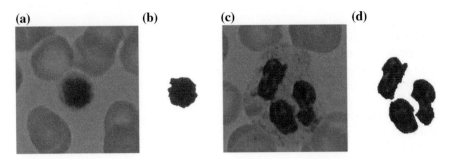

Fig. 10 Benchmark images for comparison with ground-truth. **a** Oiginal image with a single cell, **b** ground-truth, **c** original image with multiple cells and **d** ground-truth

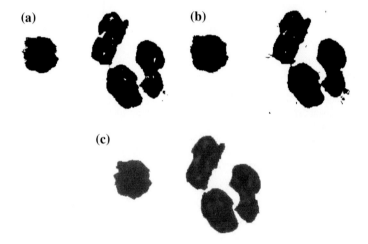

Fig. 11 Some results of the Otsu method and the AO-based algorithm. **a** AO-based. **b** Otsu. **c** Ground-truth images

Table 4 Hausdorff distance of AO against Otsu method

Method	Hausdorff distance	
	μ	σ
AO	2.1364	0.6535
Otsu	2.4655	2.9779×10^{-15}

5 Conclusions

In this chapter, a multi-thresholding segmentation algorithm based on a new evolutionary algorithm called Allostatic Optimization (AO) has been proposed. In the approach, the capacity of the Cauchy distribution to model complex problems (in presence of outliers) is exploited. Our approach assumes that the segmentation process is considered as an optimization problem by approximating the 1-D histogram of a given image in terms of a Cauchy mixture (CM) model, whose parameters are calculated through the AO algorithm.

In AO, the searcher agents emulate different body conditions which interact to each other by using operators based on the biological principles of the allostasis mechanism. The proposed approach encodes the parameters of the CM as an individual. An objective function by using the Hellinger distance evaluates the matching quality between the CM candidate and the original histogram. Guided by the values of this objective function, the set of encoded candidate mixtures are evolved using the operators defined by AO so that they can fit into the original histogram.

References

1. Arora, S., Acharya, J., Verma, A., Panigrahi, P.K.: Multilevel thresholding for image segmentation through a fast statistical recursive algorithm. Pattern Recogn. Lett. **29**(2), 119–125 (2008)
2. Naz, S., Majeed, H., Irshad, H.: Image segmentation using fuzzy clustering: a survey. In: 2010 6th International Conference on Emerging Technologies, pp. 181–186. IEEE (2010)
3. Janev, M., Pekar, D., Jakovljevic, N., Delic, V.: Eigenvalues driven gaussian selection in continuous speech recognition using hmms with full covariance matrices. Appl. Intell. **33**(2), 107–116 (2010)
4. Sezgin, M., Sankur, B.: Survey over image thresholding techniques and quantitative performance evaluation. J. Electron. Imaging **13**(1), 146–168 (2004)
5. Beevi, S.Z., Sathik, M.M., Senthamaraikannan, K., Yasmin, J.H.J.: A robust fuzzy clustering technique with spatial neighborhood information for effective medical image segmentation: an efficient variants of fuzzy clustering technique with spatial information for effective noisy medical image segmentation. In: 2010 Second International Conference on Computing, Communication and Networking Technologies, pp. 1–8. IEEE (2010)
6. Chitsaz, M., Seng, W.C.: A multi-agent system approach for medical image segmentation. In: 2009 International Conference on Future Computer and Communication, pp. 408–411. IEEE (2009)
7. Halim, N., Mashor, M., Abdul Nasir, A., Mokhtar, N., Rosline, H.: Nucleus segmentation technique for acute Leukemia. In: 2011 IEEE 7th International Colloquium on Signal Processing and Its Applications, pp. 192–197. IEEE (2011)
8. Mohapatra, S., Patra, D.: Automated cell nucleus segmentation and acute leukemia detection in blood microscopic images. In: 2010 International Conference on Systems in Medicine and Biology, pp. 49–54. IEEE (2010)
9. Mohapatra, S., Patra, D., Kumar, K.: Blood microscopic image segmentation using rough sets. In: 2011 International Conference on Image Information Processing, pp. 1–6. IEEE (2011)
10. Mohapatra, S., Samanta, S.S., Patra, D., Satpathi, S.: Fuzzy based blood image segmentation for automated leukemia detection. In: 2011 International Conference on Computers and Devices for Communication, pp. 1–5. IEEE (2011)
11. Nor Hazlyna, H., Mashor, M., Mokhtar, N., Aimi Salihah, A., Hassan, R., Raof, R., Osman, M.: Comparison of acute leukemia Image segmentation using HSI and RGB color space. In: 10th Internaional Conference on Information Science, Signal Processing and their Applications (ISSPA 2010), pp. 749–752. IEEE (2010)
12. Yang, G., Chen, K., Zhou, M., Xu, Z., Chen, Y.: Study on statistics iterative thresholding segmentation based on aviation image. In: Eighth ACIS International Conference on Software Engineering, Artificial Intelligence, Networking and Parallel/Distributed Computing (SNPD 2007), vol. 2, pp. 187–188. IEEE (2007)

13. Li, X., Ramachandran, R., He, M., Rushing, J., Graves, S., Lyatsky, W., Germany, G.: Comparing different thresholding algorithms for segmenting auroras. In: International Conference on Information Technology: Coding And Computing 2004. Proceedings. ITCC 2004, vol. 2, pp. 594–601. IEEE (2004)

14. Li, Z., Liu, C., Liu, G., Cheng, Y., Yang, X., Zhao, C.: A novel statistical image thresholding method. AEU Int. J. Electron. Commun. **64**(12), 1137–1147 (2010)

15. Akay, B.: A study on particle swarm optimization and artificial bee colony algorithms for multilevel thresholding. Appl. Soft Comput. **13**(6), 3066–3091 (2013)

16. Cuevas, E., Zaldivar, D., Pérez-Cisneros, M.: A novel multi-threshold segmentation approach based on differential evolution optimization. Expert Syst. Appl. **37**(7), 5265–5271 (2010)

17. Hammouche, K., Diaf, M., Siarry, P.: A comparative study of various meta-heuristic techniques applied to the multilevel thresholding problem. Eng. Appl. Artif. Intell. **23**(5), 676–688 (2010)

18. Horng, M.H.: A multilevel image thresholding using the honey bee mating optimization. Appl. Math. Comput. **215**(9), 3302–3310 (2010)

19. Sathya, P., Kayalvizhi, R.: Modified bacterial foraging algorithm based multilevel thresholding for image segmentation. Eng. Appl. Artif. Intell. **24**(4), 595–615 (2011)

20. Sathya, P., Kayalvizhi, R.: Optimal multilevel thresholding using bacterial foraging algorithm. Expert Syst. Appl. **38**(12), 15549–15564 (2011)

21. Otsu, N.: A threshold selection method from gray-level histograms. IEEE Trans. Syst. Man Cybern. **9**(1)

22. Al-Hussaini, E., Ateya, S.: Parametric estimation under a class of multivariate distributions. Stat. Pap. **46**(3), 321–338 (2005)

23. Liu, T., Zhang, P., Dai, W.S., Xie, M.: An intermediate distribution between Gaussian and Cauchy distributions. Phys. A Stat. Mech. Appl. **391**(22), 5411–5421 (2012)

24. Ateya, S., Madhagi, E.: On multivariate truncated generalized cauchy distribution. Stat. Pap. **54**(3), 879–897 (2013)

25. Zhang, J.: A highly efficient l-estimator for the location parameter of the cauchy distribution. Comput. Stat. **25**(1), 97–105 (2010)

26. Pander, T., Przybya, T.: Impulsive noise cancelation with simplified Cauchy-based p-norm filter. Signal Process. **92**(9), 2187–2198 (2012)

27. Gao, Q., Lu, Y., Sun, D., Sun, Z.L., Zhang, D.: Directionlet-based denoising of SAR images using a Cauchy model. Signal Process. **93**(5), 1056–1063 (2013)

28. Guozhong, C., Xingzhao, L.: Cauchy pdf modelling and its application to SAR image despeckling. J. Syst. Eng. Electron. **19**(4), 717–721 (2008)

29. Kocsor, A., Tth, L.: Application of kernel-based feature space transformations and learning methods to phoneme classification. Appl. Intell. **21**(2), 129–142 (2004)

30. Olsson, R.K., Petersen, K.B., Lehn-Schiøler, T.: State-space models: from the EM algorithm to a gradient approach. Neural Comput. **19**(4), 1097–1111 (2007)

31. Park, H., Amari, S.I., Fukumizu, K.: Adaptive natural gradient learning algorithms for various stochastic models. Neural Netw. **13**(7), 755–764 (2000)

32. Park, H., Ozeki, T.: Singularity and slow convergence of the EM algorithm for gaussian mixtures. Neural Process. Lett. **29**(1), 45–59 (2009)

33. I Abdul-Moniem, Y.M.S.: Tl-moments and l-moments estimation for the generalized pareto distribution. Appl. Math. Sci. **3**(1)

34. Reeds, J.A.: Asymptotic number of roots of cauchy location likelihood equations. Ann. Stat. **13**

35. Barnett, V.D.: Order statistics estimators of the location of the cauchy distribution. J. Am. Stat. Assoc. **61**

36. Fogel, L.J., Owens, A.J., Walsh, M.J.: Artificial Intelligence through Simulated Evolution. Wiley, Chichester (1966)

37. De Jong, K.A.: An analysis of the behavior of a class of genetic adaptive systems. Ph.D. Thesis, Ann Arbor, MI, USA (1975)

38. Holland, J.H.: Adaptation in Natural and Artificial Systems. MIT Press, Cambridge (1992)

39. De Castro, L.N., Von Zuben, F.J.: Artificial immune systems: Part i-basic theory and applications. Universidade Estadual de Campinas, Dezembro de, Technical Report 210 (1999)
40. Kennedy, J., Eberhart, R.: Particle swarm optimization. In: Proceedings of ICNN'95 - International Conference on Neural Networks, vol. 4, pp. 1942–1948. IEEE (1995)
41. Karaboga, D.: An idea based on honey bee swarm for numerical optimization. Technical Report TR06, Engineering Faculty, Computer Engineering Department, Erciyes University (2005)
42. Encyclopedia of Human Body Systems. Julie McDowell, Greenwood (2011)
43. Cannon, W.: Bodily Changes in Pain, Hunger, Fear and Rage: An Account of Recent Researchers into the Function of Emotional Excitement. Appleton, New York (1929)
44. Cannon, W.: The Wisdom of the Body. Norton (1932)
45. Gross, C.G.: Claude bernard and the constancy of the internal environment. Neuroscientist **4**
46. Fletcher, J.M.: Homeostasis as an explanatory principle in psychology. Psychol. Rev. **49**(1)
47. McEwen, B.: Allostasis and allostatic load: implications for neuropsychopharmacology. Neuropsychopharmacology **22**(2)
48. McEwen B.S., Wingfield, J.C.: The concept of allostasis in biology and biomedicine. Horm. Behav. **43**(1)
49. Romero, L.M., Dickens, M.J., Cyr, N.E.: The reactive scope model a new model integrating homeostasis, allostasis, and stress. Horm. Behav. **55**(3), 375–389 (2009)
50. Schulkin, J.: Allostasis: a neural behavioral perspective. Horm. Behav. **43**(1)
51. Sterling, P.: Allostasis: a model of predictive regulation. Physiol. Behav. **106**(1)
52. Martinez-Lavin, M., Vargas, A.: Complex adaptive systems allostasis in fibromyalgia. Rheum. Dis. Clin. North Am. **35**(2), 285–98 (2009)
53. Karunamuni, R., Wu, J.: Minimum Hellinger distance estimation in a nonparametric mixture model. J. Stat. Plan. Infer. **139**(3), 1118–1133 (2009)
54. Labati, R.D., Piuri, V., Scotti, F.: All-IDB: The acute lymphoblastic leukemia image database for image processing. In: 2011 18th IEEE International Conference on Image Processing, pp. 2045–2048. IEEE (2011)
55. Martin, D., Fowlkes, C., Tal, D., Malik, J.: A database of human segmented natural images and its application to evaluating segmentation algorithms and measuring ecological statistics. In: Proceedings of Eighth IEEE International Conference on Compututer Vision. ICCV 2001, vol. 2, pp. 416–423. IEEE Computer Society (2001)

Image Analysis and Coding Based on Ordinal Data Representation

Simina Emerich, Eugen Lupu, Bogdan Belean and Septimiu Crisan

Abstract With the use of computers and Internet in every major activity of our society, security is increasingly important. Biometric recognition is not only challenging but also computationally demanding. This chapter aims develop an iris biometric system. The iris has the advantages of uniqueness, stableness, anti-spoof, non-invasiveness and efficiency and could be applied in almost every area (banking, forensics, access control, etc.). The performance of a biometric classification system is largely depending on the techniques used for feature extraction. Inspired by the biological plausibility of ordinal measures, we propose their employment for iris representation and recognition. Qualitative measurement, associated to the relative ordering of different characteristics, is defined as ordinal measurement. Besides the proposing of a novel, fast and robust, ordinal based feature extraction method, the chapter also considers the problem of designing the decision making model so as to obtain an efficient and effective biometric system. In the literature, there are different approaches for iris recognition, nevertheless, there are still challenging open problems in improving the accuracy, robustness, security and ergonomics of biometric systems.

Keywords Ordinal measurements · Feature vector · Iris recognition · Biometrics · Wavelets

S. Emerich (✉) · E. Lupu
Faculty of Electronics, Telecommunications and Information Technology,
Technical University of Cluj-Napoca, Str. Baritiu 26-28, 400027 Cluj-Napoca, Romania
e-mail: Simina.Emerich@com.utcluj.ro

E. Lupu
e-mail: Eugen.Lupu@com.utcluj.ro

B. Belean
National Institute for Research and Development of Isotopic and Molecular
Technologies, Str. Donat, Nr. 67-103, 400293 Cluj-Napoca, Romania
e-mail: Bogdan.Belean@itim-cj.ro

S. Crisan
Faculty of Electrical Engineering, Technical University of Cluj-Napoca,
Str. Baritiu 26-28, 400027 Cluj-Napoca, Romania
e-mail: Septimiu.Crisan@ethm.utcluj.ro

© Springer International Publishing Switzerland 2016
A.I. Awad and M. Hassaballah (eds.), *Image Feature Detectors and Descriptors*,
Studies in Computational Intelligence 630, DOI 10.1007/978-3-319-28854-3_11

1 Introduction

In machine learning, feature vector extraction involves simplifying the amount of resources required to accurately describe a large set of data. A proper feature vector should be highly informative, invariant to a given set of transformations (such as rotation, scale etc.). In pattern recognition the feature extraction step is considered to be the most important step for achieving a robust and efficient system. The purpose of this framework is to guide the development of a compact, relevant and consistent set of features for the classification task [1] by choosing well-designed ordinal measures for image representation.

The measurements used in science can be classified into four types of scales: nominal, ordinal, interval and ratio. Nominal scales are used for labeling variables, without any quantitative value (ex: gender (male, female), hair color (dark, brown, blonde and grey)). Ordinal measurements describe order, but not relative size or degree of difference between the items measured. Interval scales are numeric scales in which we know not only the order, but also the exact differences between the values. The classic example of an interval scale is celsius temperature because the difference between each value is the same. Ratio scales provides the exact value between units, and they also have an absolute zero. Good examples of ratio variables include height and weight (Table 1).

Biological and psychological measurement usually operates on ordinal scales. Computer vision researchers prefer interval or ratio measures for object description and pattern recognition. As the lowest level of measurement, nominal scale is too weak for classification. But the power of ordinal measures for feature representation has been largely underestimated [2, 3].

Ordinal features come from a straightforward concept that we often use: one could easily rank or order the heights or weights of two persons, but it is hard to answer their precise differences. For computer vision, the absolute intensity information associated with an image can vary because it changes under various illumination settings. However, ordinal relationships among neighborhood pixels or regions present some stability with such changes and reflect the intrinsic natures of the object [4]. Ordinal features are efficient to compute and encode an ordinal relationship between two concepts. Less than and greater than are meaningful terms with ordinal variables. Figure 1 gives an example in which the average intensities between regions A and B are compared to give the ordinal code of 1 or 0.

Object recognition is a fundamental task that humans perform many times each day without noticeable effort. Furthermore, one can identify an object despite different changes that affect its appearance, including illumination, viewing direction, occlusion etc.

Inspired by the biological plausibility of ordinal measures, we propose to use them for object recognition. Ordinal data would use non-parametric statistics, including: median and mode rank order correlation to measure the strength of the associations between two variables, non-parametric analysis of variance etc.

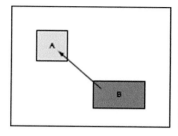
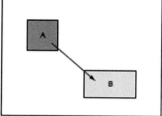

Fig. 1 Ordinal measure of relationship between two regions. *Region A* is brighter than *B*, A < B; *region A* is darker than *B*, A > B [5]

Table 1 Primary scales of measurement

Scale	Description	Examples	Permissible statistics
Nominal	Assign different names to objects ($=$, $\neq$)	Zip codes, eye colour, sex (male, female)	Mode, entropy, correlation, Chi-square
Ordinal	Indicates the relative position of objects($<$, $>$)	Ranking of teams in a tournament, grades (good, better, best)	Median, percentile, rank order correlation, sign test
Interval	Indicates the differences between objects ($+$, $-$) Zero point is arbitrary	Calendar dates, temperature	Range, mean, standard deviation, Pearson's correlation
Ratio	Ratio of scale values can be compared ($*$, $/$) Zero point is fixed	Age, mass, length, income	Geometric mean, harmonic mean, percent variation

The rest of this chapter is organized as follows. Section 2 summarizes different ordinal measures based methods proposed in literature for pattern recognition, particularly with applications in biometrics. In Sect. 3, a general algorithm for image analysis and coding, based on ordinal measures is presented. Section 4 discusses how to explore effective ordinal features for iris recognition. We conclude the chapter in Sect. 5.

2 Ordinal Measures, a Key Issue in Pattern Recognition

The advantages of ordinal measures for image representation have already been verified by some researchers. Sinha [6] was probably the first to introduce ordinal measures to computer-based vision systems. Based on the fact that several ordinal measures on facial images, such as eye-forehead and mouth-cheek, are invariant to individuals and imaging conditions, Sinha developed a ratio template for face detection [3]. He proposed a representation that is a collection of several pair-wise ordinal contrast relationships across facial regions. For instance, the average brightness of

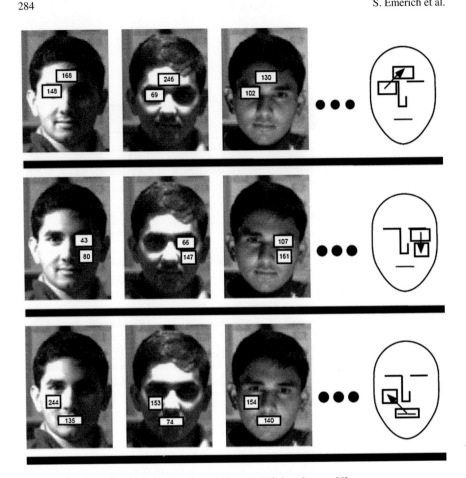

Fig. 2 Pair-wise ordinal relationships invariant under lighting changes [6]

the left eye is always less than that of the forehead, regardless of the lighting conditions. The relative magnitudes of the two brightness values may change, but the sign of the inequality does not (Fig. 2).

In other words, the ordinal relationship between the average brightness of the <left-eye, forehead> pair is invariant under lighting changes. Starting from the idea that the human visual system is far better at making relative brightness judgments than absolute ones, he suggested the structure presented in Fig. 3, for detecting faces under significant illumination variations [6].

Assuming that the ordinal relationship between neighboring image regions is stable and robust, several researchers proposed different multi-lobe differential filters (MLDFs) for ordinal image analysis and coding [3, 7]. MLDF can encode ordinal measures of multiple image regions with a flexible parameter configuration. Some of the variable parameters that can be used are presented in Fig. 4: the number of

Fig. 3 Invariant ordinal
structure of the image
brightness on a human face
under widely varying
illumination conditions [6]

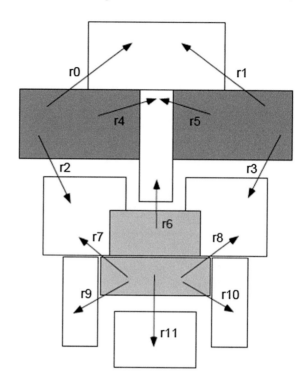

positive and negative lobes, orientation, scale and location of each lobe, inter-lobe
distance, etc.

Sun et al. proposes proposed in [3] multi-lobe differential filters for ordinal iris
feature extraction. An MLDF operator is passed across the normalized iris image
and each comparison is further encoded as one bit: 1 or 0 according to the sign
of the filtering result. For classification, the Hamming distance was employed. For
example, the ordinal measures for a group of two lobes, may denote point, line, edge,
corner, ridge, slope, etc., as shown in Fig. 5.

An effective scheme for matching noisy iris images under visible lighting is
described in [8]. For feature representation and matching, multiple cues, including ordinal measures (Fig. 6), color histogram, text on representation, and semantic
information are employed [8].

In [9] Tan introduced OM for iris, face and palmprint representation by using
a Multi-lobe Ordinal Filter (MLOF) with different parameters, such as distance,
orientation, scale and location. Each biometric region is binary encoded according
on the sign of the filtering results (Fig. 7). Their experiments have demonstrated that
the method achieves significantly higher accuracy than the state-of-the-art systems
with lower computational cost.

Orthogonal Line Ordinal Features are proposed in [4] for palm print representation. Gabor Ordinal Measures were also used for hand vein recognition [10] and

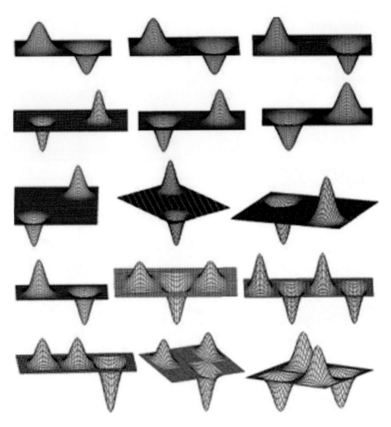

Fig. 4 Some typical multi-lobe differential filters [3]

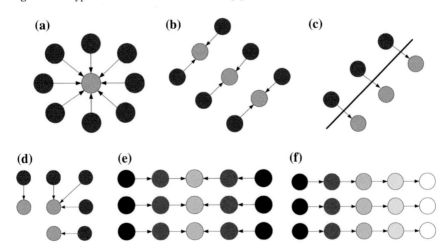

Fig. 5 Ordinal measures and their visual meanings. **a** Point. **b** Line. **c** Edge. **d** Corner. **e** Ridge. **f** Slope [3]

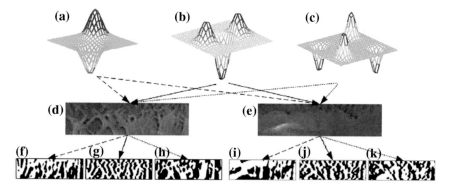

Fig. 6 **a–c** Ordinal filters; **d–e** iris sub-regions; **f–k** ordinal code [8]

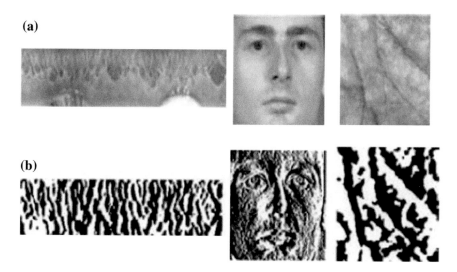

Fig. 7 Feature extraction with MLOF **a** Input images; **b** ordinal code [9]

for face recognition [5, 11, 12]. Local Ordinal Contrast was employed for lip-based speaker authentication in [13]. A robust ear recognition system is proposed in [14] using gradient ordinal relationship pattern. Based on the previous studies it can be stated that the ordinal measurements represent a feasible solution for the personal identification requirements. The existing architectures also provide directions for developing new and improved algorithms for image analysis and coding based on ordinal representations.

Object classification is a natural task for human visual system: we can classify a novel object, without effort, based on its appearance. It is therefore natural to study the biological mechanisms used for object classification and to propose similar approaches for computer vision systems.

The biological plausibility of visual ordinal measures has been verified by several neuroscience researchers. DeAngelis et al. [15] found that many striate cortical neurons' visual responses saturate rapidly regarding the magnitude of contrast as the input, which tells us the determining factor of visual cognition is not the absolute value of contrast, but the contrast polarity. Rullen et al. [16] suggested that temporal order coding might form a rank-based image representation in the visual cortex [3]. Inspired by the human visual system, Ullman et al. proposes in [17] a part-based method for pattern recognition. On their approach objects within a class are represented in terms of image fragments. The classification is based on a direct grey-level comparison between stored fragments and the input image. The method measures the qualitative shape similarity (using the ordinal ranking of the pixels in the regions) and the orientation difference (using gradient amplitude and direction) [17].

Based on ordinal pattern analysis, a mutual information technique is proposed in [18] to describe correlations of electromyogram signals during hand open/hand close states.

Several arguments in favor of ordinal measures applications are further presented:

- features based on high-level measurements are useful for image reconstruction but unnecessary for object recognition;
- high-level measures involve complex and time-consuming computations;
- ordinal measures are simple to implement and compact in feature template [3].

3 A New Algorithm for Image Analysis and Coding, Based on Ordinal Measures

The aim of this chapter is to describe a novel and robust image feature vector extraction method based on ordinal measures. Standing from the idea that a 'machine learning experimenter' needs to address three questions: (i) what to measure, (ii) how to measure it, and (iii) how to interpret it, the proposed algorithm follows the next sequence of processing steps (Fig. 8).

Fig. 8 Image feature vector extraction based on isolines

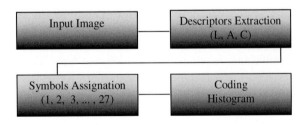

Fig. 9 Input image, viewed as a matrix

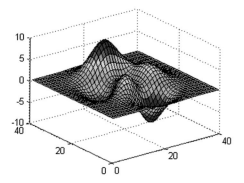

The input image is viewed as a matrix of pixels, where pixels are interpreted as heights with respect to the $x - y$ plane as shown in Fig. 9.

Firstly, the isolines of the matrix are computed where an isoline (also called a contour line) is a curve in the xy plane along which the function $f(x, y)$ has a constant value, Cj. These curves are defined by the following equation:

$$f(x, y) = Cj, j = 1, 2, \ldots, N \qquad (1)$$

where N is the number of detected contours. A local maxima or minima is surrounded by several contour lines as presented in Fig. 10. The number of the contour lines could be chosen automatically based on the minimum and maximum values of the matrix or a prefixed number of levels could be selected by the user. In our example, for the input matrix, four surfaces are delimited:

- S2 and S4 correspond to local maxima;
- S1 and S3 correspond to local minima (Fig. 10).

Fig. 10 Surfaces delimited by isolines

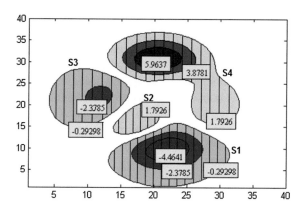

Three descriptors were associated to each surface:

- L (Level) which describes the depth of the surface by counting the number of internal isolines;
- A (Area) characterizes the area of the surface;
- C (Contour) defines the constant value related to external contour line of the surface.

The proposed new coding procedure supposes that pairs of surfaces are further compared and a symbol is provided (there are 27 different symbols: $3 \times 3 \times 3$), indicating the differences between the individual L, A and C traits of the two regions involved in the ordinal comparison ($>$, $<$ or $=$).

Each ordinal relation by itself is not reliable enough and is too coarse to provide a good discriminant function to distinguish between different classes therefore it is necessary to consider many of the relations together to achieve satisfactory performances.

The order of surfaces to be compared may be performed in different ways. The following two procedures are proposed. The first one assumes center of gravity determinations for each surface and establishing the order according to their distance from the origin of the coordinate system. A second method supposes that the delimited surfaces are compared two by two. In this way the computational cost is higher but a rotation invariant feature vector is obtained. In many practical computer vision applications, rotation invariance is a crucial issue.

A histogram of the 27 possible combinations is further produced, forming a feature vector with fixed dimension 1×27, which carries information about symbols frequency. Because of their fixed length, the resulted feature vectors are ideal form of inputs for many classifiers (Fig. 11).

A detailed description of the entire process is presented in Tables 2 and 3. By scaling the input matrix (from 40×40 pixels to 10×10) (Fig. 12) we get the contour plot given by Fig. 13.

The histogram of the obtained symbols is the same for both matrices (original and scaled). This remark indicates that the proposed coding procedure is robust to scale changes.

In a period shorter than 20 years, wavelets have imposed themselves as a fruitful tool for both signal and image processing. The theory behind wavelets has been developed independently by mathematicians, scientists and engineers. A privileged area of applications where wavelet methods have been found to be relevant is pattern recognition. Ordinal measurements could be combined with DWT2 (Bi-dimensional Discrete Wavelet Transform). DWT2 leads, after one level decomposition, into four components: the approximation coefficients (cA), and the details coefficients in the three orientations, respectively the horizontal, the vertical, and the diagonal (cH, cV, cD) [19], as shown in Fig. 14.

Figure 15 presents the surfaces obtained according to the previously presented method. The feature vector derived from approximation coefficients is the same with the one obtained from the input image, and together with corresponding vectors resulted from detail coefficients, ensures a highly informative final feature vector.

Fig. 11 Symbols assignment

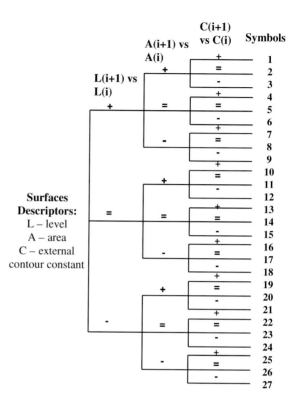

Surfaces
Descriptors:
L – level
A – area
C – external
contour constant

Table 2 Corresponding surfaces and descriptors for the input matrix

	S1	S2	S3	S4
Level (L)	3	1	2	3
Area (A)	189.10	32.40	120.07	210.12
Constant (C)	−0.2929	1.7926	−0.2929	1.7926

Table 3 Symbols assignation for the input matrix

	i = 1 (S2 vs. S1)	i = 2 (S3 vs. S2)	i = 3 (S4 vs. S3)
L(i + 1) versus L(i)	−	+	+
A(i + 1) versus A(i)	−	+	+
C(i + 1) versus C(i)	+	−	+
Symbol	25	3	1

The presented method generates simple data structures which are both compact and of known size so that, limited memory resources in embodiments such as Smart Cards can be employed efficiently. This aspect has important benefits for data storage and transfer operations.

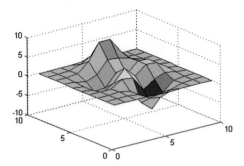

Fig. 12 Scaled version of the input matrix

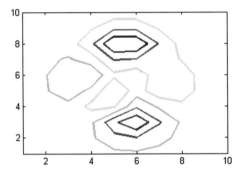

Fig. 13 Surfaces delimited by isoline

The advantages offered by the proposed method recommend it for several areas of application including biomedical image analysis, biometry (iris recognition), etc.

For one-dimensional signals a similar procedure TESPAR DZ (Time Encoded Signal Processing and Recognition) was proposed by King [20]. The TESPAR DZ method is based on an approximation model employing the zeros theory, such that the signal is divided in periods between successive zero crossing of the waveform. Duration (D), Shape (S) and Amplitude (A) are used as descriptors for each epoch. Then, pairs of epochs are compared and a symbol is provided [21] (Fig. 16).

4   Ordinal Representation for Iris Recognition

4.1   *Biometrics*

Nowadays, computers and Internet are used in every major function of our society, so security is increasingly important. For many years, passwords, PINs or identity cards have been used for person's identification. The advantage is that these modalities do

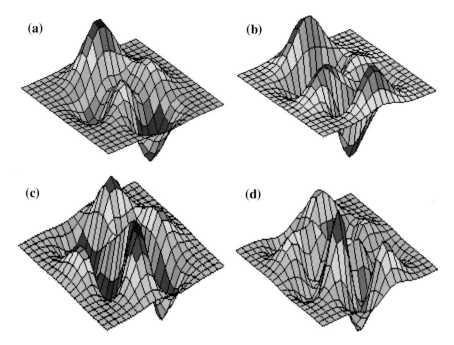

Fig. 14 Resulted matrix after DWT2 decomposition of input image **a** cA; **b** cH; **c** cV; **d** cD

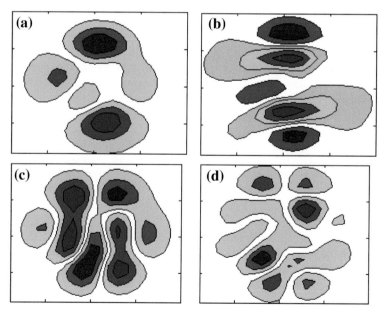

Fig. 15 Surfaces delimited by isolines for wavelet coefficients **a** cA; **b** cH; **c** cV; **d** cD

not change over time and they are not affected by the environment, but they cannot avoid problems such as being forgotten or guessed by others. Biometrics provides a promising solution for reliable personal identification [22]. The availability of faster computers and improved sensing technology coupled with significant advances in pattern recognition afford researchers to develop robust biometric systems. The goal of pattern recognition is to classify the objects in different categories or classes. In biometry the objects could be images (iris, palm vein) or signal waveforms (voice, signature, etc.).

Biometry has changed and will continue to change the way many activities are carried out by each of us. Biometric applications concern a variety of areas: civil and criminal identification, surveillance and screening, health care, eCommerce, eGovernment, physical and logical access. Biometric systems are used in airports, financial service institutions, banking ATMs, houses, etc. Many computers, laptops and smart phones have incorporated webcams, microphones and even fingerprint scanners, offering to the users the possibility to embrace biometric authentication based on fingerprint, iris, face or voice. Although several biometric modalities (i.e. fingerprint, voice and face) have already been used in large-scale deployments, there are many other attractive and "new" modalities in various stages of development and assessment (hand vein, iris, DNA etc.). However, no biometric modality is impeccable and each one has advantages and disadvantages for a given use case.

Table 4 compares several biometric technologies with each other against seven criteria.

- *Universality* describes how commonly a biometric is found in each individual.
- *Uniqueness* is how well the biometric separates one individual from another.
- *Permanence* measures how well a biometric resists aging.
- *Collectability* explains how easy it is to acquire a biometric for measurement.
- *Performance* indicates the accuracy, speed, and robustness of the system capturing the biometric.
- *Acceptability* indicates the degree of approval of a technology by the public in everyday life.
- *Circumvention* is how easy it is to fool the authentication system [23].

It is known that, from all biometric techniques, iris based biometric systems are the most promising for high security environments. Although diverse iris recognition methods have been proposed, the fundamentals of this biometrics have not a unified answer.

Though the theory behind iris recognition was studied as early as the 19th century, most research has been done in the last few decades. At the moment, two prototypes of iris recognition systems had been developed, by Daugman [25] and Wildes et al. [26]. Promising results were obtained by Boles and Boashash [27] using wavelet transform, and by Sanchez-Reillo and Sanchez-Avila in [28], where Gabor filters are employed.

Iris is a colored pigmented tissue (usually blue, brown or green), found outside the pupil to regulate it from incoming rays of light. It has an extraordinary structure and

Table 4 Comparison of various biometric technologies [24]

Biometrics	Universality	Uniqueness	Permanence	Collectability	Performance	Acceptability	Circumvention
Face	+	−	0	+	−	+	−
Facial thermogram	+	+	−	+	0	+	−
Fingerprint	0	+	+	0	+	0	+
Hand geometry	0	0	0	+	0	0	0
Hand vein	0	0	0	0	0	0	+
Iris	+	+	+	0	+	−	+
Keystroke dynamics	−	−	−	0	−	0	0
Retina	+	+	0	−	+	−	+
Signature	−	−	−	+	−	+	−
Voice	0	−	−	0	−	+	−
DNA	+	+	+	−	+	−	−
Gait	0	−	−	+	−	+	0
Ear	0	0	+	0	0	+	0
Odor	+	+	+	−	−	0	−

+ = High, 0 = Medium, − = Low

provides many interlacing minute characteristics such as freckles, coronas, stripes, furrows, crypts and so on.

The human iris has several benefits when compared with other biometrics methods: is very stable over a long period of time. In addition, the inherent isolation from the external environment and the impossibility of surgically modifying it without high risk of damaging the user's health provides strong immunity to forgery [29]. Some important factors that may affect the performances of iris recognition system are [2]:

(a) Registration: Although the rotation difference between two images can be solved by a brute force registration process, the large computational cost makes it not preferred in real time application.
(b) Normalization is expected to provide scale and position invariance of input images. Complex normalization methods were developed by researchers, but they proved to be computationally intense and for real time applications a very fast digital image processing hardware is required.
(c) Contrast variation
(d) Noise may disturb the precision of traditional algorithms. But the richness of inter-region sharp intensity differences provides a good source of ordinal measures for iris coding. Qualitative relationships across distinct iris regions can be insensitive to the contrast variations.

A proper iris recognition algorithm should be tolerant to the drawbacks mentioned above and should also encode efficiently image properties [2].

Iris scan technology has been traditionally used for surveillance and security purpose. In present iris based systems are used in airports for passenger authentication process, in financial service institutions, for ATM access usage etc.

The proposed system intends to argue that ordinal image representation provides a better trade-off for biometric recognition between accuracy, robustness and efficiency.

4.2   The Proposed Iris Based Biometric System

The processing flow used to implement the iris recognition system is presented in Fig. 17.

The qualities of the image acquisition, segmentation, normalization, feature extraction and the used classifier define the performance of the system.

Database: For experiments a public database was used. It contains 3×128 iris images (i.e. 3×64 left and 3×64 right). The features of the images are: 24 bit-RGB, 576×768 pixels, file format: PNG [30].

Iris Segmentation and Normalization: Segmentation requires a proper detection of the inner and outer boundaries of the iris texture. The specular reflections inside the pupil area are contained in the images from the database.

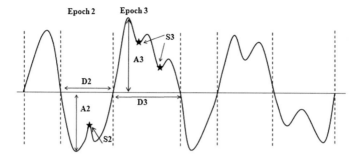

Fig. 16 TESPAR DZ coding procedure

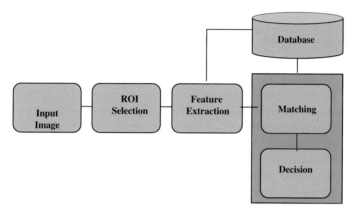

Fig. 17 Biometric system block diagram

Fig. 18 Iris, segmented image (ROI selection)

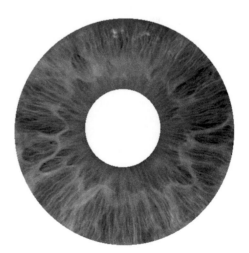

Fig. 19 Unwrapped iris (ROI transformation into polar coordinates)

At the beginning, a Canny edge detection is employed in order to generate an edge map. Then, the circular Hough transform establishes the radius and the centre coordinates of the iris region within the whole edge image which was previously obtained. Further, the pupil's radius and centre coordinates are searched within a crop of the image that contains only the region of interest (ROI) [31].

The detected iris region Fig. 18 is unwrapped by remapping each point to a pair of polar coordinates (r, θ) using the Cartesian to polar reference transform suggested by Daugman [32]. Hence, we obtain a rectangular representation for the iris after a normalization process, Fig. 19.

Feature Extraction: The feature vectors are extracted from the polar representation of the iris. Firstly, the bidirectional discrete wavelet transform (DWT-2D) was applied, by using different mother functions. Researchers are faced with an ever increasing variety of wavelets to choose from and the choice of the best wavelet is application-dependent. We selected several well-known wavelet functions such as Daubechies of order 1 and 3 (Db1, Db3), Battle–Lemarie of order 1 and 2 (Lem1, Lem2), Biorthogonal of order 1.3 (Bior1.3), Reverse Biorthogonal of order 2.4 (Rbio2.4) and Coiflet of order 1 (Coif1). Comparative studies seem to be very useful for the selection of a particular wavelet function. After one level decomposition, there are four components: approximations (cA), and details in the three orientations—horizontal, vertical and diagonal (cH, cV respectively cD).

After several experiments we conclude that for approximations 13 contour levels should be firstly computed and after that 6 of them to be selected (the first and the last three ones) (Figs. 20 and 21). For details 3 contour levels are considered Fig. 22. As presented in the previous chapter, three descriptors are associated to each surface: L (Level), A (Area) and C (Contour value).

Further, pairs of surfaces are compared and a symbol indicates the differences between the individual traits of the two surfaces being compared. The symbol stream is then condensed into fixed-size feature vectors by simply counting how many times each symbol occur. Individual vectors resulted from approximations respectively details are than fused into a final feature vector. The feature vector length has 4×27 coefficients for all irises. Thus the image is transformed into an encoded stream of discrete numerical symbols. This compact iris code greatly facilitates the matching process.

Classification: To perform training and classification tasks for identification experiments, WEKA toolkit was used. WEKA is a data mining workbench that allows comparison between many different machine learning algorithms. The first step was to represent our learning problem using an *.arff* file, where each instance is represented as a feature vector. The header of this file identifies the types of the features

Fig. 20 Surfaces delimited by isolines for unwrapped iris, approximation coefficients (full size image)

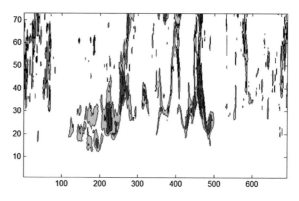

Fig. 21 Surfaces delimited by isolines for unwrapped iris, approximation coefficients (zoomed area)

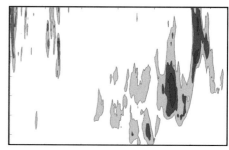

Fig. 22 Surfaces delimited by isolines for unwrapped iris, diagonal detail coefficients (zoomed area)

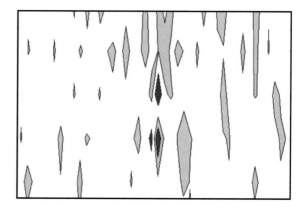

and the classes being predicted. WEKA summarizes the classification results by generating a confusion matrix [33]. In the present study, different classifiers were used, based on the following algorithms: Bayes Net (BN), Naive Bayes (NB), Radial Basis Function Neural Network (RBF), k-Nearest Neighbor ($1 - NN$ for $k = 1$ and $3 - NN$ for $k = 3$) and Support Vector Machine (SVM).

Fig. 23 The Performances
of different classifiers in
term of accuracy rates

A Bayes Classifier is a simple probabilistic classifier based on applying Bayes' theorem with strong (naive) independence assumptions. In simple terms, a naive Bayes classifier assumes that the presence (or absence) of a particular feature of a class is unrelated to the presence (or absence) of any other feature [34].

In the RBF Network, class implements a normalized Gaussian radial basis function network. It uses the k-means clustering algorithm to provide the basis functions.

The Nearest Neighbor Classifier (NNC) uses normalized Euclidean distance to find the training instance closest to the given test instance, and predicts the same class as this training instance. If multiple instances have the same (smallest) distance to the test instance, the first one found is used.

Support Vector Machines are based on the statistical learning theory of structural risk management. They are built by mapping the training patterns into a higher dimensional feature space where the points can be separated by using a hyper plane [35]. WLSVM software toolbox was employed for this classifier (Weka LibSVM—Integrating LibSVM into Weka) and may be seen as a form of implemented LibSVM working in Weka [36]. The main advantage is that LibSVM works considerably faster than WEKA. There are four kernels available for the SVM classifier: linear, polynomial, radial basis function and sigmoid. Optimal values for the SVM kernel's parameters will found by performing a grid search on the training data.

Experiments: Identification experiments were carried out, using two iris images/person for training (left and right eye) and four iris images/person for testing (two for the left and two for the right eye).

The SVM classification performances were tested for all available kernels. A proper choice of parameters is crucial for SVM to achieve good recognition rates. In the experiments the simple grid-search approach was used: parameters were varied with a fixed step-size through a wide range of values and the performance of every combination was measured. A 10-fold cross validation technique was also employed. The training data was randomly split into 10 sets, 9 of which were used in training and the 10th for validations. Then iteratively another nine were picked and so forth.

The best classification rates issued by identification experiments are presented in the Fig. 23. Comparative studies seem to be very useful for the selection of a particular classifier. The best results were obtained by using Db1 wavelet function, which is actually the Haar wavelet, together with the polynomial kernel ($C = 100$ and $d = 7$).

5 Conclusions

This chapter handles concepts of pattern recognition and their applications to solving real life problems such as biometric identification. Biometrics offers the promise of much stronger identity verification, and identity management is becoming ever more important to economic and social life. The need for enhanced security persists more than ever in a more electronically dependent and interconnected world. The traditional authentication methods are neither secure enough nor convenient for many automatic identification systems.

Firstly, a new idea for image feature representation by using ordinal measures was presented. The proposed algorithm is invariant to rotation and scale. Moreover, it provides fixed size descriptors regardless of the image dimension thus being ideal for many classifiers. Besides fixed dimension, the small number of coefficients used to store and characterize an image should be mentioned. The ordinal measures ensure the image representation to be robust to different intra-class variations (such as illumination), signal noises, misalignment or nonlinear deformations. The new method is considered to be suitable for portable applications due to its computationally low costs.

The proposed algorithm was further integrated to implement a biometric system, based on iris. The performances of several classification algorithms were investigated. The Support Vector Machines classifier was tested for linear, polynomial, radial basis function and sigmoid kernels, using 10-fold cross-validation on the training set. The obtained best parameters were then used for testing. The ability of several wavelets was also investigated, results showing that Haar functions seem to be the best suited in this case.

Acknowledgments "This work was supported by a grant of the Romanian National Authority for Scientific Research and Innovation, CNCS – UEFISCDI, project number PN-II-RU-TE-2014-4-2080".

References

1. Apatean, A., Rogozan, A., Bensrhair, A.: Visible-infrared fusion schemes for road obstacle classification. J. Transp. Res. Part C Emerg. Technol. **35**, 180–192 (2013)
2. Sun, Z., Tan, T., Wang, Y.: Robust encoding of local ordinal measures: a general framework of iris recognition. In: Biometric Authentication, pp. 270–282. Springer, Heidelberg (2004)
3. Sun, Z., Tan, T.: Ordinal measures for iris recognition. IEEE Trans. Pattern Anal. Mach. Intell. **31**(12), 2211–2226 (2009)

4. Srinivasan, B.: Performance of biometric palm print personal identification security system using ordinal measures. Int. J. Comput. Sci. Bus. Inform. **7**(1) (2013)
5. Liao, S.C., et al.: Face recognition using ordinal features. Advances in Biometrics, pp. 40–46. Springer, Heidelberg (2005)
6. Sinha, P.: Biologically Motivated Computer Vision. Qualitative Representations for Recognition. Springer, Heidelberg (2002)
7. Chai, Z., et al.: Gabor ordinal measures for face recognition. IEEE Trans. Inf. Forensics Secur. **9**(1), 14–26 (2014)
8. Tan, T., et al.: Noisy iris image matching by using multiple cues. Pattern Recogn. Lett. **33**(8), 970–977 (2012)
9. Tan, T., Sun, Z.: Ordinal representations for biometrics recognition. In: Proceedings of the Fifteenth European Conference on Signal Processing. Poznan, Poland (2007)
10. Meng, Z., Gu, X.: Hand vein identification using local Gabor ordinal measure. J. Electron. Imaging **23**(5) (2014)
11. Chai, Z.: Gabor ordinal measures for face recognition. IEEE Trans. Inf. Forensics Secur. **9**(1) (2014)
12. Chai, Z.: Histograms of Gabor ordinal measures for face representation and recognition. In: 5th IAPR International Conference on Biometrics (2012)
13. Chan, C.H., et al.: Local ordinal contrast pattern histograms for spatiotemporal, lip-based speaker authentication. IEEE Trans. Inf. Forensics Secur. **7** (2012)
14. Nigam, A., Gupta, P.: Robust ear recognition using gradient ordinal relationship pattern. In: Computer Vision-ACCV 2014 Workshops. Springer (2014)
15. DeAngelis, G.C., Ohzawa, I., Freeman, R.D.: Spatiotemporal organization of simple-cell receptive fields in the cat's striate cortex. I. General characteristics and postnatal development. J. Neurophysiol. **69**(4), 1091–1117 (1993)
16. Van Rullen, R., Thorpe, S.J.: Rate coding versus temporal order coding: what the retinal ganglion cells tell the visual cortex. Neural Comput. **13**(6), 1255–1283 (2001)
17. Ullman, S., Sali, E., Vidal-Naquet, M.: A fragment-based approach to object representation and classification. In: Visual Form 2001, pp 85–100. Springer, Heidelberg (2001)
18. Ouyang, G., Ju, Z., Liu, H.: Mutual information analysis with ordinal pattern for EMG based hand motion recognition. In: Intelligent Robotics and Applications, pp. 499–506. Springer, Heidelberg (2012)
19. Misiti, M., Misiti, Y., Oppenheim, G.: Wavelets and their application. In: Digital Signal and Image Processing Series. ISTE (2007)
20. King, R.A: Waveform coding method. U.S. patent No. 6748354B1, June 8 (2004)
21. Emerich, S., Lupu, E., Rusu, C.: A new set of features for a bimodal system based on on-line signature and speech. Digit. Signal Process. **23**(3), 928–940 (2013)
22. Li, S.Z., Jain, A.K.: Encyclopedia of Biometrics. Springer, London (2009)
23. Cheng, Qi.: User habitation in keystroke dynamics based authentication. ProQuest (2007)
24. Lupu, E., Pop, G.P.: An overview of biometrics. Acta Technica Napocensis Electron. Telecommun. **47**(2), 42–56 (2006)
25. Daugman, J.: Biometric personal identification system based on iris analysis. U.S. Patent No. 5291560 (1994)
26. Wildes, R., et al.: A system for automated iris recognition. In: Second IEEE Workshop on Applications of Computer Vision, pp. 121–128 (1994)
27. Boles, W.W., Boashash, B.: A human identification technique using images of the iris and wavelet transform. IEEE Trans. Signal Process. **46**(4), 1185–1188 (1998)
28. Sanchez-Reillo, R., Sanchez-Avila, C.: Processing of the human iris pattern for biometric identification. In: Eighth International Conference on Information Processing and Management of Uncertainty in Knowledge Based Systems, pp. 653–656. Spain (2000)
29. Sanchez-Avila, C., Sanchez-Reillo, R., de Martin-Roche, D.: Iris-based biometric recognition using dyadic wavelet transform. IEEE Aerosp. Electron. Syst. Mag. **17** (2002)
30. Dobeš, M., Machala, L.: Iris database. http://www.inf.upol.cz/iris/

31. Masek, L.: Recognition of human iris patterns for biometric identification. PhD Thesis, University of Western Australia (2003)
32. Daugman, J.G.: How iris recognition works. IEEE Trans. Circuits Syst. Video Technol. **14**, 21–30 (2004)
33. Hall, M., Frank, E., Holmes, G., Pfahringer, B., Reutemann, P., Witten, I.H.: The WEKA data mining software: an update. SIGKDD Explor. **11**(1) (2009). www.cs.waikato.ac.nz/ml/weka
34. Hossain, E., Chetty, G.: Combination of physiological and behavioral biometric for human identification. Mach. Learn. Data Min. Pattern Recogn. Lect. Notes Comput. Sci. **7376**, 380–393 (2012)
35. Vapnik, V.: The Nature of Statistical Learning Theory. Springer, New York (1995)
36. EL-Manzalawy, Y., Honavar, V.: WLSVM integrating LibSVM into Weka environment (2005). http://www.cs.iastate.edu/~yasser/wlsvm

Intelligent Detection of Foveal Zone from Colored Fundus Images of Human Retina Through a Robust Combination of Fuzzy-Logic and Active Contour Model

Rezwanur Rahman, S.M. Raiyan Kabir and Anita Quadir

Abstract Detection of the center of a fovea and its boundary from a retinal image is a challenging task due to the irregularity of the avascular foveal region. Several attempts have been made before in order to detect the foveal region and its boundary from fluorescein angiographic images. The irregularity and large variation in the human retinal images made the task increasingly difficult. In this current work funds images were considered instead of fluorescein angiographic images in order to ensure the applicability of the proposed algorithm in both biomedical and biometric analysis. A robust *fuzzy-rule* based image segmentation algorithm has been developed in order to extract the foveal region from a wide variety of images from different persons. Detection of foveal region comprised of *locating the geometric center* and *extracting the boundary*. The geometric center was evaluated by weighted averaging the grey scale intensities obtained from implementing the current algorithm. This was followed by applying gradient vector flow (GVF) based active contour technique in order to extract the boundary of the foveal region. The algorithm was applied on a several retinal images acquired from different persons with a very good success rate. The present work is considered to be an important contribution in intelligent image analysis of human retina since it incorporates a robust "fuzzy-rule" in extracting foveal region. Similar approach has not been adopted in the literature. The proposed algorithm is seen to be versatile in analyzing a wide range of retinal images.

R. Rahman (✉)
Department of Petroleum and Geosystems Engineering,
The University of Texas at Austin, 200 E. Dean Keeton St.,
Stop C0300, Austin, TX 78712, USA
e-mail: rezwanur.rahman@utexas.edu

S.M. Raiyan Kabir · A. Quadir
Department of Electrical and Electonic Engineering,
City University of London, London EC1V 0HB, UK
e-mail: Raiyan.Kabir.1@city.ac.uk

A. Quadir
e-mail: Anita.Quadir.1@city.ac.uk

© Springer International Publishing Switzerland 2016
A.I. Awad and M. Hassaballah (eds.), *Image Feature Detectors and Descriptors*,
Studies in Computational Intelligence 630, DOI 10.1007/978-3-319-28854-3_12

1   Introduction

Digital signal and image processing techniques opened the door for automation in biometric and biomedical image analysis. Biometric systems analyze images of different parts of human body to identify/distinguish different person. Biomedical image processing analyzes images in oder to diagnose different kinds of diseases. Retina is the inner part of human eye which senses the outside illumination. It has distinctive features based on the spatial arrangement of blood vessels, dark central part (fovea) and bright spot (optic nerve). These features are important for the purpose of biometric authentication and diagnosing disease (i.e. diabetes) for biomedical applications. In 1935, Dr. Carleton Simon and Dr. Isodore Goldstein published a paper where they discovered that every retina possesses a unique and different blood vessel pattern. Another study conducted by Dr. Paul Tower in 1950s showed that identical twins also have unique retina [6]. A statistical analysis was carried out on the blood vessel patterns in retinal images by Sánchez and co-workers in order to diagnose diabetic patients [18]. In another work by Bernhard and co-authors, statistical classification techniques were applied on retinal images on order to extract the feature related to an individual's diagnose for diabetes [4]. For either biomedical or biometric applications, it is required to have a reliable image acquisition method. In common practice retinal images are acquired in two major ways. One option is *Fluorescein angiogram* of the retinal image which is mainly used for biomedical purposes [26]. This method is inconvenient in biomedical purpose because it involves injecting sodium fluorescein in human body. Biometric retinal authentication systems are designed for frequent daily use. So, fluorescein angiography may not be recommended for this application either. On the other hand, *color fundus* images are popular due to easier image acquisition method by high resolution RGB digital cameras. Since no chemicals are required, this image acquisition method is more feasible for both biometric and biomedical applications [15, 21]. One caveat with the colored funds images is the challenge in detecting fovea and its boundary from an arbitrary image. The brightness and contrast in the fundus images may vary with an individual's eye. Previously few attempts were made in order to extract the foveal region [5, 19, 26]. However, it has been a difficult task to detect the fovea region and its boundary simultaneously for wide range of dataset. As mentioned earlier, fovea is a small, slightly concave area without retinal capillaries. It has no precise definition in terms of geometric parameters [5, 26]. It is the most accuracy vision zone of the retina. **Foveal Avascular Zone (FAZ)** refers to the dark area surrounding the fovea [5]. In the context of various applications it is often required to have a reasonably precise location of fovea and optic nerve in order to define a global axis of the images because different fundus images from a same person might not have translational or rotational consistency [15, 17]. Hence, in addition to estimating the *FAZ*, it's geometric centroid and shape are ought to be detected [5, 26]. This demands for a robust and computationally efficient algorithm which is the overall goal of this work.

2   Background

In a typical color fundus retinal image showed in Fig. 1, retinal image can be divided into three different and unique parts.

1. The solid brightest part of the retinal image. It is the **optic disk**. At optic disk, optic nerve joins the retina and acts as the transmission line between the brain and the eye.
2. The red non-linear fragmented lines all over the retinal image are called the **blood vessels**.
3. The center of the image is called the **fovea**. It is the darkest part of the retina.

In previous works, *fixed thresholding* technique was applied based on observation in order to determine the pixels belonging to the FAZ and the centroid of these pixels was taken as the center of fovea in [15, 17]. As the gray level of FAZ is different in different retinal images, this method was proven to be highly exclusive to specific images and hence became less accurate. The ambiguity in the color distribution is considered to be a primary reason. It is clearly seen in Fig. 1 that the obscure nature of the blood vessels around the FAZ leads to challenge in detecting this region and its center. Hence the current work addresses the problem through *fuzzy-rule* based analysis [29]. Previously, fuzzy logic based technique was applied on blood vessel segmentation [1]. However, detecting FAZ using the fuzzy classification technique needs entirely different attention. In addition, active contour model based on Gradient Vector Flow (GVF) is used to precisely detect the boundary of the fovea, using some parameters estimated from the fuzzy rule based analysis [5, 7, 16, 20, 22, 28, 29].

Fig. 1 A color fundus retinal image

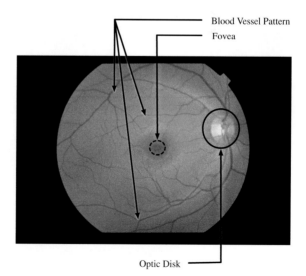

Blood Vessel Pattern

Fovea

Optic Disk

3 Preprocessing of Retinal Image

The acquired fundus images were of high resolution. So, any filtration was not applied in order to pre-process them. As it is seen in Fig. 2, optic disk in the red layer is more detectable, blood vessel pattern is clearly visible in the green layer and FAZ is significantly darker in the blue layer. Hence, based on this observation Red, Green or Blue layers were considered in further analysis in order to extract different components from the retinal images.

The optic disk is the brightest region in the retinal image. It typically occupies approximately $14 \sim 15\%$ of the entire image. The appearance of the optic disk can be characterized by a relatively rapid variation in the color intensity in red layer [19]. In this layer the disk looks like a solid oval shaped region. Blood vessels overlapping the optic disk almost disappear. Figure 3a gives a clear view of the optic disk and its surroundings in red layer and Fig. 3b shows the histogram of the grey scale values. It is observed that the median of the histogram resides within the higher color range. Based on this observation, a threshold value I_{th_1} is selected as initial guess which includes the median of the histogram and I_{th_1} remains in the higher color range. It is also observed that for all retinal images the histograms near the optic disk region are similar. The color centroid of the pixels over the threshold limit was taken based on Eq. 1 [15, 17]. The detection of optic disk was refined by considering the new threshold to be $I_{th_2} = I_{th_1} - \Delta$. Δ is the level of confidence (e.g. 90%). In Fig. 3b, there are several peaks and valleys in the histogram. In order to consider all

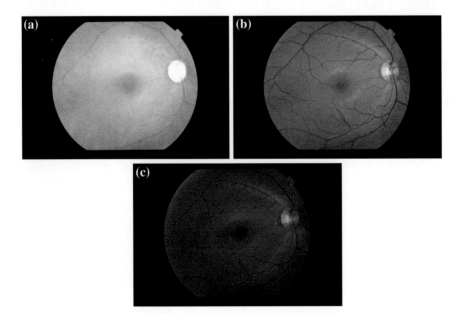

Fig. 2 **a** *Red layer* of retinal image. **b** *Green layer* of retinal image. **c** *Blue layer* of retinal image

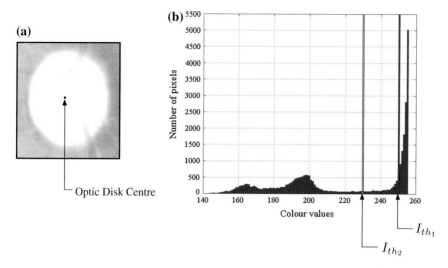

Fig. 3 **a** Optic disk in *red layer* of a retinal image, **b** The histogram of the optic disk in *red layer* with threshold values I_{th_1} and I_{th_2}

pixels corresponding to the optic disk, the value of Δ is chosen so that the threshold I_{th_2} resides between the highest peak and the peak ahead to the highest peak [11]. Afterwards, the boundary of the aperture is detected by edge detection with canny method [2]. The location for optic disc center is defined as follows:

$$x_{oc^1} = \frac{\sum M_1(x, y).x}{\sum M_1(x, y)}, \quad y_{oc^1} = \frac{\sum M_1(x, y).y}{\sum M_1(x, y)},$$

$$M_1(x, y) = \begin{cases} I(x, y) & I(x, y) \geqslant I_{th_1}. \\ 0 & otherwise \end{cases}$$

(1)

Here, $I(x, y)$ = intensity of the pixel at position (x, y) and I_{th_1} is the threshold value for step one. (x_{oc^1}, y_{oc^1}) is the coordinate position of primary optic disk center.

The foveal region (FAZ) in retina is defined as an area of circle. The geometric center of FAZ is located at $\approx 2DD$ (DD = optic disk diameter) far from the optic disk center along the line connecting the center of optic disk and the center of the fovea [9]. It is observed that the FAZ resides around the line connecting the optic disk center and its nearest point on the boundary. Figure 4a shows the straight line through the optic disk center and nearest point on the boundary around it. As the aperture of the retina is circular, farthest point from the optic disk will be on the straight line. The mid point between the nearest and farthest point from optic disk center is taken as the *initial guess for the center* of the retina. It is noticed that the FAZ remains in the nearest neighborhood zone of the initial center of the retina. Prior to applying any pattern analysis method the FAZ needs to be rotationally and translationally invariant. The translational normalization is done by moving the initial center of the

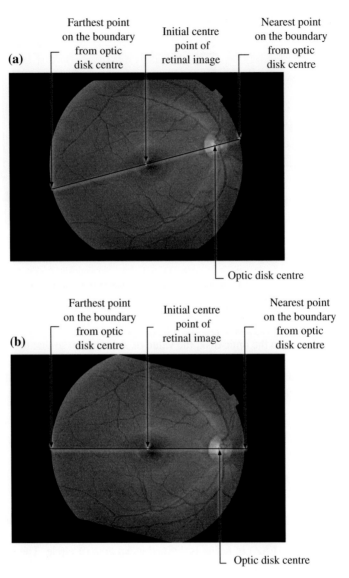

Fig. 4 a Optic disk center, nearest and farthest point on boundary from optic disk center and initial center point on the connecting *straight line*, **b** Image in the *blue layer* after translational and rotational normalization

retina on the global center of the image. The rotational normalization is performed by considering line through the initial center and the optic disk center. The image in blue layer was considered for further analysis and normalization due to the fact that FAZ is the most significant in this layer [15, 17]. Output from the normalization is shown in Fig. 4b.

3.1 Detection of Region of Interest

As mentioned in [10], the most likely geometric description for FAZ is: circular with a radius of $\approx 1DD$. Figure 5 shows the schematic diagram of the position of the candidate region for FAZ. Initially, the center of the retinal image is assumed to be coinciding on the fovea center. The distance between the optic disk center and the initially guessed center for retina was assumed to be $\approx 2DD$. The possibility of finding fovea increases with the increase in distance towards the center of the retina. The chance of detecting fovea decreases with the increase in distance from the center of the retina. As a result, an ellipse can be defined which contains the fovea. The horizontal axis of the image can be taken as the major axis. Twice the distance between the optic disk center and the initial center of retina can be taken as the length of the major axis. As the radius of the probable zone is half of the distance between the optic disk center and initial center of retina [9, 10], 50 % of length of the major axis can be considered as the length of the minor axis.

A rectangle with a length of $2DD$ can be taken as the **Region of Interest (ROI)**. Figure 5 shows the schematic diagram of the ROI within the ellipse. The width of the rectangle can be taken as the vertical line segment between the intersecting points of the ellipse and a line $1DD$ away from the initial center of retina. Taking the initial center of the retina as the origin of the global coordinate system, the ROI can be explained mathematically by Eq. 2 and Fig. 6a.

Fig. 5 Schematic position of the probable zone and the region of interest in retinal image

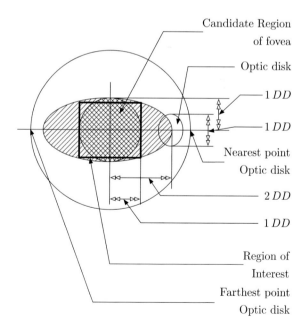

Candidate Region
of fovea

Optic disk

1 DD

1 DD

Nearest point
Optic disk

2 DD

1 DD

Region of
Interest

Farthest point
Optic disk

(a)   (b)

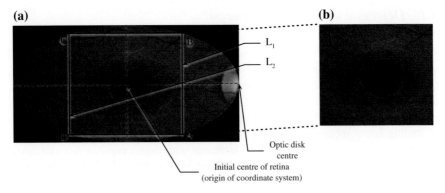

Fig. 6 **a** Elliptical section containing the fovea, **b** Rectangular section from the central part of the elliptical region

$$
\begin{aligned}
\{A, B\} &= \{(x, y)|x, y \in L_1\}, \\
\{C, D\} &= \{(x, y)|x, y \in L_2\}, \\
E &= \{(x, y)|\tfrac{x^2}{a^2} + \tfrac{y^2}{b^2} = 1\}, \\
L_1 &= \{(x, y)|x = \tfrac{w}{2}, -\infty < y < \infty\}, \\
L_2 &= \{(x, y)|x = -\tfrac{w}{2}, -\infty < y < \infty\}.
\end{aligned}
\tag{2}
$$

Here, $a = 2DD$, $b = 1DD$ and $w = 2DD$. $d(A, B)$ and $d(C, D)$ are the distances between A, B and C, D, respectively. Hence, the four vertices of the ROI rectangle are $A(w/2, h/2)$, $B(w/2, -h/2)$, $C(-w/2, -h/2)$ and $D(-w/2, h/2)$. The distance $d(A, B) = d(C, D)$ is taken as the width of the rectangular ROI. Mathematically, ROI can be defined as a set of points **ROI** with Eq. 3 and Fig. 6b.

$$
\textbf{ROI} := \left\{(x, y)| -\frac{w}{2} \leqslant x \leqslant \frac{w}{2}, -\frac{h}{2} \leqslant y \leqslant \frac{h}{2}\right\}.
\tag{3}
$$

4 Steps for Detecting the Center of the Fovea

After detecting and extracting the ROI from the retinal image, the histogram and contour plot of the ROI of different images were observed. Figure 7 shows the histogram and contour plot of the ROI of a retinal image. It is clearly seen in Fig. 7b that the FAZ is a small area and its grey scale intensity level is significantly lower than the surrounding parts in the ROI. The histograms of ROI are almost normally distributed with a mean μ and standard deviation σ (Fig. 7a). As the grey scale intensity level of FAZ is significantly lower than the mean and its area is comparatively smaller, the pixels of the FAZ reside below the mean grey scale values. Mathematically, if $X \times Y$ is a retinal image and I is the set of grey scale intensity in the image, then the intensity of $X \times Y$ can be expressed as a function in Eq. 4.

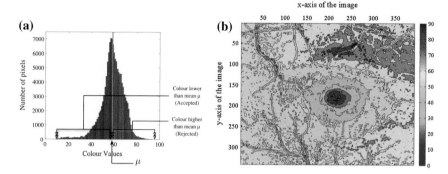

Fig. 7 a Histogram of the region of interest, **b** Filled contour plot of the region of interest **ROI**

$$I = \{i|i = 1, 2, 3 \ldots 255\},$$
$$\forall x \forall y \, \mathcal{I} : X \times Y \to I \quad \neg \exists \ \ \mathcal{I}^{-1}. \tag{4}$$

As **ROI** is a part of the retinal image, it can be expressed as **ROI** $\subset X \times Y$. So, the intensity of the *ROI* can be expressed as a function in Eq. 5.

$$\forall x \forall y \, \mathcal{I}_{ROI} : \textbf{ROI} \to I_{ROI} \quad \neg \exists \ \ \mathcal{I}_{ROI}^{-1}. \tag{5}$$

Here, $I_{ROI} \subset I$, I_{ROI} is the set of intensities in **ROI**. FAZ is a part of the ROI which is expressed in terms of **FAZ** $\subset$ **ROI**. As observed in the histogram in Fig. 7a, I_{ROI} can be divided into two intensity sets I' and I'' with Eq. 6.

$$I' = \{i \in I_{ROI} | i \leqslant \mu\},$$
$$I'' = \{i \in I_{ROI} | i \geqslant \mu\}. \tag{6}$$

Here, $I' \cup I'' = I_{ROI}$ and $I' \cap I'' = \{\phi\}$. As mentioned earlier, the points of **FAZ** have intensities lower then μ. The intensity set of the **FAZ**, $I_{FAZ} \subset I'$. So, points with intensity $i \in I'$ can be discarded and a set of points **ROI**$'$ can be defined by Eq. 7.

$$\textbf{ROI}' = \{(x, y) | i \in I'\}. \tag{7}$$

Here, $\forall x \forall y \, \mathcal{I}_{ROI'} : \textbf{ROI}' \to I' \quad \neg \exists \ \ \mathcal{I}_{ROI'}^{-1}$ and **ROI**$'$ $\subset$ **ROI**.

4.1 Clusterization of Region of Interest

In an image, all pixels are strongly correlated with its neighboring pixels due to the nonlocal nature in the color intensity level. If the neighborhood of a point is small with respect to the image resolution, the standard deviation of the grey scale intensity within that neighborhood is expected to be small. In order to reduce the computational

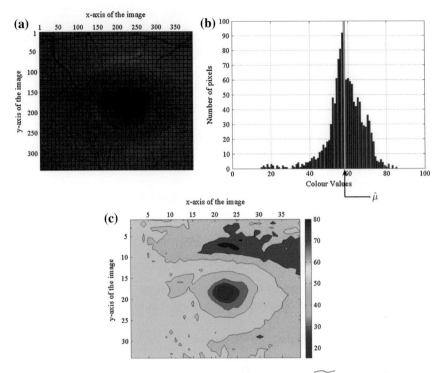

Fig. 8 **a** ROI after clusterization with 10×10 cluster, **b** Histogram of $\widetilde{\text{ROI}}$ with mean $\hat{\mu}$, **c** Contour plot of $\widetilde{\text{ROI}}$

time, the image can be divided into small clusters[1] and the mean of intensity of the clusters can be considered to be the homogenized grey scale level of an individual cluster. Hence, the image is divided into several small square clusters with dimension $\mathcal{D} = 10 \times 10$. After clusterization, **ROI** can be expressed as an equivalent class of $\mathcal{D}$ by Eq. 8 [3]. The value of each point of $\widetilde{\text{ROI}}$ can be considered based on Eq. 9. The clusters of the **ROI** and the contour plot of the $\widetilde{\text{ROI}}$ are shown in Fig. 8a and Fig. 8c, respectively.

$$\begin{aligned}
&\widetilde{\text{ROI}} := \text{ROI}/\mathcal{D}, \\
&\forall x \forall y \, \Phi : \text{ROI} \to \widetilde{\text{ROI}} \;\exists\; \Phi^{-1}.
\end{aligned} \tag{8}$$

And,

$$\begin{aligned}
&\hat{I}(m, n) = \frac{\sum_{i=1}^{N} I(x, y)}{N}, \\
&I_{\widetilde{ROI}} := \{\hat{I}(m, n) | 0 \leqslant \hat{I}(m, n) \leqslant 255\}, \\
&\forall x \forall y \, \mathcal{I}_{\widetilde{ROI}} : \widetilde{\text{ROI}} \to I_{\widetilde{ROI}} \;\neg\exists\; \mathcal{I}_{\widetilde{ROI}}^{-1}.
\end{aligned} \tag{9}$$

---

[1] Segment of an image with very small dimension.

Here, (m, n) is coordinate position of a point in $\widetilde{\mathbf{ROI}}$, N is the number of pixels in a cluster and $I(x, y)$ is the intensity of a point in $\mathbf{ROI}$.

4.2 Homogenization of the ROI

Figure 8b shows the histogram of $\widetilde{\mathbf{ROI}}$. It is observed from Fig. 7a and Fig. 8b that the distribution of the histogram of $\mathbf{ROI}$ and $\widetilde{\mathbf{ROI}}$ are similar. The mean of grey level intensity is defined by $\mu \approx \hat{\mu}$. It is seen from Fig. 7b and Fig. 8c that the intensity deviation of FAZ with respect to the surrounding spatial regions of $\widetilde{\mathbf{ROI}}$ is similar to the one from $\mathbf{ROI}$. As a result, the histogram of $\widetilde{\mathbf{ROI}}$ was divided into disjoint two sets $L'_{\widetilde{ROI}}$ and $L''_{\widetilde{ROI}}$ (Eq. 10). Hence, expectedly FAZ will reside within $\widetilde{\mathbf{ROI}}'$ which has lower intensity level than $\hat{\mu}$. Figure 9 shows the contour plot of $\widetilde{\mathbf{ROI}}'$ after discarding the pixels with higher intensity than $\hat{\mu}$.

$$
\begin{aligned}
L'_{\widetilde{ROI}} &= \{i \in I_{\widetilde{ROI}} | i \leqslant \hat{\mu}\}, \\
L''_{\widetilde{ROI}} &= \{i \in I_{\widetilde{ROI}} | i \geqslant \hat{\mu}\}, \\
\widetilde{\mathbf{ROI}}' &= \{(x, y) | i \in L'_{\widetilde{ROI}}\}.
\end{aligned}
\tag{10}
$$

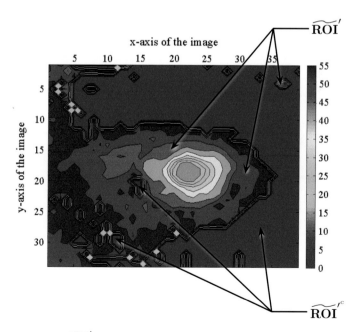

Fig. 9 Contour plot of $\widetilde{\mathbf{ROI}}'$

Here, $I'_{\widetilde{ROI}} \cup I''_{\widetilde{ROI}} = I_{\widetilde{ROI}}$, $I'_{\widetilde{ROI}} \cap I''_{\widetilde{ROI}} = \{\phi\}$, $\forall x \forall y\, \mathcal{I}_{\widetilde{ROI}} : \widetilde{\mathbf{ROI}}' \to I'_{\widetilde{ROI}} \neg\exists\, \mathcal{I}^{-1}_{\widetilde{ROI}}$
and $\widetilde{\mathbf{ROI}}' \subset \widetilde{\mathbf{ROI}}$.

4.3 Extracting the Candidate Region

According to Fig. 9, $\widetilde{\mathbf{ROI}}'$ is a multiply connected region [8]. FAZ is a simply connected region with the simply connected part $\widetilde{\mathbf{ROI}}'' \subset \widetilde{\mathbf{ROI}}'$, which contains the FAZ should be extracted as the candidate region. At first, $(\tilde{x}_c, \tilde{y}_c)$, i.e. the location of the global minima of $I_{\widetilde{ROI}}$ in $\widetilde{\mathbf{ROI}}'$ (boundary $\partial\widetilde{\mathbf{ROI}}'$) is calculated. $\widetilde{\mathbf{ROI}}'$ is scanned parallel to x-axis. As shown in Fig. 10a for $y = a$ two lines $L_{H1} \times a$ and $L_{H2} \times a$ can be found. The projection of line $L_{H1} \times a$, A_1 and line $L_{H2} \times a$, A_2 on x-axis were evaluated. If $\tilde{x}_c \in A_1$ then a set $S_{(j,y)_H}$ will be formed with $L_{H1} \times a$. Otherwise if $\tilde{x}_c \in A_2$, $S_{(j,y)_H}$ will be formed with $L_{H2} \times a$. If none of them contain $\tilde{x}_c$, $S_{(j,y)_H} = \{\phi\}$ i.e. both $L_{H1} \times a$ and $L_{H2} \times a$ will be rejected. According to Fig. 10a, $\tilde{x}_c \in L_{H2} \times a$ leads to

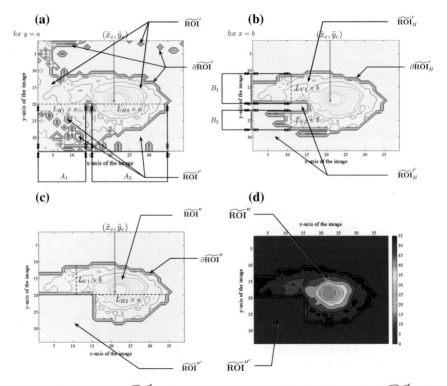

Fig. 10 **a** Contour plot of $\widetilde{\mathbf{ROI}}'$ before the algorithm was applied, **b** Contour plot of $\widetilde{\mathbf{ROI}}'_H$ after application of the first part of the algorithm, **c** Selection of the horizontal and vertical line containing $(\tilde{x}_c, \tilde{y}_c)$, **d** Contour plot of $\widetilde{\mathbf{ROI}}''$ after application of the complete algorithm

Fig. 11 Four circles
containing the FAZ in **ROI**

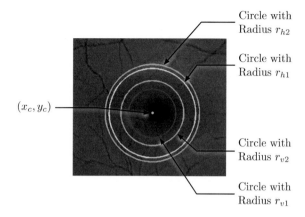

Circle with
Radius r_{h2}

Circle with
Radius r_{h1}

(x_c, y_c)

Circle with
Radius r_{v2}

Circle with
Radius r_{v1}

$S_{(2,a)_H} = \{L_{H2} \times a\}$. Figure 10c shows selected $L_{H2} \times a$ in $\widetilde{\mathbf{ROI}}''$. The process mentioned above is applied for all $y \in Y$. $\widetilde{\mathbf{ROI}}'_H$ (Fig. 10b) is formed with the product of all $S_{(j,y)_H}$. Now, $\widetilde{\mathbf{ROI}}'_H$ (boundary $\partial \widetilde{\mathbf{ROI}}'_H$) is scanned parallel to y-axis. As seen in Fig. 10b, for $x = b$, the projection of line $L_{V1} \times b$ and $L_{V2} \times b$ on the y-axis B_1 and B_2, respectively were considered. If $\tilde{y}_c \in B_1$ then a set $S_{(j,x)_V}$ will be formed with $L_{V1} \times b$. Otherwise if $\tilde{y}_c \in B_2$, $S_{(j,x)_V}$ will be formed with $L_{V2} \times b$. If none of them contain $\tilde{y}_c$, $S_{(j,x)_V} = \{\phi\}$ i.e. both $L_{V1} \times b$ and $L_{V2} \times b$ will be rejected. In Fig. 10a, $\tilde{y}_c \in L_{V1} \times b$. As a result, $S_{(1,b)_V} = \{L_{V1} \times a\}$. Figure 10c shows selected $L_{V1} \times b$ in $\widetilde{\mathbf{ROI}}''$. The process mentioned above is applied for all $x \in X$. The $\widetilde{\mathbf{ROI}}''$ (Fig. 10c) is formed with the product of all $S_{(j,x)_V}$. Figure 10d shows the finally extracted $\widetilde{\mathbf{ROI}}''$.

4.4 Determination of Initial Foveal Radius

As mentioned in Sects. 1 and 3.1, the FAZ can be defined as a slightly concave and circular zone which was show in Fig. 12a. According to Figs. 10b and c, $\widetilde{\mathbf{ROI}}''$ is an irregular shaped region. In order to define this region a group of radii from concentric circles containing FAZ and the grey scale intensity distribution of the $\widetilde{\mathbf{ROI}}$ can be observed (Fig. 11). FAZ certainly resides within the concave part of intensity distribution of $\widetilde{\mathbf{ROI}}$ (Fig. 12a). As $(\tilde{x}_c, \tilde{y}_c)$ stays almost at the concave part (Fig. 12b), a set of lines thought $(\tilde{x}_c, \tilde{y}_c)$ can be considered. In order reduce computational time, only the vertical and horizontal lines through $(\tilde{x}_c, \tilde{y}_c)$ are evaluated. Initial set of radii for FAZ is referred as $\widetilde{\mathcal{R}}_{if} = \{\tilde{r}_i | \tilde{r}_i \in \mathbb{R}, i \in \mathbb{N}\}$ where, $\tilde{r}_i = \sqrt{(\tilde{x}_c - \tilde{x})^2 + (\tilde{y}_c - \tilde{y})^2}$, $\tilde{x}, \tilde{y} \in \widetilde{\mathbf{ROI}}''$, $\mathbb{R}$ is the set of real numbers and $\mathbb{N}$ is the set of natural numbers. As the vertical and horizontal lines were considered, two sets $\widetilde{\mathcal{R}}_{if_v} = \{\tilde{r}_{vi} | \tilde{r}_{vi} \in \mathbb{R}, i \in \{1, 2\}\}$ and $\widetilde{\mathcal{R}}_{if_h} = \{\tilde{r}_{hi} | \tilde{r}_{hi} \in \mathbb{R}, i \in \{1, 2\}\}$ were obtained from observing the concave behavior of the intensity along the lines $L_{Vi} \times \tilde{x}_c$ and $L_{Hi} \times \tilde{y}_c$. According to Fig. 12c,

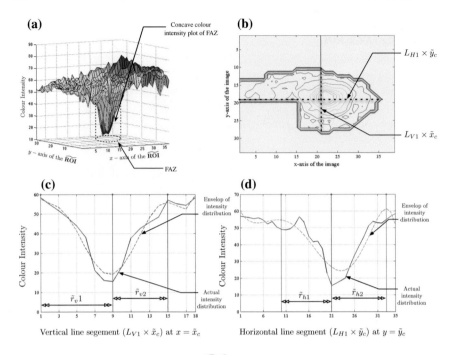

Fig. 12 **a** 3-D surface plot of intensity of $\widetilde{\mathrm{ROI}}$, **b** Central horizontal and vertical lines $L_{V1} \times \tilde{x}_c$ and $L_{H1} \times \tilde{y}_c$ on $\widetilde{\mathrm{ROI}}''$, **c** intensity plot of $L_{V1} \times \tilde{x}_c$ and its polynomial envelop, **d** intensity plot of $L_{H1} \times \tilde{y}_c$ and its polynomial envelop

d the intensity increases monotonically with the increase in distance from $(\tilde{x}_c, \tilde{y}_c)$. The intensity of the $\widetilde{\mathrm{ROI}}''$ is defined as Eq. 11.

$$\forall x \forall y \, \mathcal{I}_{\widetilde{ROI}''} : \widetilde{\mathrm{ROI}}'' \rightarrow I_{\widetilde{ROI}''} \quad \neg \exists \quad \mathcal{I}_{\widetilde{ROI}''}^{-1}. \tag{11}$$

The relation between the intensity function and the distances from $\tilde{y}_c$ and $\tilde{x}_c$ along the lines $L_{Vi} \times \tilde{x}_c$ and $L_{Hi} \times \tilde{y}_c$ can be defined by Eq. 12 and Eq. 13 respectively

$$\begin{aligned} \mathbf{d_v} &= \{|\tilde{y}_c - y| : \tilde{r}_{v1} \leqslant y \leqslant \tilde{r}_{v2}\}, \\ \mathbf{d_v} &\propto (I_{(L_{Vi} \times \tilde{x}_c)})^{n_1}. \end{aligned} \tag{12}$$

Here, $I_{(L_{Vi} \times \tilde{x}_c)} \subset I_{\widetilde{ROI}''}$, $L_{Vi} \times \tilde{x}_c \subset \widetilde{\mathrm{ROI}}''$ and $n_1 \in \mathbb{R}$.

$$\begin{aligned} \mathbf{d_h} &= \{|\tilde{x}_c - x| : \tilde{r}_{h1} \leqslant x \leqslant \tilde{r}_{h2}\}, \\ \mathbf{d_h} &\propto (I_{(L_{Hi} \times \tilde{y}_c)})^{n_2}. \end{aligned} \tag{13}$$

Here, $I_{(L_{Hi} \times \tilde{y}_c)} \subset I_{\widetilde{ROI}''}$, $L_{Hi} \times \tilde{y}_c \subset \widetilde{\mathrm{ROI}}''$ and $n_2 \in \mathbb{R}$.

It was seen in Fig. 12c, d the actual intensity distribution is highly erratic in nature. For this reason, the intensity distribution is non-differentiable. In order to obtain a differentiable envelop, it can be approximated with a polynomial which sustains the fundamental profile of intensity. The distances from $\tilde{y}_c$ and $\tilde{x}_c$ are proportional to a higher order of the intensities within the initial radii (Eqs. 12 and 13) and the intensity is monotonically increasing. The two maxima m_{v1}, m_{v2} and m_{h1}, m_{h2} respectively in each polynomial surrounding the global minima are traced which leads to find out the radii. The initial radii are chosen with Eq. 14.

$$
\begin{aligned}
\widetilde{\mathcal{R}}_{if_v} &= \{\tilde{r}_{vi} = |\tilde{y}_c - \tilde{y}_{m_{vi}}| : \tilde{r}_{vi} \in \mathbb{R}, \ i \in \{1, 2\}\}, \\
\widetilde{\mathcal{R}}_{if_h} &= \{\tilde{r}_{hi} = |\tilde{x}_c - \tilde{x}_{m_{hi}}| : \tilde{r}_{hi} \in \mathbb{R}, \ i \in \{1, 2\}\}.
\end{aligned}
\tag{14}
$$

Here, $(\tilde{x}_c, \tilde{y}_{m_{vi}})$ is the position of m_{vi} and $(\tilde{x}_{m_{hi}}, \tilde{y}_c)$ is the position of m_{hi} in $\widetilde{\mathbf{ROI}}''$. $\widetilde{\mathcal{R}}_{if_v}$ and $\widetilde{\mathcal{R}}_{if_h}$ contains the radii in $\widetilde{\mathbf{ROI}}''$. The location of points $(\tilde{x}_c, \tilde{y}_{m_{v1}})$, $(\tilde{x}_c, \tilde{y}_{m_{v2}})$, $(\tilde{x}_{m_{h1}}, \tilde{y}_c)$, $(\tilde{x}_{m_{h2}}, \tilde{y}_c)$ and $(\tilde{x}_c, \tilde{y}_c)$ in $\mathbf{ROI}$ was calculated through the inverse mapping given in Eq. 8. Equation 15 shows the location of the points in $\mathbf{ROI}$.

$$
\begin{aligned}
\Phi^{-1}((\tilde{x}_c, \tilde{y}_{m_{v1}})) &= (x_c, y_{m_{v1}}), \\
\Phi^{-1}((\tilde{x}_c, \tilde{y}_{m_{v2}})) &= (x_c, y_{m_{v2}}), \\
\Phi^{-1}((\tilde{x}_{m_{h1}}, \tilde{y}_c)) &= (x_{m_{h1}}, y_c), \\
\Phi^{-1}((\tilde{x}_{m_{h2}}, \tilde{y}_c)) &= (x_{m_{h2}}, y_c), \\
\Phi^{-1}((\tilde{x}_c, \tilde{y}_c)) &= (x_c, y_c).
\end{aligned}
\tag{15}
$$

The radii sets $\mathcal{R}_{if_v}$ and $\mathcal{R}_{if_h}$ in $\mathbf{ROI}$ were defined with Eq. 16.

$$
\begin{aligned}
\mathcal{R}_{if_v} &:= \{r_{vi} = |y_c - y_{m_{vi}}| : r_{vi} \in \mathbb{R}, \ i \in \{1, 2\}\}, \\
\mathcal{R}_{if_h} &:= \{r_{hi} = |x_c - x_{m_{hi}}| : r_{hi} \in \mathbb{R}, \ i \in \{1, 2\}\}.
\end{aligned}
\tag{16}
$$

A combined set of radii $\mathcal{R}_{if_{vh}} = \mathcal{R}_{if_v} \cup \mathcal{R}_{if_h}$ has been defined. Figure 11 shows four circles with the radii from $\mathcal{R}_{if_{vh}}$ in $\mathbf{ROI}$. After sorting the elements of $\mathcal{R}_{if_{vh}}$, a cardinally equivalent set with monotonically increasing order $\mathcal{R}_{if_{vhs}}$ was with Eq. 17.

$$
\begin{aligned}
f &: \mathcal{R}_{if_{vh}} \to \mathcal{R}_{if_{vhs}}, \\
\mathcal{R}_{if_{vhs}} &= \{r_i | r_i < r_{i+1}, \ i \in \{1, 2, 3, 4\}\}, \\
\mathcal{R}_{if_{vh}} \setminus \mathcal{R}_{if_{vhs}} &= \{\phi\}.
\end{aligned}
\tag{17}
$$

4.5   Fuzzy-Rule Based Scheme and Extraction of the Fovea Center

It was mentioned earlier that the intensity level and boundary of the FAZ vary in every retinal image. Therefore, the threshold should be selected according to the intensity distribution of FAZ and its surrounding. As it is seen in Fig. 11, the FAZ is obscure region which does not have well defined edge encapsulating the FAZ region [5]. In order to overcome the difficulty a *fuzzy-rule* based segmentation technique was incorporated. Fuzzy set theory and fuzzy logic provide powerful tools to manage this type of issues [22]. From thorough observation, two linguistic variables can be found.

1. Distance from the center
2. Intensity of the pixel

The possibility of a pixel to be a member of the FAZ decreases with the increase in distance from the center of fovea. The intensity is darker in the FAZ than in its neighborhood. The possibility of a pixel in order to be a member of the FAZ is also dependent on its grey scale intensity. Darker pixels are more likely to be a member of the FAZ than brighter pixels. As a result, a rule based fuzzy system was designed in order to analyze and detect the center of the FAZ [22]. Fuzzy systems use normalized versions of membership functions. There are many types of fuzzy membership functions such as, triangles, trapezoids, bell curves, Gaussian, Sigmoidal functions etc [7]. Due to simple formulas and computational efficiency, both triangular and trapezoidal membership functions are popular in different fuzzy logic based applications. In the current work *trapezoidal membership functions* were applied in developing the fuzzification rule.

4.5.1 Classifcation Based on Radial Distance

Prior to applying fuzzy analysis **ROI** was divided into two disjoint regions based on Eq. 18.

$$\mathbf{ROI}_1 := \{(x, y) | (x - x_c)^2 + (y - y_c)^2 \leqslant r_4\},$$
$$r_4 \in \mathcal{R}_{if_{vhs}},$$
$$\mathbf{ROI}_1^c := \mathbf{ROI} \backslash \mathbf{ROI}_1. \tag{18}$$

Due to the possibility of a point $(x, y) \in \mathbf{ROI}_1^c \subset FAZ$ is trivial, $\mathbf{ROI}_1^c$ was discarded for further analysis and treated as a background. The four radii in $\mathcal{R}_{if_{vhs}}$, $\mathbf{ROI}_1$ were divided into four different regions: central region C, first annular region A_1, second annular region A_2 and third annular region A_3 (Eq. 19). Their intensity sets I_C, I_{A_1}, I_{A_2} and I_{A_3} respectively are defined by Eq. 20. Figure 13 shows, all four regions of $\mathbf{ROI}_1$.

$$C = \{(x, y) | (x - x_c)^2 + (y - y_c)^2 \leqslant r_1^2\}, A_1 = \{(x, y) | r_1^2 < (x - x_c)^2 + (y - y_c)^2 \leqslant r_2^2\},$$
$$A_2 = \{(x, y) | r_2^2 < (x - x_c)^2 + (y - y_c)^2 \leqslant r_3^2\}, A_3 = \{(x, y) | r_3^2 < (x - x_c)^2 + (y - y_c)^2 \leqslant r_4^2\}. \tag{19}$$

$$\forall x \forall y \, \mathcal{I}_C : C \to I_C \quad \neg\exists \ \mathcal{I}_C^{-1}, \forall x \forall y \, \mathcal{I}_{A_1} : A_1 \to I_{A_1} \quad \neg\exists \ \mathcal{I}_{A_1}^{-1},$$
$$\forall x \forall y \, \mathcal{I}_{A_2} : A_2 \to I_{A_2} \quad \neg\exists \ \mathcal{I}_{A_2}^{-1}, \forall x \forall y \, \mathcal{I}_{A_3} : A_3 \to I_{A_3} \quad \neg\exists \ \mathcal{I}_{A_3}^{-1}. \tag{20}$$

Fig. 13 Four circular regions C, A_1, A_2 and A_3 of **ROI**$_1$ with **ROI**$_1^c$

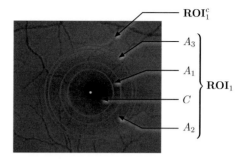

4.5.2 Fuzzification of the Boundaries of Radial Classes

All the four circles in Fig. 11 are co-centric and the center is (x_c, y_c) which is the inverse mapping of $(\tilde{x}_c, \tilde{y}_c)$ in **ROI** (Eq. 15). (x_c, y_c) is not the actual center of the fovea. As the linguistic variables are related to the distance from the real center of fovea, a region rather than a point should be given higher priority to be in a specific class. Again, there is an uncertainty of a point to be in a specific region if it is residing in the neighborhood of a common boundary of two regions. Hence, the trapezoidal membership function is considered. Radii of all points of **ROI**$_1$ can be calculated and mapped to a set of radii: $\mathcal{R}_{ROI_1}$ with Eq. 21. $\mathcal{R}_{ROI_1}$ was mapped to a fuzzy set using the membership function $\mu_{\mathbb{X}_i}$ in Eq. 22 [29].

$$\mathcal{R}_{ROI_1} := \left\{ \begin{array}{l} r_{ROI_1} | r_{ROI_1} = \sqrt{(x - x_c)^2 + (y - y_c)^2}, \\ (x, y) \in \textbf{ROI}_1. \end{array} \right\} \tag{21}$$

$$\mu_{\mathbb{X}_i} : \mathcal{R}_{ROI_1} \to [0, 1]; \quad \mathbb{X}_1 = C_{(fuzzy)}, \quad \mathbb{X}_2 = A_{1(fuzzy)},$$
$$\mathbb{X}_3 = A_{2(fuzzy)}, \quad \mathbb{X}_4 = A_{3(fuzzy)}, \quad i \in \{1, 2, 3, 4\}$$

$$\mu_{C_{(fuzzy)}} := \begin{cases} 1, & 0 \leqslant r \leqslant r_1 - \gamma r_1, \\ 1 - \frac{r - r_1 + \gamma r_1}{\gamma r_2}, & r_1 - \gamma r_1 < r \leqslant r_1 + \gamma(r_2 - r_1), \\ 0, & r > r_1 + \gamma(r_2 - r_1), \end{cases}$$

$$\mu_{A_{1(fuzzy)}} := \begin{cases} 0, & r \leqslant r_1 + \gamma(r_2 - r_1), \\ \frac{r - r_1 + \gamma r_1}{\gamma r_2}, & r_1 - \gamma r_1 < r \leqslant r_1 + \gamma(r_2 - r_1), \\ 1, & r_1 + \gamma(r_2 - r_1) < r \leqslant r_2 - \gamma(r_2 - r_1), \\ 1 - \frac{r - r_2 + \gamma(r_2 - r_1)}{\gamma(r_3 - r_1)}, & r_2 - \gamma(r_2 - r_1) < r \leqslant r_2 + \gamma(r_3 - r_2), \\ 0, & r > r_2 + \gamma(r_3 - r_2), \end{cases} \tag{22}$$

$$\mu_{A_{2(fuzzy)}} := \begin{cases} 0, & r \leqslant r_2 + \gamma(r_3 - r_2), \\ \frac{r - r_2 + \gamma(r_2 - r_1)}{\gamma(r_3 - r_1)}, & r_2 - \gamma(r_2 - r_1) < r \leqslant r_2 + \gamma(r_3 - r_2), \\ 1, & r_2 + \gamma(r_3 - r_2) < r \leqslant r_3 - \gamma(r_3 - r_2), \\ 1 - \frac{r - r_3 + \gamma(r_3 - r_2)}{\gamma(r_4 - r_2)}, & r_3 - \gamma(r_3 - r_2) < r \leqslant r_3 + \gamma(r_4 - r_3), \\ 0, & r > r_3 + \gamma(r_4 - r_3), \end{cases}$$

Fig. 14 Membership grade
profile for the fuzzification
of C, A_1, A_2 and A_3 based on
radial distance from (x_c, y_c)

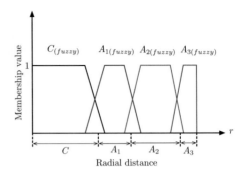

$$
\mu_{A_{3(fuzzy)}} := \begin{cases} 0, & r \leqslant r_3 + \gamma(r_4 - r_3), \\ \frac{r - r_3 + \gamma(r_3 - r_2)}{\gamma(r_4 - r_2)}, & r_3 - \gamma(r_3 - r_2) < r \leqslant r_3 + \gamma(r_4 - r_3), \\ 1, & r \leqslant r_4. \end{cases}
$$

Here, γ is the overlapping ratio, r is the radial distance from (x_c, y_c) and $r_1, r_2, r_3, r_4 \in \mathcal{R}_{if_{vhs}}$.

Using membership functions of Eq. 22 four crisp sets from Eq. 19 were fuzzyfied and four fuzzy sets $C_{(fuzzy)}$, $A_{1(fuzzy)}$, $A_{2(fuzzy)}$ and $A_{3(fuzzy)}$ were deduced. Figure 14 shows the membership profile of the fuzzy regions of $\mathcal{R}_{ROI_1}$ based on radial distance. If $\gamma = 0$, $C_{(fuzzy)}$, $A_{1(fuzzy)}$, $A_{2(fuzzy)}$ and $A_{3(fuzzy)}$ will be converted into crisp sets of Eq. 19.

4.5.3 Fuzzification Based on Intensity

The intensity of pixels belong to **ROI**$_1$ can be classified into three intensity regions dark, gray and bright. As a result, the set of intensity of **ROI**$_1$ can be defined with Eq. 23.

$$
\forall x \forall y \, \mathcal{I}_{ROI_1} : \mathbf{ROI}_1 \rightarrow I_{ROI_1} \quad \neg \exists \, \mathcal{I}_{ROI_1},
$$
$$
I_{ROI_1} = I_D \cup I_G \cup I_B. \tag{23}
$$

Here, I_D, I_G and I_B are the sets of intensities of dark, gray and bright regions respectively. The maximum and minimum intensity of **ROI**$_1$, $\max\{I_{ROI_1}\}$ and $\min\{I_{ROI_1}\}$ respectively, were obtained from I_{ROI_1}. $\max\{I_{ROI_1}\}$ and $\min\{I_{ROI_1}\}$ were defined as the supremum of I_B and infimum of I_D, respectively. The supremum of I_D, infimum and supremum of I_G and infimum of I_B are not defined. As most of the members of set C will be in the FAZ, initially the supremum of I_D and infimum of I_G can be taken as $\sup I_D = \inf I_G = \bar{I}_C$, where, $\bar{I}_C$ is the weighted average of I_C which can be calculated with Eq. 24.

$$
\bar{I}_C = \frac{\sum_{j=1}^{|I_C|} f_j i_{Cj}}{\sum_{j=1}^{|I_C|} f_j}. \tag{24}
$$

Here, f_j is the frequency of intensity i_{Cj} in C and $i_{Cj} \in I_C$. The intensity of **ROI**$_1$ increases with radial distance from the center of fovea. As a result, the supremum of I_G and infimum of I_B should be within A_1 and A_2. Initially, the supremum of I_G and infimum of I_B is talken as $\sup I_G = \inf I_B = \bar{I}_{A_1 \cup A_2}$, where, $\bar{I}_{A_1 \cup A_2}$ is the weighted average of $I_{A_1} \cup I_{A_2}$ which was calculated with Eq. 25.

$$\bar{I}_{A_1 \cup A_2} = \frac{\sum_{j=1}^{|A|} f_j i_{Aj}}{\sum_{j=1}^{|A|} f_j},$$
$$\mathbb{A} = A_1 \cup A_2. \tag{25}$$

Here, f_j is the frequency of intensity i_{Aj} in $\mathbb{A}$ and $i_{Aj} \in I_\mathbb{A}$. As $\sup I_D$, $\inf I_G$, $\sup I_G$ and $\inf I_B$ are not known a priori, I_D, I_G and I_B should be considered to be fuzzy sets instead of crisp sets. So, I_{ROI_1} can be fuzzyfied using the membership function $\mu_{\mathbb{Y}_i}$ defined in Eq. 26. After fuzzification of I_{ROI_1}, three fuzzy sets $I_{D(fuzzy)}$, $I_{G(fuzzy)}$ and $I_{B(fuzzy)}$ were obtained. Figure 15 shows the membership grade profile of the fuzzy regions of I_{ROI_1} based on intensity.

$$\mu_{\mathbb{Y}_i} : I_{ROI_1} \rightarrow [0, 1]; \quad \mathbb{Y}_1 = I_{D(fuzzy)}, \ \mathbb{Y}_2 = I_{G(fuzzy)},$$
$$\mathbb{Y}_3 = I_{B(fuzzy)}, \ i \in \{1, 2, 3\},$$

$$\mu_{I_{D(fuzzy)}} := \begin{cases} 1, & \min\{I_{ROI_1}\} \leqslant i \leqslant \bar{I}_C \\ & -\gamma(\bar{I}_C - \min\{I_{ROI_1}\}), \\ 1 - \dfrac{i - \bar{I}_C + \gamma(\bar{I}_C - \min\{I_{ROI_1}\})}{\gamma(\bar{I}_{A_1 \cup A_2} - \min\{I_{ROI_1}\})}, & \bar{I}_C - \gamma(\bar{I}_C - \min\{I_{ROI_1}\}) \\ & < i \leqslant \bar{I}_C + \gamma(\bar{I}_{A_1 \cup A_2} - \bar{I}_C), \\ 0, & i > \bar{I}_C + \gamma(\bar{I}_{A_1 \cup A_2} - \bar{I}_C), \end{cases}$$

$$\mu_{I_{G(fuzzy)}} := \begin{cases} 0, & i \leqslant \bar{I}_C - \gamma(\bar{I}_C - \min\{I_{ROI_1}\}), \\[2mm] \dfrac{i - \bar{I}_C + \gamma(\bar{I}_C - \min\{I_{ROI_1}\})}{\gamma(\bar{I}_{A_1 \cup A_2} - \min\{I_{ROI_1}\})}, & \bar{I}_C - \gamma(\bar{I}_C - \min\{I_{ROI_1}\}) \\ & < i \leqslant \bar{I}_C + \gamma(\bar{I}_{A_1 \cup A_2} - \bar{I}_C), \\[2mm] 1, & \bar{I}_C + \gamma(\bar{I}_{A_1 \cup A_2} - \bar{I}_C) \\ & < i \leqslant \bar{I}_{A_1 \cup A_2} \\ & -\gamma(\bar{I}_{A_1 \cup A_2} - \bar{I}_C), \\[2mm] 1 - \dfrac{i - \bar{I}_{A_1 \cup A_2} - \gamma(\bar{I}_{A_1 \cup A_2} - \bar{I}_C)}{\gamma(\max\{I_{ROI_1}\} - \bar{I}_C)}, & \bar{I}_{A_1 \cup A_2} - \gamma(\bar{I}_{A_1 \cup A_2} - \bar{I}_C) \\ & < i \leqslant \bar{I}_{A_1 \cup A_2} \\ & +\gamma(\max\{I_{ROI_1}\} - \bar{I}_{A_1 \cup A_2}), \\[2mm] 0, & i > \bar{I}_{A_1 \cup A_2} \\ & +\gamma(\max\{I_{ROI_1}\} - \bar{I}_{A_1 \cup A_2}), \end{cases} \tag{26}$$

Fig. 15 Membership grade
profile for fuzzification of I_D,
I_G and I_B based on intensity

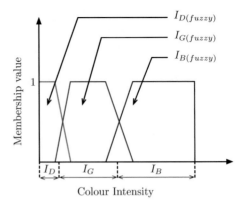

$$
\mu_{I_{B(fuzzy)}} := \begin{cases}
0, & i \leqslant \bar{I}_{A_1 \cup A_2} \\
& + \gamma(\max\{I_{ROI_1}\} - \bar{I}_{A_1 \cup A_2}), \\[2ex]
\dfrac{i - \bar{I}_{A_1 \cup A_2} - \gamma(\bar{I}_{A_1 \cup A_2} - \bar{I}_C)}{\gamma(\max\{I_{ROI_1}\} - \bar{I}_C)}, & \bar{I}_{A_1 \cup A_2} - \gamma(\bar{I}_{A_1 \cup A_2} - \bar{I}_C) \\
& < i \leqslant \bar{I}_{A_1 \cup A_2} \\
& + \gamma(\max\{I_{ROI_1}\} - \bar{I}_{A_1 \cup A_2}), \\[2ex]
1, & \bar{I}_{A_1 \cup A_2} \\
& + \gamma(\max\{I_{ROI_1}\} - \bar{I}_{A_1 \cup A_2}) \\
& < i \leqslant \max\{I_{ROI_1}\}.
\end{cases}
$$

Here, γ is the overlap ratio and $i \in I_{ROI_1}$.

4.5.4 Fuzzy Rule Based Classification

ROI$_1$ was fuzzyfied according to Eq. 22 and Eq. 26 based on two linguistic variables;
radial distance and intensity, respectively. As the fovea is the darkest region in the
neighborhood of (x_c, y_c), these two linguistic variables should be combined together
in order to detect the pixels of fovea. So, the linguistic variables, fuzzy rules should
be extracted and applied on **ROI**$_1$. During classifying foveal and non-foveal pixels,
a region based classification should be carried out [20]. The classified **ROI**$_1$ was
expressed with **ROI**$_{1(classified)}$ with a bijective mapping given in Eq. 27.

$$
\forall x \forall y \, \exists \zeta : \mathbf{ROI}_1 \rightarrow \mathbf{ROI}_{1(classified)}. \tag{27}
$$

In Sect. 4.4, the four radii of **ROI**$_1$ were selected according to Eq. 16 and Eq. 15;
where, radii were the distances between $(\tilde{x}_c, \tilde{y}_c)$ and maximum points on the polynomial envelop of the horizontal and vertical lines through $(\tilde{x}_c, \tilde{y}_c)$. As a result, FAZ

should be residing within C. The boundary of FAZ remains inside the inflectional region of the polynomial envelop (Fig. 12). It is observed that for all retinal images, $r_1 > \frac{r_4}{3}$ and FAZ mostly stays within the circle with radius $r_4/3$. As seen in Fig. 12, the intensity pattern increases with the distance to a certain high intensity level. So, $\mathbf{ROI}_1$ was divided into three circular regions based on the distance and the grey scale intensity. Using these information we can divide the $\mathbf{ROI}_1$ into following three regions:

Fovea: Fovea is the region nearest to (x_c, y_c) with lowest intensity.

Semi Fovea: Semi fovea is the region surrounding the foveal region. Its intensity exists within the inflection region of the intensity distribution of $\mathbf{ROI}_1$ (Fig. 12).

Outer Fovea: Outer Fovea is the outer most region of the $\mathbf{ROI}_1$ surrounding both the semi foveal and foveal region. The intensity of the outer fovea is highest in $\mathbf{ROI}_1$.

$$\mathbf{ROI}_1 = \mathbf{F} \cup \mathbf{SF} \cup \mathbf{OF}. \tag{28}$$

Mathematically, the regions can be expressed as Eq. 28. Here, $\mathbf{F}$, $\mathbf{SF}$ and $\mathbf{OF}$ are the sets for Fovea, Semi Fovea and Outer Fovea, respectively. There is no well defined boundary of these regions. Initially, they can be defined with Eq. 29. Figure 16 shows the three initial regions.

$$
\begin{aligned}
\mathbf{F} &= \{(x, y) | r \leqslant (r_4/3)\}, \\
\mathbf{SF} &= \{(x, y) | (r_4/3) < r \leqslant (2r_4/3)\}, \\
\mathbf{OF} &= \{(x, y) | (2r_4/3) < r \leqslant r_4\},
\end{aligned}
\tag{29}
$$

Here, $r = \sqrt{(x - x_c)^2 + (y - y_c)^2}$. As the boundaries among $\mathbf{F}$, $\mathbf{SF}$ and $\mathbf{OF}$ are not determined based on any prior information from $\mathbf{ROI}_1$, these boundaries in between two sets may overlap. Mathematically, $\sup \mathbf{F} \cap \inf \mathbf{SF} \neq \{\phi\}$ and $\sup \mathbf{SF} \cap \inf \mathbf{OF} \neq \{\phi\}$. So, $\mathbf{F}$, $\mathbf{SF}$ and $\mathbf{OF}$ should be treated as fuzzy sets $\mathbf{F}_{(fuzzy)}$, $\mathbf{SF}_{(fuzzy)}$ and $\mathbf{OF}_{(fuzzy)}$ with the membership function $\mu_{\mathbb{D}_i}$ (Eq. 30). Figure 17 shows the membership grade profile of three destination fuzzy sets.

Fig. 16 Initial $\mathbf{F}$, $\mathbf{SF}$ and $\mathbf{OF}$ on $\mathbf{ROI}$

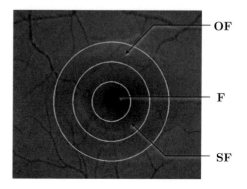

Fig. 17 Membership grade profile for fuzzification of destination sets **F**, **SF** and **OF**

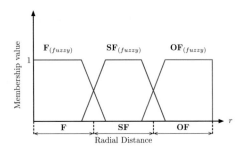

$$\mu_{\mathbb{D}_i} : \mathbf{ROI}_1 \to [0, 1]; \quad \mathbb{D}_1 = \mathbf{F}_{(fuzzy)}, \; \mathbb{D}_2 = \mathbf{SF}_{(fuzzy)}, \; \mathbb{D}_3 = \mathbf{OF}_{(fuzzy)}, \; i \in \{1, 2, 3\},$$

$$\mu_{\mathbf{F}_{(fuzzy)}} := \begin{cases} 1, & 0 \leqslant r \leqslant \frac{r_4}{3} - \gamma\frac{r_4}{3}, \\ 1 - \frac{r - \frac{r_4}{3} + \gamma\frac{r_4}{3}}{2\gamma\frac{r_4}{3}}, & \frac{r_4}{3} - \gamma\frac{r_4}{3} \leqslant r \leqslant \frac{r_4}{3} + \gamma\frac{r_4}{3}, \\ 0, & r > \frac{r_4}{3} + \gamma\frac{r_4}{3}. \end{cases}$$

$$\mu_{\mathbf{SF}_{(fuzzy)}} := \begin{cases} 0, & r \leqslant \frac{r_4}{3} + \gamma\frac{r_4}{3}, \\ \frac{r - \frac{r_4}{3} + \gamma\frac{r_4}{3}}{2\gamma\frac{r_4}{3}}, & \frac{r_4}{3} - \gamma\frac{r_4}{3} \leqslant r \leqslant \frac{r_4}{3} + \gamma\frac{r_4}{3}, \\ 1, & \frac{r_4}{3} + \gamma\frac{r_4}{3} < r \leqslant \frac{2r_4}{3} - \gamma\frac{r_4}{3}, \\ 1 - \frac{r - \frac{2r_4}{3} + \gamma\frac{r_4}{3}}{2\gamma\frac{r_4}{3}}, & \frac{2r_4}{3} - \gamma\frac{r_4}{3} < r \leqslant \frac{2r_4}{3} + \gamma\frac{r_4}{3}, \\ 0, & r > \frac{2r_4}{3} + \gamma\frac{r_4}{3}, \end{cases}$$ (30)

$$\mu_{\mathbf{OF}_{(fuzzy)}} := \begin{cases} 0, & r \leqslant \frac{2r_4}{3} + \gamma\frac{r_4}{3}, \\ \frac{r - \frac{2r_4}{3} + \gamma\frac{r_4}{3}}{2\gamma\frac{r_4}{3}}, & \frac{2r_4}{3} - \gamma\frac{r_4}{3} < r \leqslant \frac{2r_4}{3} + \gamma\frac{r_4}{3}, \\ 1, & \frac{2r_4}{3} + \gamma\frac{r_4}{3} < r \leqslant r_4. \end{cases}$$

Here, γ is the overlap ratio.

Foveal pixels should be closer to the center of the fovea and the intensity should be lower than any other parts of the retinal image. As (x_c, y_c) is not at the center of fovea, rather, it should stay inside the neighborhood region. There is a possibility of a pixel to be a member of **F**, which actually resides in **SF** and vice versa. Similar possibility exists for a pixel in **OF** to be a member of **SF** and vice versa. The uncertainty of a pixel to be a member of a single fuzzy set can be eliminated by applying a fuzzy *if-then* rule based classification algorithm. As the classification of pixels can be based on both the radial distance and intensity, fuzzy inference given in Table 1 were applied.

The above mentioned algorithm contains following steps [7],

- Fuzzy Matching
- Inference
- Combination
- Defuzzification

Table 1 Inference table for detecting the center of fovea

	$C_{(fuzzy)}$	$A_{1(fuzzy)}$	$A_{2(fuzzy)}$	$A_{3(fuzzy)}$
$I_{D(fuzzy)}$	F	F	SF	OF
$I_{G(fuzzy)}$	SF	SF	OF	OF
$I_{B(fuzzy)}$	SF	OF	OF	OF

Fuzzy Matching

In fuzzy matching, membership values of both $\mu_{\mathbb{X}_i}$ (radial distance) and $\mu_{\mathbb{Y}_i}$ (intensity) of every pixel were considered. The supremum minimum of $\mu_{\mathbb{X}_i}$ and $\mu_{\mathbb{Y}_i}$ of a particular pixel is taken as the membership function of that pixel in a specific class of $\mathbf{ROI}_{1(classified)}$ [14]. Mathematically, it can be expressed in terms of Eq. 31.

$$\mu_{\mathbb{D}_{ik}} = \sup \min\{\mu_{\mathbb{X}_{ik}}, \mu_{\mathbb{Y}_{ik}}\},$$
$$= \mu_{\mathbb{X}_{ik}} \wedge \mu_{\mathbb{Y}_{ik}}. \tag{31}$$

Here, $\mu_{\mathbb{X}_{ik}}$ and $\mu_{\mathbb{Y}_{ik}}$ are the membership functions of the pixel at $(x_k, y_k) \in \mathbf{ROI}_1$ obtained from Eqs. 22 and 26. "$\wedge$" is the "min" operator on fuzzy sets. If $\mu_{\mathbb{X}_{ik}}$ and $\mu_{\mathbb{Y}_{ik}}$ have more then one values, all combinations of $(\mu_{\mathbb{X}_{ik}}, \mu_{\mathbb{Y}_{ik}})$ are taken into consideration.

Inference

As there are more than one values of membership functions in each combination for every pixel, inference step is invoked in order to produce a conclusion based on the values of membership functions in a combination and to assign the membership value to a destination set [7]. In this step the inferencing in clipping method was used. A collection of rules were applied to perform the inference step. The pattern of the rules is shown in Eq. 32.

$$\text{IF } r_k \text{ in } \mathbb{X}_i \text{ and } i_k \text{ in } \mathbb{Y}_i \text{ THEN } (x_k, y_k) \text{ in } \mathbb{D}_i,$$
$$\text{Symbolically, } \mathbb{X}_i \text{ and } \mathbb{Y}_i \to \mathbb{D}_i. \tag{32}$$

Here, $r_k \in \mathcal{R}_{ROI_1}$ is the radial distance of a point (x_k, y_k) from (x_c, y_c), $i_k \in I_{ROI_1}$ is the intensity of the point (x_k, y_k) and $\mathbb{D}_i \subset \mathbf{ROI}_{1(classified)}$. So, the t-norm based proposition $P(r_k, i_k)$ is given by Eq. 33 [14].

$$P(r_k, i_k) = \mathbb{X}_i(r_k) \mathbf{t} \mathbb{Y}_i(i_k),$$
$$P \qquad = \mathbb{X}_i \times \mathbb{Y}_i. \tag{33}$$

Using this proposition in Eq. 33, $\mathbb{D}'_i \subset \mathbb{D}_i$ can be obtained from Eq. 34.

$$\begin{aligned} \mathbb{D}'_i &:= (\mathbb{X}_i \times \mathbb{Y}_i) \circ (\mathbb{X}_i \text{ and } \mathbb{Y}_i \to \mathbb{D}_i), \\ &:= (\mathbb{X}_i \times \mathbb{Y}_i) \circ f_c. \end{aligned} \tag{34}$$

Here, f_c is the fuzzy conjunction proposed by Mamdani [13]. f_c was expressed as Eq. 35.

$$\begin{aligned} f_c(\mathbb{X}_i(r_k), \mathbb{Y}_i(i_k)) &:= \mathbb{X}_i(r_k) \wedge \mathbb{Y}_i(i_k), \\ \forall \, (r_k, y_k) &\in \mathcal{R}_{ROI_1} \times I_{ROI_1}. \end{aligned} \tag{35}$$

Alternatively, $\mathbb{D}'_{ik}$ for a point (x_k, y_k) was expressed with Eq. 36.

$$\mathbb{D}'_{ik} := \mu_{\mathbb{D}_{ik}} \wedge \mathbb{D}_i. \tag{36}$$

All $\mathbb{D}'_i$s were obtained using *if-then* rules based on the inferences of Table 1.

Combination

As the boundaries of the fuzzy sets are overlapped, rule based fuzzy system consists of a set of rules for a point in the overlapping region. A particular input of a system often satisfies multiple fuzzy rules. As a result, the combination step is required. For a particular pixel at (x_k, y_k) more than one values of one $\mathbb{D}'_{ik}$ can be obtained. So, every value of $\mathbb{D}'_{ik}$ was expressed in terms of $\mathbb{D}'_{ijk}$. Here, j is the index of the value. If p is the number of values for corresponding $\mathbb{D}'_{ik}$, the combination set $\mathbf{D}_k$ can be calculated using Eq. 37.

$$\mathbf{D}_k = \bigcup_{i=1}^{3} \bigcup_{j=1}^{p} \mathbb{D}'_{ijk}. \tag{37}$$

Defuzzification

Defuzzification step is performed to obtain a crisp decision using the result of the combination step. To obtain a new position for a pixel based on its position and intensity in $\mathbf{ROI}_1$, the defuzzification is performed. In this step center of area (COA) or centroid method is used to calculate the a new radius for every pixel. The center of the area can be calculated using Eq. 38 [7].

$$r_{k(new)} = \frac{\sum_{m=0}^{r_4} \mu_{\mathbf{D}_k} \times r_m}{\sum_{m=0}^{r_4} \mu_{\mathbf{D}_k}}. \tag{38}$$

Here, $\mu_{\mathbf{D}_k}$ is the membership function of $\mathbf{D}_k$ and $0 \leqslant r_m \leqslant r_4$. After defuzzification, the pixel (x_k, y_k) was moved to a new position $(x_{k(new)}, y_{k(new)})$ with the conditions given in Eq. 39 in $\mathbf{ROI}_{1(classified)}$.

$$\sqrt{(x_{k(new)} - x_c)^2 + (y_{k(new)} - y_c)^2} = r_{k(new)},$$

$$\tan^{-1} \frac{y_k - y_c}{x_k - x_c} = \tan^{-1} \frac{y_{k(new)} - y_c}{x_{k(new)} - x_c}.$$

(39)

For example, if the radial distance of a pixels at (x_k, y_k) from (x_c, y_c) is r_k and its intensity is i_k, in Fig. 19a, the membership function $\mu_{\mathbb{X}_{ik}}$ has values $\mu_{C_{r_k}}, \mu_{A_{1r_k}}$. In Fig. 19b, the membership function $\mu_{\mathbb{Y}_{ik}}$ has values $\mu_{I_{Di_k}}, \mu_{I_{Gi_k}}$. As a result, four different combinations $(\mu_{C_{r_k}}, \mu_{I_{Di_k}})$, $(\mu_{C_{r_k}}, \mu_{I_{Gi_k}})$, $(\mu_{A_{1r_k}}, \mu_{I_{Di_k}})$ and $(\mu_{A_{1r_k}}, \mu_{I_{Gi_k}})$ can be found. For combination $(\mu_{C_{r_k}}, \mu_{I_{Di_k}})$ the supremum minimum $\mu_{I_{Di_k}} = \mu_{C_{r_k}} \wedge \mu_{I_{Di_k}}$ was considered as the membership value for the combination. Similarly, the membership values for other combinations are calculated. To assign the membership value of a combination to a destination set, a collection of rules can be extracted from the inference table given in Table 1. For the pixel at (x_k, y_k) the collection of rules is given in Table 2. For combination $(\mu_{C_{r_k}}, \mu_{I_{Di_k}})$, as $r_k \in C_{(fuzzy)}$ and $i_k \in I_{D(fuzzy)}$, the pixels $(x_k, y_k) \in \mathbf{F}_{(fuzzy)}$ with the membership value $\mu_{\mathbf{D}_{1k}} = \mu_{\mathbf{F}_k} = \mu_{C_{r_k}} \wedge \mu_{I_{Di_k}}$. Using the clipping method, $\mathbb{D}'_{11k} = \mu_{\mathbf{D}_{1k}} \wedge \mathbb{D}_1$ is calculated. Figure 19c shows $\mathbb{D}'_{11k}$. Similarly, the membership values were assigned to destination sets for other combinations mentioned in Table 2. Figure 19d, Fig. 19e and Fig. 19f show $\mathbb{D}'_{21k}$, $\mathbb{D}'_{12k}$ and $\mathbb{D}'_{22k}$, respectively.

To obtain the combination set $\mathbf{D}_k$, union of $\mathbb{D}'_{11k}$, $\mathbb{D}'_{12k}$, $\mathbb{D}'_{21k}$ and $\mathbb{D}'_{22k}$ are taken. Figure 19g shows the union of all the $\mathbb{D}_{ijk}$s. Figure 19h shows the set $\mathbf{D}_k$. During defuzzification the centroid of $\mathbf{D}_k$ was calculated. Figure 19h shows the center of area $r_{k(new)}$ of $\mathbf{D}_k$. After fuzzy inference and defuzzification were applied on $\mathbf{ROI}_1$, all pixels of fovea and its neighborhood became classified into three distinguishable regions, $\mathbf{R}_{(fovea)}$, $\mathbf{R}_{(semi\,fovea)}$ and $\mathbf{R}_{(outer\,fovea)}$ for foveal pixels, semi foveal pixels and outer foveal pixels respectively; where, $\mathbf{R}_{(fovea)} \cup \mathbf{R}_{(semi\,fovea)} \cup \mathbf{R}_{(outer\,fovea)} = \mathbf{ROI}_{1(classified)}$. Figure 18 shows three distinguishable regions of $\mathbf{ROI}_{1(classified)}$.

Table 2 Collection of rules extracted from inference for (x_k, y_k)

Combination	IF	THEN	WITH	$\mathbb{D}'_{ijk}$
$(\mu_{C_{r_k}}, \mu_{I_{Di_k}})$	$r_k \in C_{(fuzzy)}$ and $i_k \in I_{D(fuzzy)}$	$(x_k, y_k) \in \mathbf{F}_{(fuzzy)}$	$\mu_{\mathbf{D}_{1k}} = \mu_{\mathbf{F}_k} = \mu_{C_{r_k}} \wedge \mu_{I_{Di_k}}$	$\mu_{\mathbf{D}_{1k}} \wedge \mathbb{D}_1 = \mathbb{D}'_{11k}$
$(\mu_{C_{r_k}}, \mu_{I_{Gi_k}})$	$r_k \in C_{(fuzzy)}$ and $i_k \in I_{G(fuzzy)}$	$(x_k, y_k) \in \mathbf{SF}_{(fuzzy)}$	$\mu_{\mathbf{D}_{2k}} = \mu_{\mathbf{SF}_k} = \mu_{C_{r_k}} \wedge \mu_{I_{Gi_k}}$	$\mu_{\mathbf{D}_{2k}} \wedge \mathbb{D}_2 = \mathbb{D}'_{21k}$
$(\mu_{A_{1r_k}}, \mu_{I_{Di_k}})$	$r_k \in A_{1(fuzzy)}$ and $i_k \in I_{D(fuzzy)}$	$(x_k, y_k) \in \mathbf{F}_{(fuzzy)}$	$\mu_{\mathbf{D}_{1k}} = \mu_{\mathbf{F}_k} = \mu_{A_{1r_k}} \wedge \mu_{I_{Di_k}}$	$\mu_{\mathbf{D}_{1k}} \wedge \mathbb{D}_1 = \mathbb{D}'_{12k}$
$(\mu_{A_{1r_k}}, \mu_{I_{Gi_k}})$	$r_k \in A_{1(fuzzy)}$ and $i_k \in I_{G(fuzzy)}$	$(x_k, y_k) \in \mathbf{SF}_{(fuzzy)}$	$\mu_{\mathbf{D}_{2k}} = \mu_{\mathbf{SF}_k} = \mu_{A_{1r_k}} \wedge \mu_{I_{Gi_k}}$	$\mu_{\mathbf{D}_{2k}} \wedge \mathbb{D}_2 = \mathbb{D}'_{22k}$

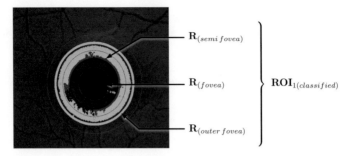

Fig. 18 $\mathbf{ROI}_{1(classified)}$ pixels with three distinguishable regions $\mathbf{R}_{(fovea)}$, $\mathbf{R}_{(semi\,fovea)}$ and $\mathbf{R}_{(outer\,fovea)}$ plotted on **ROI**

4.6   Calculation of the Center of Fovea

Members of $\mathbf{R}_{(fovea)}$ reside within the neighborhood of (x_c, y_c) and also have lowest intensities in $\mathbf{ROI}_{1(classified)}$ and $\mathbf{ROI}_1$. So, in order to detect the center of fovea, only $\mathbf{R}_{(fovea)}$ was considered. All pixels in $\mathbf{R}_{(fovea)}$ were inversely mapped to their previous position and these points can be taken as the set of foveal pixels $\mathbf{F}_{(actual)}$ with Eq. 40. Figure 20a shows the $\mathbf{F}_{(actual)}$ on a white background.

$$\forall (x_{k(new)}, y_{k(new)}) \in \mathbf{R}_{(fovea)}, \\ \zeta^{-1} : \mathbf{R}_{(fovea)} \rightarrow \mathbf{F}_{(actual)}. \tag{40}$$

Here, $\mathbf{R}_{(fovea)} \subset \mathbf{ROI}_{1(classified)}$, $\mathbf{F}_{(actual)} \subset \mathbf{ROI}_1$. The center of fovea (x_{fo}, y_{fo}) can be calculated by calculating the intensity centroid of $\mathbf{F}_{(actual)}$ [15, 17]. Figure 20a shows (x_{fo}, y_{fo}) in F(actual) and Fig. 20b shows (x_{fo}, y_{fo}) in **ROI**. The grey scale intensity of $\mathbf{F}_{(actual)}$ can be expressed with Eq. 41.

$$\forall x \forall y \, \mathcal{I}_{\mathbf{F}_{(actual)}} : \mathbf{F}_{(actual)} \rightarrow I_{\mathbf{F}_{(actual)}} \quad \neg \exists \, \mathcal{I}_{\mathbf{F}_{(actual)}}^{-1}. \tag{41}$$

5   Detection of the Boundary of FAZ

Figure 21 shows center of fovea and detected foveal region in three different retinal images. As seen in Fig. 21, foveal region $\mathbf{R}_{(fovea)}$ is a non-compact set. As a result, a closed boundary cannot be defined. The fovea is typically a circular region containing all foveal pixels. The FAZ has a very smooth transition from central region towards outward direction. Therefore, a scheme is needed to detect the region where FAZ fades out to blood vessels.

Gradient Vector Flow based active contour model (GVF snake) uses a *gradient vector flow* as its external force which is a dense vector fields derived from images by minimizing a certain energy functional in a variational framework. The minimization

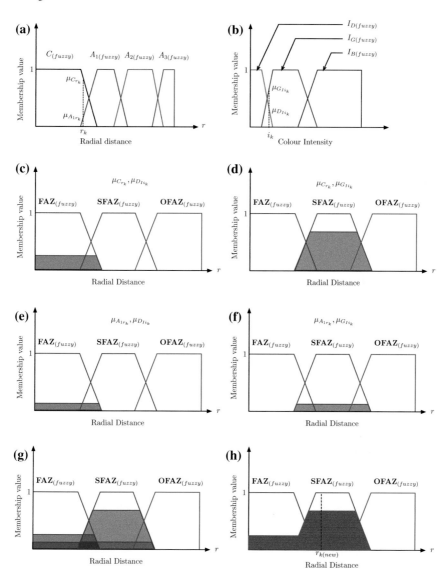

Fig. 19 Steps of fuzzy analysis to determine center of fovea

is achieved by solving a pair of decoupled linear partial differential equations that diffuses the gradient vectors of a gray-level or binary edge map computed from the image [28]. The advantages of GVF snake are its invariance to the initial contour and its ability to converge towards the boundary concavity whether the initial snake is taken inside; outside or across the object boundary. The range of the GVF snake is also larger. So, GVF based snake is chosen to find the boundary of FAZ on fundus

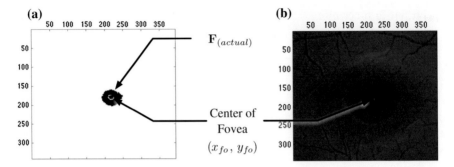

Fig. 20 **a** $F_{(actual)}$ and center of fovea (x_{fo}, y_{fo}) on a white background, **b** Center of fovea (x_{fo}, y_{fo}) on **ROI**

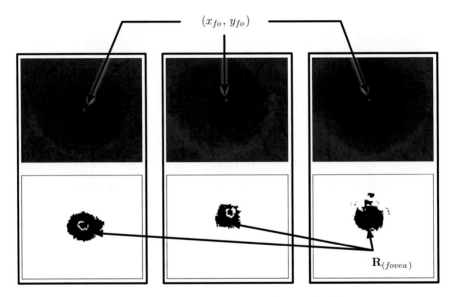

Fig. 21 Detected $F_{(actual)}$ and (x_{fo}, y_{fo}) on ROI of three different retinal images

images. As the FAZ is usually a compact circular region, a circle is taken as the initial snake. Pixels on the circle can be expressed as $\mathbf{z}(s) = [x(s), y(s)], s \in [0, 1]$. The dynamic energy functional of the snake traveling through the spatial domain of the **ROI** can be expressed with Eq. 42 [28].

$$E = \int_0^1 \frac{1}{2}[\alpha|\mathbf{z}'(s, t)|^2 + \beta|\mathbf{z}''(s, t)|^2] + E_{ext}(\mathbf{z}(s))ds. \tag{42}$$

Here, α and β are the parameters for controlling the tension and rigidity, respectively. $E_{ext}(\mathbf{z}(s))$ is the external energy function. For GVF based snake, $E_{ext}(\mathbf{z}(s)) = \bar{\mathbf{v}}(x, y)$. $\bar{\mathbf{v}}(x, y)$ is the *gradient vector flow* (GVF), which can be extracted from

$I_{ROI}(x, y)$. In order to extract $\bar{\mathbf{v}}(x, y)$ from $I_{ROI}(x, y)$, equilibrium solution of the system of partial differential equation in Eq. 43 was considered as $\bar{\mathbf{v}}(x, y)$ [27].

$$\frac{\partial \mathbf{v}}{\partial t} = \mu \nabla^2 \mathbf{v} - (\mathbf{v} - \nabla f)|\nabla f|^2. \tag{43}$$

Here, μ is the regularization parameter [28] and f is an edge map derived from $I_{ROI}(x, y)$ which is larger near the image edge. I_{ROI} is the intensity of f which can be calculated using Eq. 44 [27]. As, the intensity value in the FAZ is lower, the vector field diverge for the FAZ. The initial snake towards FAZ was considered in the inverted image $[255 - I_{ROI}(x, y)]$.

$$\pm G_\sigma(x, y) * [255 - I_{ROI}(x, y)]. \tag{44}$$

Here, $G_\sigma(x, y)$ is a gaussian window with standard deviation σ. Minimizied E should satisfy the Euler equation given in Eq. 45 [28].

$$\alpha \mathbf{z}''(s, t) - \beta \mathbf{z}'''(s, t) - \nabla E_{ext}(\mathbf{z}(s)) = 0. \tag{45}$$

For GVF based snake, Eq. 45 can be expressed as Eq. 46 [12, 27].

$$\alpha \mathbf{z}''(s, t) - \beta \mathbf{z}'''(s, t) + \kappa \bar{\mathbf{v}}(x, y) = 0. \tag{46}$$

Here, κ is the weight of $\bar{\mathbf{v}}(x, y)$. Every control point on the initial snake is guided with $\bar{\mathbf{v}}(x, y)$. In every iteration all points on the snake is updated to minimize the energy [5]. This process is performed until a stopping criteria is reached. To determine the boundary of FAZ, a circle $\mathbb{C}_0 = \{\mathbf{z}(s)|\mathbf{z}(s) = [x(s), y(s)], s \in [0, 1]\}$ with center at (x_{fo}, y_{fo}) was considered in the semi foveal region as the initial snake. Figure 22a shows initial snake on **ROI**. In every iteration, a contour $\mathbb{C}_i = \{\mathbf{z}(s)|\mathbf{z}(s) = [x(s), y(s)], s \in [0, 1]\}$ is obtained which encloses an area $\mathbb{S}_i \subset$ **ROI**; where, $\mathbb{S}_i$ is a closed set and $\mathbb{C}_i \rightarrow \mathbb{S}_i$. Figure 22b shows snakes after several iteration. The intensity of $\mathbb{S}_i$ can be given by Eq. 47.

$$\forall \mathbf{z}(s) \, \mathcal{I}_{\mathbb{S}_i} : \mathbb{S}_i \rightarrow I_{\mathbb{S}_i} \quad \neg \exists I_{\mathbb{S}_i}. \tag{47}$$

During extracting FAZ, $\mu_{I_{F_{(actual)}}}$ i.e. the average of $I_{F_{(actual)}}$ was considered as the stopping criteria. As the initial snake was placed outside the **F**, the average of $I_{\mathbb{S}_0}$, $\mu_{I_{\mathbb{S}_0}} > \mu_{I_{F_{(actual)}}}$. At every iteration, the snake $\mathbb{C}_i$ approaches FAZ and $\mu_{I_{\mathbb{S}_i}}$, mean of $I_{\mathbb{S}_i}$ approached to $\mu_{I_{F_{(actual)}}}$. Whenever, $\mu_{I_{\mathbb{S}_i}} \leq \mu_{I_{F_{(actual)}}}$, the snake $\mathbb{C}_i$ was considered to be the boundary of **FAZ** and enclosed area $\mathbb{S}_i$ was considered to be **FAZ**. Figure 22c shows final area of **FAZ**.

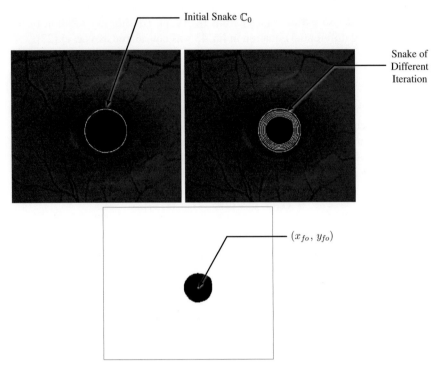

Fig. 22 **a** Initial snake on **ROI**, **b** Snakes after several iterations, **c** Final area of **FAZ** with center of fovea (x_{fo}, y_{fo})

6 Result Analysis

The above mentioned process is applied on 20 fundus images. All of the images are of same resolution. The algorithm was implemented with MATLAB.

6.1 Results of the Center of Fovea Detection

In the process of detecting the center of fovea, overlap ratio γ (please see: Eqs. 22, 26, 30), may be varied to change the overlapping regions of the membership functions and the detection can be more precise. The value of γ for precise detection of the center of fovea varies from person to person. It is observed in all image that the center of fovea is correctly detected when the ratio of overlap for all membership functions are equal. So, the overlap ratios for detecting the center of fovea were taken equal for all the membership functions. The fuzzy detection process was applied for different values of γ on all images.

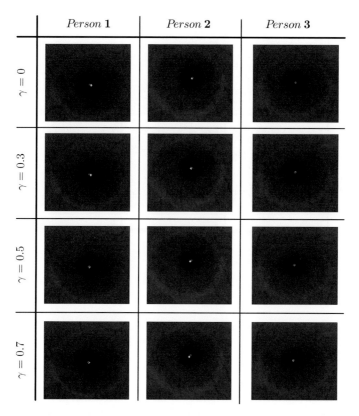

Fig. 23 Variation of the position of the center of fovea (x_{fo}, y_{fo}) with overlap ratio γ of three different person shown in *yellow dots*. The *red dots* show the manually detected centers of foveae

Figure 23 shows variation in the position of (x_{fo}, y_{fo}) (yellow dots) with γ in retinal image of three different persons. It also exhibits manually detected fovea center on every image (red dots). The manually detected points helps us to validate the performance of the proposed method. In the case of *Person* 1, the center of fovea converges towards the actual center with increase of γ. Difference between position of the center of fovea for $\gamma = 0.5$ and $\gamma = 0.7$ seems to be insignificant. For lower γ, some of the foveal pixels are not considered for calculation of the center of fovea. For *Person* 2, the center of fovea converges towards the prescribed center with the decrease of γ. Difference among positions of the center of fovea for $\gamma = 0$, $\gamma = 0.3$ and $\gamma = 0.5$ are minimal. For this retinal image, all points of the foveal zone lies within the boundary of initial circle. When $\gamma > 0.5$ the more noisy points from semi fovea region become inevitable during calculating the center of fovea. For *Person* 3, the center of fovea is insensitive of γ. All of the foveal pixels lie within the initial foveal boundary. All pixels of the foveal zones are clearly classified. After evaluating sensitivity of γ for all sample retinal images, $0.3 \leqslant \gamma \leqslant 0.5$ is considered as the optimal range of overlapping ratio for the detection of the center

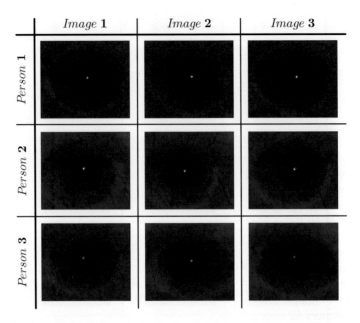

Fig. 24 Consistent center of fovea in different retinal images of same eye of same person. The *yellow dots* shows the center of fovea detected by proposed method. The *red dot* shows manually detected centers of fovea

of fovea. Figure 24 shows the consistency of the detected center of fovea in different retinal images acquired from the same eye of the same person. It was seen that in all images of *Person* 1, the center of fovea resides inside the brighter annular region visible at the central fovea. For *Person* 2, the center of fovea resides on the ring in all images. In case of *Person* 3, there is no such ring and the center of fovea stays at the center of the FAZ for the suggested range. Figure 24 also compares the calculated centers ($C_{f(calc)} = (x_{of}, y_{of})$, yellow dots) with the manually detected centers ($C_{f(manual)} = (x_{of(m)}, y_{of(m)})$, red dots). The relative difference between the calculated center and the manually detected center can be quantified using Eq. 48.

$$v = \frac{\Delta r}{2r_1}. \qquad (48)$$

Here, $\Delta r = \sqrt{(x_{of} - x_{of(m)})^2 + (y_{of} - y_{of(m)})^2}$ and $r_1 \in \mathcal{R}_{if_{vhs}}$ in (Eq. 17) is the radius of the inner fovea circle introduced in Sect. 4.4 and shown in Fig. 12.

Table 3, exhibits deviation of the calculated fovea centers from the manually detected centers in Fig. 24. The variation of the center calculated is very small compared to the resolution of the image. The r_1 considered for the normalization of the variation is the radius of the smallest which must contain the center of fovea. It also provides a very small area of variation compared to any other boundary considered during the detection process of center of fovea. The maximum possible deviation

Table 3 Measured deviation from manually detected fovea center, v

	Image 1	Image 2	Image 3
Person 1	0.019	0.026	0.026
Person 2	0.035	0.035	0.036
Person 3	0.012	0.01	0.01

(Eq. 48) is 1.0 when both the points are located on the periphery of the circle at two opposite points on the circle diameter. This should only happen when there is a very high degree of disagreement between the manually selected point and the calculated point. A complete agreement with the proposed technique will superimpose the calculated center on top of the manually detected centre and that would make value of v to be 0. Therefore, the dynamic range v is : $0 \leqslant v \leqslant 1$. It is observed that, the variation inside the smallest area was very small for all the images. Therefore, the detected centers and the manually selected centers have direct agreement for all the images considered for this study. The deviation presented in the table are also very consistent on multiple images of the same person. As a result, the location of the calculated centers for each person on multiple images are very precise.

6.2 Results of Detection of Boundary of FAZ

In order to detect the boundary of the FAZ accurately, GVF snake (active contour model based on GVF) was applied where tension α, rigidity β and weight of external GVF κ were varied precisely. The contour containing pixels with average intensity almost equal to the average intensity of the pixels of $\mathbf{F}_{(actual)}$ was considered to be **FAZ**. As a result, boundary of FAZ will also vary with the changing γ. As the boundary of the fovea should be dependent on the image of the fovea and β, the rigidity parameter reduces the sensitivity of the external force and κ was chosen to be 1. Effect of other parameters are explained in the following sections.

6.2.1 Effect of γ on FAZ

Figure 25 shows variation of **FAZ** with γ; keeping α, β and κ fixed. As seen in the images, generally area of **FAZ** increases with the increase in overlap ratio γ. With the increase in γ, more pixels with higher intensity become members of $\mathbf{F}_{(actual)}$. As a result, $\mu_{I_{\mathbf{F}_{(actual)}}}$ increases. The area of **FAZ** increases with the stopping criteria $\mu_{I_{S_i}} \lesssim \mu_{I_{\mathbf{F}_{(actual)}}}$ for detecting **FAZ**. For *Person* 1 and *Person* 2 the area of **FAZ** increases with $\gamma < 0.6$. If $\gamma > 0.6$ the shape of **FAZ** gets distorted. For *Person* 3, **FAZ** does not get distorted for $\gamma > 0.6$ but the area increased gradually. In Fig. 25 for *Person* 1, *Person* 2 and *Person* 3 with $\gamma < 0.3$ **FAZ** does not enclose completely

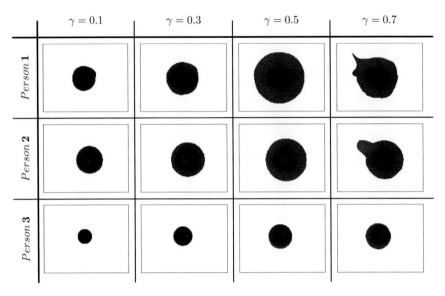

Fig. 25 Variation of γ in different retinal images with $\alpha = 0$, $\beta = 0$ and $\kappa = 1$

the fovea region. For $\gamma = 0.3$ the **FAZ** encloses complete foveal region. For $\gamma > 0.3$, **FAZ** encloses complete foveal region as well as some of its surrounding area. From observing all samples it was found that $0.2 < \gamma < 0.4$ leads to the best result for detection of **FAZ**.

6.2.2 Effect of α on FAZ

Figure 26 shows the variation in the contours with α (yellow contour) and the manually detected FAZ boundary (red contour). It is clearly seen that with increasing α, the first snake $\mathbb{C}_1$ (given the initial snake $\mathbb{C}_0$), comes closer to the center of fovea (x_{of}, y_{of}). As α is the drift towards (x_{of}, y_{of}), $\mathbb{C}_1$ moves towards (x_{of}, y_{of}) with the increment of α. So, it will take less number of iteration to determine the **FAZ** if **FAZ** $\subset \mathbb{S}_1$. If α is too high and $\mathbb{C}_1$ moves too close to (x_{of}, y_{of}), i.e. **FAZ** $\supset \mathbb{S}_1$. As a result, **FAZ** will enclose a part of the actual Foveal Avascular Zone. In all these cases for $0 \leqslant \alpha \lesssim 5$, $\mathbb{C}_1$ was seen to converge towards the manually detected **FAZ** boundaries.

Figure 27, shows the effect of high α on different retinal images. It is seen that, **FAZ** becomes smaller as α increases. It was observed that, for $0 \leqslant \alpha \lesssim 5$, $\mathbb{C}_1$ always encloses **FAZ** (See Fig. 26). For $7 > \alpha > 5$, either **FAZ** $\subset \mathbb{S}_1$ or **FAZ** $\supset \mathbb{S}_1$ for all the images. For $\alpha > 7$, **FAZ** $\supset \mathbb{S}_1$. So, **FAZ** is smaller. Due to noise in GVF in semi foveal region, shape of the boundary of **FAZ** became distorted. α can be used to obtain correct shape of the **FAZ**. Figure 28 shows, effect of α on the shape of the boundary of **FAZ**. As seen in Fig. 28, the shape of **FAZ** becomes more circular with

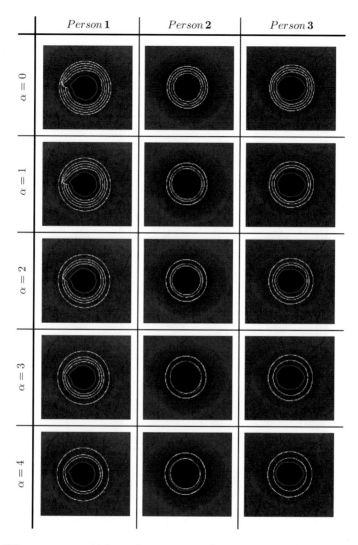

Fig. 26 Effect on contour with increasing α; $\gamma = 0.3$, $\beta = 0$ and $\kappa = 1$. The *yellow* contours are the snakes generated by GVF enclosing the FAZ. The *red* contour is the manually detected FAZ

the increase in α. For $\alpha > 5$, the size of **FAZ** shrinks. In order to avoid distortion and retain the proper area of **FAZ**, the optimum interval for α is $3 \leqslant \alpha \leqslant 5$.

6.2.3 Effect of β on FAZ

In most of the cases, $3 \leqslant \alpha \leqslant 5$ solves the issue of distortion of the boundary of **FAZ**. If the GVF is too noisy, the rigidity β of the initial has to be increased. So that

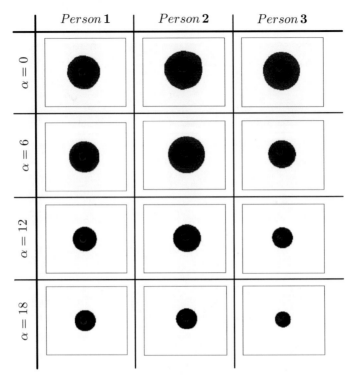

Fig. 27 Effect of increase in α on area of **FAZ** with $\gamma = 0.3$, $\beta = 0$ and $\kappa = 1$

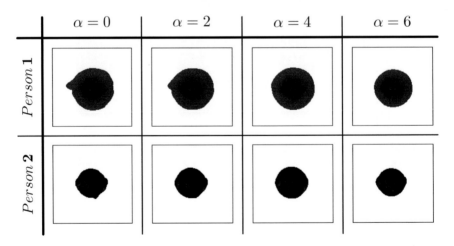

Fig. 28 Correction of distorted boundary of FAZ by changing α; $\gamma = 0.3$, $\beta = 0$ and $\kappa = 1$

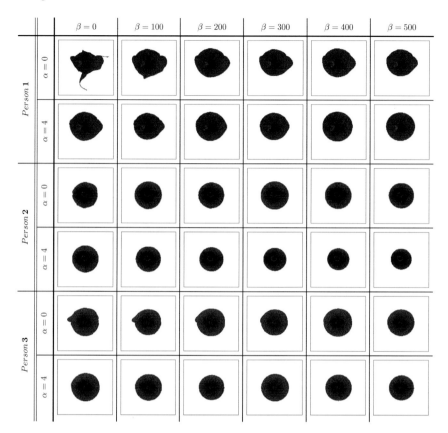

Fig. 29 Effect of increase in β on FAZ with $\alpha = 0$ & 4, $\gamma = 0.3$ and $\kappa = 1$

the snake will be less sensitive to the noisy GVF. In Fig. 29, Effect of increase in β with $\alpha = 0$ and $\alpha = 4$ in three different images of three different persons are shown.

The image of *Person 1* has high noise in GVF. As a result, the detected **FAZ** with $\alpha = 0$ and $\beta = 0$ has deformed boundary. As shown in Fig. 29, for $\alpha = 4$, the boundary of **FAZ** has improved but not fully undistorted. As seen in Fig. 29, with the increase in β, the boundary of **FAZ** has further improved. For $\beta \geqslant 200$, the boundary of **FAZ** becomes more circular. For $\beta \geqslant 400$, boundary of **FAZ** become insensitive to GVF and becomes completely circular. The image of *Person 2* has one of the least noisy GVF. As a result, detected boundary of **FAZ** is almost circular with $\alpha = 0$ and $\alpha = 4$. As β increases the boundary of **FAZ** becomes more circular with $\alpha = 0$ but the enclosed area of **FAZ** decreases with increase in β for $\alpha = 4$. By observing all the images of retina it was found that the effect of increase in β does not affect all the images with low noise GVF. In some images the area of **FAZ** decreases slightly, in other images the **FAZ** is insensitive to β. It is also observed that in images with low noise GVF and sensitivity to β, the **FAZ** for $200 \leqslant \beta \leqslant 300$ contains the Foveal

Avascular Zone. The image of *Person* **3** has a noisy GVF that can be processed with only $3 \leqslant \alpha \leqslant 5$. As seen in Fig. 29, for $\alpha = 4$ and $\alpha = 0$, the boundary of FAZ is circular. Increase in β has little effect on the boundary of **FAZ**. As a result, in order to avoid distortion of the boundary of **FAZ**, the optimum range of β is $200 \leqslant \beta \leqslant 300$.

The current work does not depend on any conventional image pre-processing techniques. The foveal region is evaluated based on an intelligent image segmentation algorithm. Exact similar kind of work has not been addressed in the literature. For the purpose of comparison, the outputs from the algorithm were compared with the data obtained from rigorous inspection. The foveal region and its center were located on the test images and compared with the results from the proposed method. The reference data are 100% accurate since they were evaluated manually. From the above discussion it was clearly seen that the proposed method was able to achieve very good accuracy based on a reasonably large dataset.

7   Conclusion

In this work the major effort was given to extract the foveal zone from a retinal image. The fuzzy logic based algorithm is implemented in order to classify the candidate region for fovea. This approach helps us to detect the fovea zone as well as fovea center. Besides performing biomedical image analysis, there have been several methods developed for different biometric applications. One of the most popular is the "face detection" [23–25]. The proposed method in this paper will be a significant contribution to the field of "retinal image detection" which is another growing major field in biometrics. The proposed technique does not utilizes any image filltration although it showed very high agreement with the manually detected fovea centers and boundaries of all images it was applied on. The accuracy of the center detection technique is dependent on formation of fuzzy inference table. For most of the images the inference tables are similar. In order to capture the boundary of fovea zone, the output from fuzzy classifier is analyzed further. In this regard, gradient vector flow based boundary detection method is applied. The gradient flow algorithm traces for the boundary of fovea zone iteratively by reaching the global minimum of energy functional. This approach demands a thorough analysis of different parameters involved. In this paper we did some study on these parameters. However, the more detail analysis can be done in future works. The proposed method exhibited promising performance based on decent agreement with manually detected fovea centers and boundaries.

References

1. Ayala, G., León, T., Zapater, V.: Different averages of a fuzzy set with an application to vessel segmentation. IEEE Trans. Fuzzy Syst. **13**(3), 384–393 (2005)
2. Canny, J.: A computational approach to edge detection. IEEE Trans. Pattern Anal. Mach. Intell. **6**, 679–698 (1986)
3. Dugundji, J.: Topology, wm. c (1989)
4. Ege, B.M., Hejlesen, O.K., Larsen, O.V., Møller, K., Jennings, B., Kerr, D., Cavan, D.A.: Screening for diabetic retinopathy using computer based image analysis and statistical classification. Comput. Methods Programs Biomed. **62**(3), 165–175 (2000)
5. Gutierrez, J., Epifanio, I., de Ves, E., Ferri, E.J.: An active contour model for the automatic detection of the fovea in fluorescein angiographies. In: 15th International Conference on Pattern Recognition. Proceedings, vol. 4, pp. 312–315. IEEE (2000)
6. Jain, A., Bolle, R., Pankanti, S.: Biometrics: Personal Identification in Networked Society, vol. 479. Springer Science and Business Media (2006)
7. John, Y., Langari, R.: Fuzzy Logic: Intelligence, Control, and Information, pp. 379–383. Publish by Dorling Kindersley, India (1999)
8. Klette, R., Rosenfeld, A.: Digital Geometry: Geometric Methods for Digital Picture Analysis. Elsevier (2004)
9. Larsen, H.W.: The Ocular Fundus: A Color Atlas. WB Saunders Company (1976)
10. Li, H., Chutatape, O.: Automated feature extraction in color retinal images by a model based approach. IEEE Trans. Biomed. Eng. **51**(2), 246–254 (2004)
11. Lipschutz, S.: Schaums outline of theory and problems of set theory (1998)
12. Luo, S., Li, R., Ourselin, S.: A new deformable model using dynamic gradient vector flow and adaptive balloon forces. In: APRS Workshop on Digital Image Computing, pp. 9–14 (2003)
13. Mamdani, E.H.: Application of fuzzy algorithms for control of simple dynamic plant. In: Proceedings of the Institution of Electrical Engineers, vol. 121, pp. 1585–1588. IET (1974)
14. Pedrycz, W., Gomide, F.: An Introduction to Fuzzy Sets: Analysis and Design. MIT Press (1998)
15. Rahman, R., Chowdhury, M., Kabir, S., Khan, M.: A novel method for person identification by comparing retinal patterns. In: CAINE-2006. Proceedings: 19th International Conference on Computer Applications in Industry and Engineering: (CAINE-2006): Nov. 13–15, 2006, Las Vegas, Nevada, USA, pp. 14–17. International Society for Computers and Their Applications (2006)
16. Ross, T.J.: Fuzzy Logic with Engineering Applications. Wiley (2009)
17. Raiyan Kabir, S.M., Rahman, R., Habib, M., Rezwan Khan, M.: Person identification using retina pattern matching. In: Proceeding of 3rd International Conference on Electrical and Computer Engineering 2004, pp. 522–525 (2004)
18. Sanchez, C., Hornero, R., Lopez, M., Poza, J., et al.: Retinal image analysis to detect and quantify lesions associated with diabetic retinopathy. In: Engineering in Medicine and Biology Society. IEMBS'04. 26th Annual International Conference of the IEEE, vol. 1, pp. 1624–1627. IEEE (2004)
19. Sinthanayothin, C., Boyce, J.F., Cook, H.L., Williamson, T.H.: Automated localisation of the optic disc, fovea, and retinal blood vessels from digital colour fundus images. Br. J. Ophthalmol. **83**(8), 902–910 (1999)
20. Sonka, M., Hlavac, V., Boyle, R.: Image Processing, Analysis, and Machine Vision. Cengage Learning (2014)
21. Styles, I.B., Calcagni, A., Claridge, E., Orihuela-Espina, F., Gibson, J.M.: Quantitative analysis of multi-spectral fundus images. Med. Image Anal. **10**(4), 578–597 (2006)
22. Tizhoosh, H.R.: Fuzzy Image Processing: Introduction in Theory and Practice. Springer (1997)
23. Türkan, M., Dülek, B., Onaran, I., Enis Cetin, A.: Human face detection in video using edge projections. In: Defense and Security Symposium, pp. 624607–624607. International Society for Optics and Photonics (2006)

24. Türkan, M., Pardas, M., Enis Cetin, A.: Human eye localization using edge projections. In: VISAPP (1), pp. 410–415 (2007)
25. Turkan, M., Pardas, M., Enis Cetin, A.: Edge projections for eye localization. Opt. Eng. **47**(4), 047007–047007 (2008)
26. Victoria Ibañez, M., Simó, A.: Bayesian detection of the fovea in eye fundus angiographies. Pattern Recognit. Lett. **20**(2), 229–240 (1999)
27. Xu, C., Prince, J.L.: Generalized gradient vector flow external forces for active contours. Sig. Process. **71**(2), 131–139 (1998a)
28. Xu, C., Prince, J.L.: Snakes, shapes, and gradient vector flow. IEEE Trans. Image Process. **7**(3), 359–369 (1998)
29. Yuan, B.: Fuzzy Sets and Fuzzy Logic: Theory and Applications (1995)

Registration of Digital Terrain Images Using Nondegenerate Singular Points

A. Ben Hamza

Abstract Registration of digital elevation models is a vital step in fusing sensor data. In this chapter, we present a robust topological framework for entropic image registration using Morse singularities. The core idea behind our proposed approach is to encode a digital elevation model into a set of nondegenerate singular points, which are the maxima, minima and saddle points of the Morse height function. An information-theoretic dissimilarity measure between the Morse features of two misaligned digital elevation models is then maximized to bring the elevation data into alignment. In addition, we show that maximizing this dissimilarity measure leads to minimizing the total length of the joint minimal spanning tree of two misaligned digital elevation data models. Illustrating experimental results are presented to show the efficiency and registration accuracy of the proposed framework compared to existing entropic approaches.

Keywords Registration · Tsallis entropy · Morse theory · Minimal spanning tree

1 Introduction

Information-theoretic divergence measures [1] have been successfully applied in many areas including image retrieval [2], multimedia protection [3], text categorization [4], image edge detection [5, 6], and image registration [7–11]. The latter will be the focus of the present chapter. Image registration or alignment refers to the process of aligning images so that their details overlap accurately [7, 8]. Images are usually registered for the purpose of combining or comparing them, enabling the fusion of information in the images. Roughly speaking, the image alignment problem may be formulated as a two-step process: the first step is to define a dissimilarity measure that quantifies the quality of spatial alignment between the fixed image and the spatially transformed moving image, and the second step is to develop an efficient

A. Ben Hamza (✉)
Concordia Institute for Information Systems Engineering, Concordia University,
Montreal, QC, Canada
e-mail: hamza@ciise.concordia.ca

© Springer International Publishing Switzerland 2016
A.I. Awad and M. Hassaballah (eds.), *Image Feature Detectors and Descriptors*,
Studies in Computational Intelligence 630, DOI 10.1007/978-3-319-28854-3_13

optimization algorithm for maximizing this dissimilarity measure in order to find the optimal transformation parameters. Recently, much attention has been paid to the image registration problem due in large part to its importance in a variety of tasks including data fusion, navigation, motion detection, and clinical studies [7, 8]. A wide range of image registration techniques have been developed for many different types of applications and data, such as mean squared alignment, correlation registration, moment invariant matching, and entropic image alignment [9–12].

Multisensor data fusion technology combines data and information from multiple sensors to achieve improved accuracies and better inference about the environment that could be achieved by the use of a single sensor alone. In this chapter, we introduce a nonparametric multisensor data fusion algorithm for the registration of digital elevation models (DEMs). The goal of geo-registration is to align a fixed DEM to the geographic location of a moving DEM using global or feature-based techniques [13–17]. Our proposed method falls into the category of feature-based techniques which require that features be extracted and described before two DEMs can be registered. In [15], a robust multi-temporal DEM registration technique for detecting terrain changes was proposed using the least trimmed squares estimator with the least Z-difference algorithm. The approach yields high accuracy in both matching and change detection. In [16], a registration method for DEM pairs was introduced using the elevation difference residuals and the elevation derivatives of slope and aspect. The method represents the complete analytical solution of a three-dimensional shift vector between two DEMs. A comprehensive review of DEM registration methods can be found in [17].

The proposed framework consists of two major steps, namely topological feature extraction and rigid point feature registration. The first step involves extracting Morse singularity features [18–20] from the DEMs represented as 3D surfaces. These geometric features form the so-called Reeb graph [20], which is a topological representation of a digital elevation surface and has the advantage to be stored or transmitted with a much smaller amount of data. Inspired by the success of information-theoretic measures in image registration, we propose in the second step the use of the Jensen-Tsallis (JT) divergence [21] as a dissimilarity measure between Morse features of two misaligned DEMs. The JT divergence is defined in terms of the non-extensive Tsallis entropy, which can be estimated using the length of the minimal spanning tree (MST) over Morse feature points of a DEM. Then, the registration is performed by minimizing the length of a joint MST which spans the graph generated from the overlapping misaligned DEMs.

The rest of the chapter is organized as follows. The next section provides a brief overview of the Jensen-Tsallis dissimilarity measure and its main properties. Section 3 describes the proposed feature extraction approach for DEM data using Morse theory. In Sect. 4, we propose a topological framework for DEM registration, and discuss in more detail its most important algorithmic steps. In Sect. 5, we provide experimental results to show the robustness and the geo-registration accuracy of the proposed method. And finally, we conclude in Sect. 6.

2 Background

2.1 Jensen-Tsallis Divergence

Let X be a continuous random variable with a probability density function f defined on $\mathbb{R}$. Shannon's entropy is defined as

$$H(f) = -\int f(x) \log f(x)\, dx, \tag{1}$$

and it is a measure of uncertainty, dispersion, information and randomness. The maximum uncertainty or equivalently minimum information is achieved by the uniform distribution. Hence, we can think of the entropy as a measure of uniformity of a probability distribution. Consequently, when uncertainty is higher it becomes more difficult to predict the outcome of a draw from a probability distribution. A generalization of Shannon entropy is Rényi entropy [22] given by

$$R_\alpha(f) = \frac{1}{1-\alpha} \log \int f(x)^\alpha\, dx, \quad \alpha \in (0,1) \cup (1,\infty). \tag{2}$$

Another important generalization of Shannon entropy is Tsallis entropy [23–25] given by

$$H_\alpha(f) = \frac{1}{1-\alpha}\left(\int f(x)^\alpha\, dx - 1\right) = -\int f(x)^\alpha \log_\alpha f(x)\, dx, \tag{3}$$

where $\log_\alpha$ is the α-logarithm function defined as $\log_\alpha(x) = (1-\alpha)^{-1}(x^{1-\alpha} - 1)$ for $x > 0$. This generalized entropy was first introduced by Havrda and Charvát in [23], who were primarily interested in providing another measure of entropy. Tsallis, however, appears to have been principally responsible for investigating and popularizing the widespread physics applications of this entropy, which is referred to nowadays as Tsallis entropy [24]. More recently, there has been a concerted research effort in statistical physics to explore the properties of Tsallis entropy, leading to a statistical mechanics that satisfies many of the properties of the standard theory [24]. It is worth noting that for $\alpha \in (0, 1]$, Rényi and Tsallis entropies are both concave functions; and for $\alpha > 1$ Tsallis entropy is also concave, but Rényi entropy is neither concave nor convex. Furthermore, both entropies reduce to Shannon entropy as $\alpha \to 1$, and are related by

$$H_\alpha(f) = \frac{1}{1-\alpha}[\exp\{(1-\alpha)R_\alpha(f)\} - 1]. \tag{4}$$

Figure 1 depicts Tsallis entropy of a Bernoulli distribution $\mathbf{p} = (p, 1-p)$, for different values of the entropic index. As illustrated in Fig. 1, the measure of uncertainty is at a minimum when Shannon entropy is used, and for $\alpha \geq 1$ it decreases as the

Fig. 1 Tsallis entropy $H_\alpha(\mathbf{p})$ of a Bernoulli distribution $\mathbf{p} = (p, 1 - p)$ for different values of α

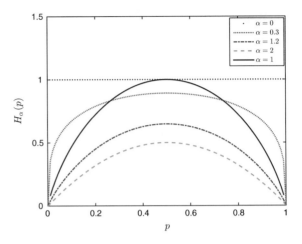

parameter α increases. Furthermore, Tsallis entropy attains a maximum uncertainty when its exponential order α is equal to zero.

For $x, y > 0$, the α-logarithm function satisfies the following property

$$\log_\alpha(xy) = \log_\alpha(x) + \log_\alpha(y) + (\alpha - 1)\log_\alpha(x)\log_\alpha(y).$$

If we consider that a physical system can be decomposed in two statistical independent subsystems with probability density functions f_1 and f_2, then using Eq. (5) it can be shown that the joint Tsallis entropy is pseudo-additive

$$H_\alpha(f_1, f_2) = H_\alpha(f_1) + H_\alpha(f_2) + (1 - \alpha)H_\alpha(f_1)H_\alpha(f_2), \qquad (5)$$

whereas the joint Shannon and Rényi entropies satisfy the additivity property:

$$H(f_1, f_2) = H(f_1) + H(f_2), \qquad (6)$$

and

$$R_\alpha(f_1, f_2) = R_\alpha(f_1) + R_\alpha(f_2). \qquad (7)$$

The pseudo-additivity property implies that Tsallis entropy has a nonextensive property for statistical independent systems, whereas Shannon and Rényi entropies have the extensive property (i.e. additivity). Furthermore, standard thermodynamics is extensive because of the short-range nature of the interaction between subsystems of a composite system. In other words, when a system is composed of two statistically independent subsystems, then the entropy of the composite system is just the sum of entropies of the individual systems, and hence the correlations between the subsystems are not accounted for. Tsallis entropy, however, does take into account these correlations due to its pseudo-additivity property. Furthermore, many objects in nature interact through long-range interactions such as gravitational or unscreened Coulomb forces. Therefore the property of additivity is very often violated, and consequently the use of a nonextensive entropy is more suitable for real-world applications.

Definition 1 Let $f_1, f_2, \ldots, f_n$ be n probability density functions. The JT divergence is defined as

$$D(f_1, \ldots, f_n) = H_\alpha \left(\sum_{i=1}^{n} \lambda_i f_i \right) - \sum_{i=1}^{n} \lambda_i H_\alpha(f_i), \tag{8}$$

where $H_\alpha(\cdot)$ is Tsallis entropy, and $\lambda = (\lambda_1, \ldots, \lambda_n)$ be a weight vector such that $\sum_{i=1}^{n} \lambda_i = 1$ and $\lambda_i \geq 0$.

Using the Jensen inequality, it is easy to check that the JT divergence is nonnegative for $\alpha > 0$. It is also symmetric and vanishes if and only if the probability density functions $f_1, f_2, \ldots, f_n$ are equal, for all $\alpha > 0$. Note that the Jensen-Shannon divergence [26] is a limiting case of the JT divergence when $\alpha \to 1$.

Unlike other entropy-based divergence measures such as the Kullback-Leibler divergence, the JT divergence has the advantage of being symmetric and generalizable to any arbitrary number of probability density functions or data sets, with a possibility of assigning weights to these density functions [21]. The following result establishes the convexity of the JT divergence of a set of probability density functions [25].

Proposition 1 *For $\alpha \in [1, 2]$, the JT divergence is a convex function of $f_1, \ldots, f_n$.*

In the sequel, we will restrict $\alpha \in [1,2]$, unless specified otherwise. In addition to its convexity property, the JT divergence is an adapted measure of disparity among n probability density functions as shown in the next result.

Proposition 2 *The JT divergence achieves its maximum value when $f_1, \ldots, f_n$ are degenerate density functions, that is f_i is a Dirac delta function.*

Proof The domain of the JT divergence is a convex polytope in which the vertices are degenerate probability density functions. That is, the maximum value of the JT divergence occurs at one of the extreme points which are the degenerate density functions.   ∎

3 Feature Extraction

A digital elevation model (DEM) is a raster of elevation values, and consists of an array of points of elevations, sampled systematically at equally spaced intervals [27]. We may represent a DEM as an image $I : \Omega \subset \mathbb{R}^2 \to \mathbb{R}$ (see Fig. 2left), where Ω is

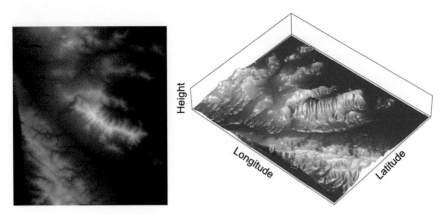

Fig. 2 Representation of a digital elevation model in 2D and 3D

a bounded set (usually a rectangle) and each image location $I(x, y)$ denotes a height value. DEMs are usually constructed from aerial photographs and require at least two images of a scene.

3.1 Morse Singular Points

A digital elevation model $I : \Omega \subset \mathbb{R}^2 \to \mathbb{R}$ may be viewed as a 3D surface $\mathbb{M} \subseteq \mathbb{R}^3$ defined as $\mathbb{M} = \{(x, y, z) : z = I(x, y)\}$ where the coordinates x, y are the latitude and longitude of a DEM, respectively, and $z = I(x, y)$ is the height value on the DEM domain Ω, as illustrated in Fig. 2 right. The surface $\mathbb{M}$ is given by the Monge patch defined by $\mathbf{r} : \Omega \to \mathbb{M}$ such that $\mathbf{r}(x, y) = (x, y, I(x, y))$. Note that the patch $\mathbf{r}$ covers all $\mathbb{M}$ (i.e. $\mathbf{r}(\Omega) = \mathbb{M}$), and it is regular, i.e. $\mathbf{r}_x \times \mathbf{r}_y \neq 0$ or equivalently, the Jacobian matrix of $\mathbf{r}$ has rank two.

Surfaces consist of geometric and topological data, and their compact representation is an important step towards a variety of computer vision and medical imaging applications. Geometry deals with measuring and computing geometric concepts, whereas topology is concerned with those features of geometry which remain unchanged after twisting, stretching or other deformations of a geometrical space. Surface singularities are prominent landmarks and their detection, recognition, and classification is a crucial step in computer vision and 3D graphics [18].

Height Function Let $h : \mathbb{M} \to \mathbb{R}$ be a height function defined by $h(\mathbf{p}) = I(x, y)$ for all $\mathbf{p} = (x, y, I(x, y)) \in \mathbb{M}$, i.e. h is the orthogonal projection with respect to the z-axis.. A point $\mathbf{p}_0$ on $\mathbb{M}$ is a singularity or critical point of h if $\mathbf{p}_0 = \mathbf{r}(x_0, y_0) = (x_0, y_0, I(x_0, y_0))$, for some $(x_0, y_0) \in \Omega$, and the gradient of $h \circ \mathbf{r}$ at (x_0, y_0) vanishes, i.e. $\nabla(h \circ \mathbf{r}(x_0, y_0)) = 0$. A singularity $\mathbf{p}_0$ is nondegenerate if the Hessian matrix $\nabla^2(h \circ \mathbf{r}(x_0, y_0))$ is nonsingular. We say that the height function is a Morse

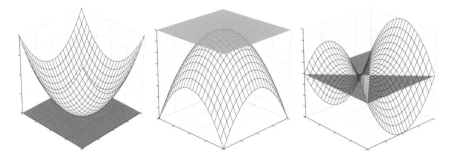

Fig. 3 Nondegenerate singular points: minimum (*left*), maximum (*center*) and saddle (*right*)

function if all its singular points are nondegenerate, that is minimum, maximum and saddle points as depicted in Fig. 3. The main theory about nondegenerate singularities is Morse theory that describes the topology changes of the level sets of Morse functions at those singularities [18, 28]. Moreover, Morse theorem is an important result which says that a small, smooth perturbation of a Morse function yields another Morse function. The density means that there is a Morse function arbitrarily close to any non-Morse function [28].

Level Sets Around Morse Critical Points   The level sets of the height function are the intersections of the surface $\mathbb{M}$ with planes orthogonal to the z-axis, as shown in Fig. 4. The original manifold may then be reconstructed if we know all its sections by these parallel planes (i.e. the surface is the union of these planes). Clearly, the level sets of the height function may have isolated points, or curves, or may contain an open subset of the plane. Furthermore, the level sets may be connected or disconnected, and the curves may have complicated Fig. 4 singularities.

Nondegenerate singularities are isolated, i.e. there cannot be a sequence of nondegenerate singularities converging to a nondegenerate singularity $\mathbf{x} \in \mathbb{M}$. A *level*

Fig. 4 Cross-sections of the height function on a torus

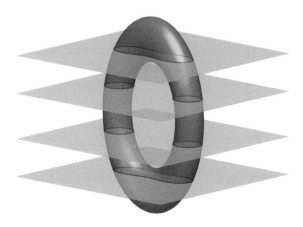

Fig. 5 Level curve Γ_a (*left*); Subsurface $\mathbb{M}_a$ (*center*); Subsurface and level curve (*right*)

Fig. 6 Evolution of the subsurface $\mathbb{M}_a$ as a changes

set $f^{-1}(a)$ of a smooth, real-valued function f on $\mathbb{M}$ at a value a may be composed of one or many connected components. Morse deformation lemma states that if no critical points exist between two level sets of f, then these two level sets are topologically equivalent and can be deformed onto one another [18]. In particular, they consist of the same number of connected components. Moreover, Morse theory implies that topological changes on the level sets occur only at critical points. This property can be illustrated by considering the subsurface $\mathbb{M}_a$ consisting of all points at which f takes values less than or equal to a real number a:

$$\mathbb{M}_a = \{\mathbf{x} \in \mathbb{M} : f(\mathbf{x}) \leq a\}. \tag{9}$$

Denote by $\Gamma_a = f^{-1}(a)$ the set of points where the value of f is exactly a. Note that when a is a regular value, the set Γ_a is a smooth curve of $\mathbb{M}$ and is the boundary of $\mathbb{M}_a$, as illustrated in Fig. 5.

Figure 6 shows the evolution of the subsurface $\mathbb{M}_a$ as a changes, when f is a height function. If $a < \min_{\mathbf{x} \in \mathbb{M}}\{f(\mathbf{x})\}$, then $\mathbb{M}_a$ is the empty set. And as we increase the parameter a, the subsurface $\mathbb{M}_a$ changes until it covers the entire surface $\mathbb{M}$.

4 Proposed Framework

In this section, a topological framework for DEM registration is presented. The proposed feature-based approach encodes a DEM into a set of Morse singularities. Then the JT dissimilarity measure is optimized in an effort to quantify the difference between two misaligned DEMs. We show that maximizing the JT divergence is tantamount to minimizing the total edge length of the joint minimal spanning tree that spans the graph generated from the Morse singularities of two misaligned DEMs.

4.1 Problem Statement

Let $I_1, I_2 : \Omega \subset \mathbb{R}^2 \to \mathbb{R}$ be two misaligned digital elevation models. The goal of geo-registration is to align the fixed I_1 to the geographic location of the moving I_2 by maximizing a dissimilarity measure D between I_1 and $\mathcal{T}_{(\mathbf{t},\theta,s)} I_2$, where $\mathcal{T}_{(\mathbf{t},\theta,s)}$ is a Euclidean transformation with translation parameter vector $\mathbf{t} = (t_x, t_y)$, a rotation parameter θ, and a scaling parameter s. In other words, the registration problem may be succinctly stated as

$$(\mathbf{t}^\star, \theta^\star, s^\star) = \arg \max_{(\mathbf{t},\theta,s)} D(I_1, \mathcal{T}_{(\mathbf{t},\theta,s)} I_2), \qquad (10)$$

or equivalently

$$(\mathbf{t}^\star, \theta^\star, s^\star) = \arg \max_{(\mathbf{t},\theta,s)} D(\mathcal{U}, \mathcal{T}_{(\mathbf{t},\theta,s)} \mathcal{V}), \qquad (11)$$

where $\mathcal{U}$ and $\mathcal{V}$ are two sets of feature vectors extracted from I_1 and I_2 respectively. To geo-register the fixed DEM to the moving DEM, we propose in this section a feature-based approach by extracting Morse singular points from these DEMs, and then quantifying the difference between them via the JT divergence as a dissimilarity measure [29].

4.2 Algorithm

The goal of our proposed approach may be described as follows: Given two misaligned DEMs to be registered, we first extract their Morse singularity features, and we compute a joint MST connecting their Morse features, then we optimize the JT divergence to bring these DEMs into alignment. Without loss of generality, we consider the transformation $\mathcal{T}_\ell$, where $\ell = (\mathbf{t}, \theta)$, i.e. a Euclidean transformation with translation parameter vector $\mathbf{t} = (t_x, t_y)$, and a rotation parameter θ. In other words, for $\mathbf{x} = (x, y)$ we have $\mathcal{T}_\ell(\mathbf{x}) = \mathbf{R}\mathbf{x} + \mathbf{t}$, where $\mathbf{R}$ is a rotation matrix given by

$$\mathbf{R} = \begin{pmatrix} \cos\theta & \sin\theta \\ -\sin\theta & \cos\theta \end{pmatrix}. \qquad (12)$$

354   A. Ben Hamza

Let I_1 and I_2 be the fixed and moving DEMs to be registered. The proposed registration algorithm may now be concisely summarized in the following three steps:

(i) Find the Morse singularity features $\mathcal{U} = \{\mathbf{u}_1, \ldots, \mathbf{u}_k\}$ and $\mathcal{V} = \{\mathbf{v}_1, \ldots, \mathbf{v}_m\}$ of I_1 and I_2, respectively.
(ii) Transform $\mathcal{V}$ to a new set $\mathcal{W} = \mathcal{T}_\ell \mathcal{V} = \{\mathbf{w}_1, \ldots, \mathbf{w}_m\}$
(iii) Find the optimal parameter vector $\boldsymbol{\ell}^\star = (\mathbf{t}^\star, \theta)$ of the JT divergence

$$\boldsymbol{\ell}^\star = \arg \max_\ell \ D(\mathcal{U}, \mathcal{W}), \tag{13}$$

where
$$D(\mathcal{U}, \mathcal{W}) = \widehat{H}_\alpha(\mathcal{U} \cup \mathcal{W}) - \lambda \widehat{H}_\alpha(\mathcal{U}) - (1 - \lambda)\widehat{H}_\alpha(\mathcal{W}), \tag{14}$$

and $\lambda = k/(k + m)$.

Figure 7 displays the block diagram of the proposed framework.

4.3 Feature Extraction

We applied the algorithm introduced by Takahashi et al. [19] to extract Morse singular points from the fixed and moving DEMs. This algorithm extracts the singularity points that follow the criteria of topological integrity of the surface of a DEM. This means that the singular points must satisfy the Euler-Poincaré formula, which states that the number of maxima, minima and saddles satisfy the topological relation given by

$$\chi = \#\text{minima} - \#\text{saddlepoints} + \#\text{maxima} = 2. \tag{15}$$

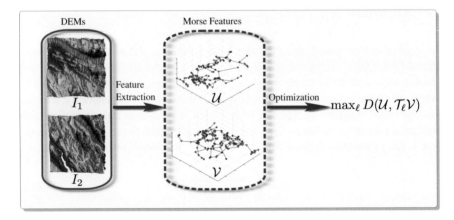

Fig. 7 Block diagram of the proposed approach

Fig. 8 The 8-neighborhood of a vertex **v**

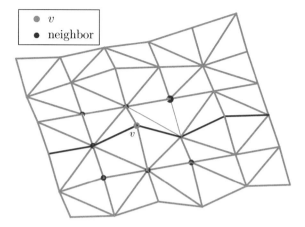

The extraction algorithm is based on the 8-neighbor method, which compares a vertex **v** with its 8-adjacent neighbors by computing the height difference between **v** and each of its neighbors, as illustrated in Fig. 8.

In addition, to ensure that the potential singular points satisfy the Euler-Poincaré formula, it is necessary to study the changes on the contours considering their heights. The classification of a feature point is based on the topological variations between the cross-sectional contours. According to this criterion, a feature point **v** is classified as a maximum point if a new contour appears at **v**. Therefore, a peak is given by a point that is higher than all other points in its neighborhood. A feature point **v** is classified as a minimum point if an existing contour disappears at **v**. Therefore, a pit is given by a point that is lower than all other points in its neighborhood. A pass occurs when a contour is divided or two contours are merged at **v**. Figure 9 displays the Morse singular points of a DEM and the corresponding MST connecting all these features.

4.4 Entropy Estimation Using Minimal Spanning Tree

Let $\mathcal{V} = \{\mathbf{v}_1, \mathbf{v}_2, \ldots, \mathbf{v}_n\}$ be a set of feature vectors (e.g., Morse singularities), where $\mathbf{v}_i \in \mathbb{R}^d$ (e.g., $d = 3$ for 3D data). A spanning tree $\mathcal{E}$ is a connected acyclic graph that passes through all features and it is specified by an ordered list of edges e_{ij} connecting certain pairs $(\mathbf{v}_i, \mathbf{v}_j)$, $i \neq j$, along with a list of edge adjacency relations. The edges e_{ij} connect all n features such that there are no paths in the graph that lead back to any given feature vector. The total length $L_{\mathcal{E}}(\mathcal{V})$ of a tree is given by

$$L_{\mathcal{E}}(\mathcal{V}) = \sum_{e_{ij} \in \mathcal{E}} \|e_{ij}\|. \tag{16}$$

The minimal spanning tree $\mathcal{E}^\star$ is the spanning tree that minimizes the total edge length $L_{\mathcal{E}}(\mathcal{V})$ among all possible spanning trees over the given features

Fig. 9 Morse singular points
of a DEM (*left*), and its
corresponding MST (*right*)

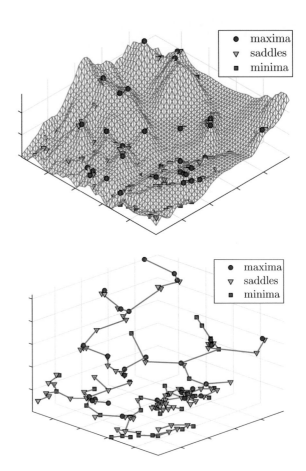

$$L^{\star}(\mathcal{V}) = \sum_{e_{ij} \in \mathcal{E}^{\star}} \|e_{ij}\| = \min_{\mathcal{E}} L_{\mathcal{E}}(\mathcal{V}). \tag{17}$$

Figure 10a, b depict an example of MSTs with 2D and 3D feature vectors, respectively.

The set $\mathcal{V}$ is called a random feature set if its elements are random variables with a probability density function f. It can be shown that

$$\lim_{n \to \infty} \frac{L^{\star}(\mathcal{V})}{n^{\alpha}} = \beta \int f(x)^{\alpha} dx \qquad a.s. \tag{18}$$

where the constant β plays a role of bias correction [11]. Hence, we may define an estimator $\widehat{H}_{\alpha}$ of Tsallis entropy as

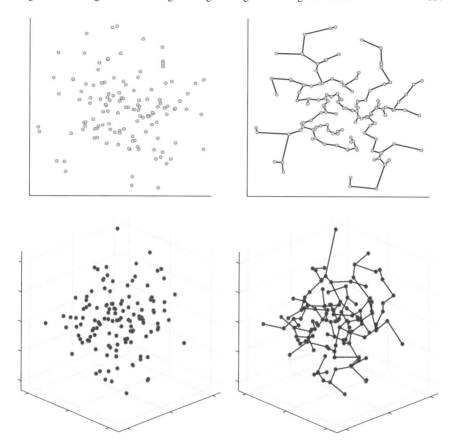

Fig. 10 MSTs in 2D and 3D using random data of 128 samples

$$\widehat{H}_\alpha(\mathcal{V}) = \frac{1}{1-\alpha} \left[\frac{L^\star(\mathcal{V})}{\beta \, n^\alpha} - 1 \right].$$ (19)

4.5 *Optimization*

Recall from Proposition 1 that the JT divergence is convex when $\alpha \in [1,2]$. Using (19) and (13), it is clear that solving

$$\boldsymbol{\ell}^\star = \arg\max_{\boldsymbol{\ell}} \; D(\mathcal{U}, \mathcal{T}_\ell \mathcal{V}),$$ (20)

may be reduced to solving

$$\ell^* = \arg\max_{\ell} \widehat{H}_\alpha(\mathcal{U} \cup \mathcal{T}_\ell V)$$

$$= \arg\max_{\ell} \frac{1}{1-\alpha} \left[\frac{L^*(\mathcal{U} \cup \mathcal{T}_\ell V)}{\beta\,(k+m)^\alpha} - 1 \right], \tag{21}$$

or equivalently solving the minimization problem

$$\ell^* = \arg\min_{\ell} L^*(\mathcal{U} \cup \mathcal{T}_\ell V)$$

$$= \arg\min_{\ell} \sum_{e_{ij} \in \mathcal{E}^*} \|e_{ij}\|, \tag{22}$$

where $\mathcal{E}^*$ is the MST of $\mathcal{U} \cup \mathcal{T}_\ell V$, and $\|e_{ij}\|$ is the edge length which depends on the transformation parameter ℓ and is given by

$$\|e_{ij}\| = \sqrt{(I_1(\mathbf{x}_i) - I_1(\mathbf{x}_j))^2 + (\mathcal{T}_\ell I_2(\mathbf{x}_i) - \mathcal{T}_\ell I_2(\mathbf{x}_j))^2}$$

$$= \sqrt{(I_1(\mathbf{x}_i) - I_1(\mathbf{x}_j))^2 + (I_2(\mathbf{R}\mathbf{x}_i + \mathbf{t}) - I_2(\mathbf{R}\mathbf{x}_j + \mathbf{t}))^2}, \tag{23}$$

where $\mathbf{x}_i = (x_i, y_i) \in \Omega$, and $\mathbf{x}_j = (x_j, y_j) \in \Omega$.

Hence, the DEM registration is performed by minimizing the total length of the minimum spanning tree which spans the joint MST generated from the overlapping feature vectors of the fixed and moving DEMs. The minimization problem given by Eq. (22) may be solved using the method of steepest descent [30]. Let ℓ^r be the value of the objective function given by Eq. (22) at the rth iteration. At each iteration, the update rule is given by

$$\ell^{r+1} = \ell^r + \alpha\, \mathbf{d}^r = \ell^r - \alpha \sum_{e_{ij} \in \mathcal{E}^*} \nabla \|e_{ij}\|, \tag{24}$$

where $\mathbf{d}^r = -\sum_{e_{ij} \in \mathcal{E}^*} \nabla \|e_{ij}\|$ is the direction vector of steepest descent that is the direction for which the objective function will decrease the fastest, and α is a scalar which determines the size of the step taken in that direction. Taking the gradient of the edge length with respect to ℓ yields

$$\nabla \|e_{ij}\| = \frac{1}{\|e_{ij}\|} \Big(I_2(\mathbf{R}\mathbf{x}_i + \mathbf{t}) - I_2(\mathbf{R}\mathbf{x}_j + \mathbf{t}) \Big)$$

$$\times \Big(\nabla I_2(\mathbf{R}\mathbf{x}_i + \mathbf{t}) \cdot J_{\mathbf{R}\mathbf{x}_i + \mathbf{t}} - \nabla I_2(\mathbf{R}\mathbf{x}_j + \mathbf{t}) \cdot J_{\mathbf{R}\mathbf{x}_j + \mathbf{t}} \Big), \tag{25}$$

where $J_{\mathbf{Rx+t}}$ denotes the Jacobian matrix of the function

$$(t_x, t_y, \theta) \rightarrow (x \cos\theta + y \sin\theta + t_x, -x \sin\theta + y \cos\theta + t_y), \tag{26}$$

which yields

$$J_{\mathbf{Rx+t}} = \begin{pmatrix} 1 & 0 & -x\sin\theta + y\cos\theta \\ 0 & 1 & -x\cos\theta - y\sin\theta \end{pmatrix} \tag{27}$$

5 Experimental Results

In this section experimental results are presented to show the much improved performance of the proposed method in DEM registration. We carried out three sets of experiments. In the first experiment, we show that Tsallis entropy gives the best results in comparison with Shannon and Rényi entropy estimators. In the second experiment, we calculate the total length of the joint MST and then estimate Tsallis entropy for two misaligned DEMs. And, in the third experiment we test the performance of the proposed gradient descent algorithm in estimating the registration parameters. In all the experiments, the tuning-parameter α of Tsallis entropy estimator was set to $\alpha = 1.5$.

5.1 Tsallis Entropy Estimator

In the first experiment we tested the performance of Tsallis entropy in comparison with Shannon and Rényi entropies, and the results are listed in Table 1. A graphical comparison of these entropies is also shown in Fig. 11, where the best performance of Tsallis entropy is clearly illustrated. We used the same $\alpha = 1.5$ for Rényi entropy. This better performance of Tsallis entropy is consistent with a variety of DEMs used for experimentation, and also with different values of the parameter α. Sample DEMs are shown in Fig. 12.

Table 1 Estimation of Shannon, Rényi and Tsallis entropies for different rotation angles

Angle θ	Shannon entropy	Rényi entropy	Tsallis entropy
15°	4.8560	4.7834	1.8171
30°	5.0303	4.9020	1.8276
45°	5.0774	4.9322	1.8302
60°	5.0863	4.9404	1.8309

Fig. 11 Graphical
comparison between
Shannon, Rényi and Tsallis
entropy estimators

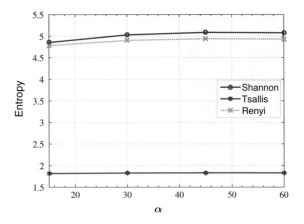

5.2   Joint MST Length and Tsallis Entropy Estimation

The second experiment deals with two misaligned DEMs I_1 and I_2 as shown in
Fig. 13left and right respectively, where for ease of visualization, only the maxima
features are displayed. We calculated the total length of the joint MST and we estimated Tsallis entopy for two misaligned DEMs. Tsallis entropy estimator for the
feature sets of the fixed and moving DEMs is given by

$$\widehat{H}_\alpha(\mathcal{U} \cup \mathcal{T}_\ell \mathcal{V}) = \frac{1}{1-\alpha}\left[\frac{L^\star(\mathcal{U} \cup \mathcal{T}_\ell \mathcal{V})}{\beta\,(k+m)^\alpha} - 1\right], \qquad (28)$$

where k and m are the numbers of Morse singular points of I_1 and I_2, respectively.
   The numerical results are depicted in Table 2. As expected, note that both the
length of the joint MST and Tsallis entropy increase as the rotation angle between
the fixed and moving DEMs increases.
   We also tested the sensitivity of the JT divergence to the entropic index α and the
bias parameter β. As can be seen in Fig. 14left, for a fixed value of α the JT divergence
is a decreasing function of the bias parameter. On the other hand, for a fixed value
of the bias parameter the JT divergence slightly increases when the entropic index
increases, but it stabilizes for larger values of α, as shown in Fig. 14right.

5.3   Estimation of the Registration Parameters

In the third experiment, we estimated the parameters corresponding to the spatial
transformation between the fixed and moving DEMs. We applied a Euclidean transformation to the moving DEM with known parameters (t_x, t_y, θ). To geo-register
these two DEMs, we then applied the iterative gradient descent algorithm described

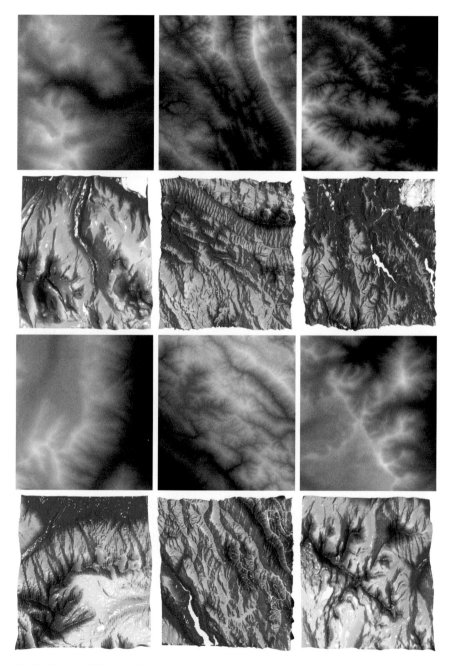

Fig. 12 Sample DEMs used for experimentation along with their surface representations

Fig. 13 Misaligned DEMs:
fixed DEM (*left*) and moving
DEM (*right*). For visual
simplicity, only maxima
features are displayed

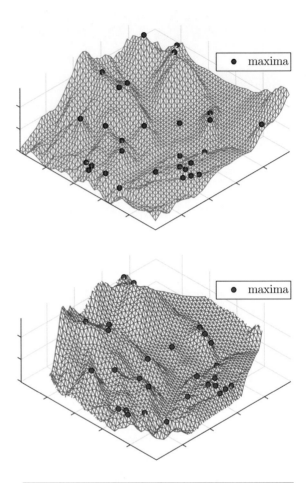

Table 2 Total length and
Tsallis entropy estimator of
the joint MST

Angle θ	$L^\star(\mathcal{U} \cup \mathcal{T}_\ell \mathcal{V})$	$\widehat{H}_\alpha(\mathcal{U} \cup \mathcal{T}_\ell \mathcal{V})$
15°	823.9646	8.9509
30°	951.1633	9.2380
60°	1064.5076	9.4632

in Sect. 3 to find the optimal parameters $t_x^\star$, $t_x^\star$ and $\theta^\star$. The registration results are shown in Table 3, where the errors between the original and the estimated transformation parameters are also listed. The estimated values indicate the effectiveness and the accuracy of the proposed algorithm in geo-registering elevation data. The joint MST of the fixed and moving are displayed in Fig. 15left and also in Fig. 15right, where the overlapping DEM surfaces are shown as well. Note that for the sake of clarity, only the maxima features are displayed.

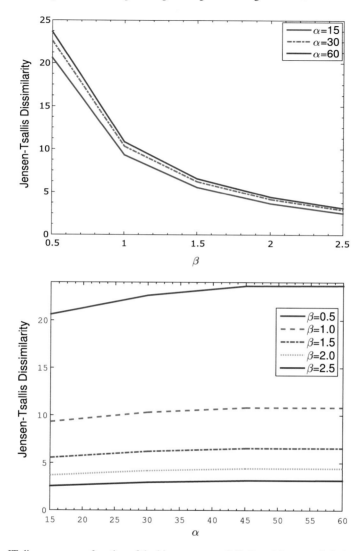

Fig. 14 JT divergence as a function of the bias parameter β (*left*) and the entropic index α (*right*)

Table 3 Registration results

ℓ			$\ell^\star$			Error $= \ell - \ell^\star$		
t_x	t_y	θ	$t_x^\star$	$t_y^\star$	$\theta^\star$	t_x^e	t_y^e	θ^e
2	4	0	2.64	3.20	1.26	0.64	0.80	1.26
2	4	5	2.02	3.65	5.70	0.02	0.35	0.70
5	10	15	5.77	9.53	15.34	0.77	0.47	0.34
4	8	20	3.46	7.38	20.87	0.54	0.62	0.87

Fig. 15 Joint MST of fixed
and moving DEMs (*left*).
Overlapping surfaces and
joint MST of fixed and
moving DEMs (*right*). For
visual simplicity, only
maxima features are
displayed

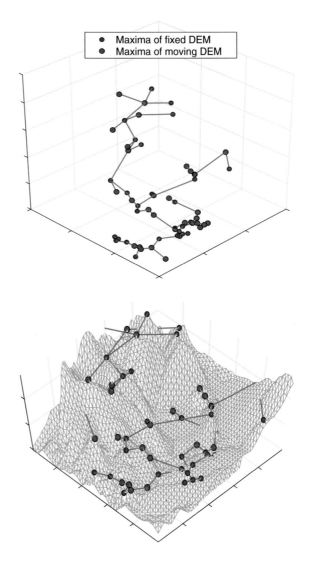

6   Conclusions

In this chapter, we proposed a topological approach for entropic registration of digital elevation models using Morse singularities. The geo-registration is achieved by minimizing the total length of the joint minimal spanning tree of the fixed and the moving elevation models. The main advantages of the proposed approach are: (i) Tsallis entropy provides a reliable data estimator, (ii) the proposed approach is simple and computationally fast, and (iii) the experimental results provide accurate registration results and clearly indicate the suitability of the proposed method for geo-registration of 3D terrain elevation data. Our future work direction is to extend the proposed singularity-based framework to 3D shape matching and retrieval by designing efficient shape signatures that will capture both the geometric and topological structure of 3D shapes.

References

1. Basseville, M.: Divergence measures for statistical data processing an annotated bibliography. Sig. Process **93**, 621–633 (2013)
2. Tang, J., Li, Z., Wangand, M., Zhao, R.: Neighborhood discriminant hashing for large-scale image retrieval. IEEE Trans. Image Process. **24**, 2827–2840 (2015)
3. Tarmissi, K., Ben Hamza, A.: Information-theoretic hashing of 3D objects using spectral graph theory. Expert Syst. Appl. **36**, 9409–9414 (2009)
4. Martins, A., Smith, N.A., Xing, E.P., Aguiar, P., Figueiredo, M.: Nonextensive information theoretic kernels on measures. J. Mach. Learn. Res. **10**, 935–975 (2009)
5. Ben Hamza, A., Krim, H.: Jensen-Rényi divergence measure: theoretical and computational perspectives. In: Proceedings of IEEE International Symposium on Information Theory (2003)
6. Xiao, Y., Cao, Z., Yuan, J.: Entropic image thresholding based on GLGM histogram. Pattern Recogn. Lett. **40**, 47–55 (2014)
7. Hajnal, J., Hill, D., Haweks, D.: Medical Image Registration. CRC Press LLC (2001)
8. Goshtasby, A.A.: 2-D and 3-D Image Registration for Medical, Remote Sensing, and Industrial Applications. Wiley Publishers (2005)
9. Viola, P., Wells, W.M.: Alignment by maximization of mutual information. Int. J. Comput. Vis. **24**, 154–173 (1997)
10. Maes, F., Collignon, A., Vandermeulen, D., Marchal, G., Suetens, P.: Multimodality image registration by maximization of mutual information. IEEE Trans. Med. Imaging **16**, 187–198 (1997)
11. Hero, A.O., Ma, B., Michel, O., Gorman, J.: Applications of entropic spanning graphs. IEEE Signal Process. Mag. **19**, 85–95 (2002)
12. Woo, J., Stone, M., Prince, J.L.: Multimodal registration via mutual information incorporating geometric and spatial context. IEEE Trans. Image Process. **24**, 757–769 (2015)
13. Demers, M.N.: Fundamentals of Geographic Information Systems. Wiley (1996)
14. Van Niel, T.G., McVicar, T.R., Li, L.T., Gallant, J.C., Yang, Q.K.: The impact of misregistration on SRTM and DEM image differences. Remote Sens. Environ. **112**, 2430–2442 (2008)
15. Zhang, T., Cen, M.: Robust DEM co-registration method for terrain changes assessment using least trimmed squares estimator. Adv. Space Res. **41**, 1827–1835 (2008)
16. Nuth, C., Kääb, A.: Co-registration and bias corrections of satellite elevation data sets for quantifying glacier thickness change. Cryosphere **5**, 271–290 (2011)
17. Ravanbakhsh, M., Fraser, C.S.: A comparative study of DEM registration approaches. J. Spat. Sci. **58**, 79–89 (2013)
18. Fomenco, A.T., Kunii, T.L.: Topological Modeling for Visualization. Springer, Tokyo (1997)
19. Takahashi, S., Ikeda, T., Shinagawa, Y., Kunii, T.L., Ueda, M.: Algorithms for extracting correct critical points and constructing topological graphs from discrete geographical elevation data. Comput. Graph. Forum **14**, 181–192 (1995)
20. Krim, H., Ben Hamza, A.: Geometric Methods in Signal and Image Analysis. Cambridge University Press (2015)
21. Khader, M., Ben Hamza, A.: An information-theoretic method for multimodality medical image registration. Expert Syst. Appl. **39**, 5548–5556 (2012)
22. Rényi, A.: On measures of entropy and information. Selected Papers of Alfréd Rényi **2**, 525–580 (1961)

23. Havrda, M.E., Charvát, F.: Quantification method of classification processes: concept of structural α-entropy. Kybernitica **3**, 30–35 (1967)
24. Tsallis, C.: Possible generalization of Boltzmann-Gibbs statistics. J. Stat. Phys. **52**, 479–487 (1988)
25. Burbea, J., Rao, C.R.: On the convexity of some divergence measures based on entropy functions. IEEE Trans. Inf. Theory **28**, 489–495 (1982)
26. Lin, J.: Divergence measures based on the Shannon entropy. IEEE Trans. Inf. Theory **37**, 145–151 (1991)
27. Wright, R., Garbeil, H., Baloga, S.M., Mouginis-Mark, P.J.: An assessment of shuttle radar topography mission digital elevation data for studies of volcano morphology. Remote Sens. Environ. **105**, 41–53 (2006)
28. Milnor, J.: Morse Theory. Princeton University Press, Princeton (1963)
29. Ben Hamza, A.: A nonextensive information-theoretic measure for image edge detection. J. Electron. Imaging **15**, 13011.1–13011.8 (2006)
30. Sabuncu, M.R., Ramadge, P.J.: Using spanning graphs for efficient image registration. IEEE Trans. Image Process. **17**, 788–797 (2008)

Visual Speech Recognition with Selected Boundary Descriptors

Preety Singh, Vijay Laxmi and Manoj Singh Gaur

Abstract Lipreading is an important research area for human-computer interaction. In this chapter, we explore relevant features for a visual speech recognition system by representing the lip movement of a person during speech, by a set of spatial points on the lip boundary, termed as *boundary descriptors*. In a real time system, minimizing the input feature vector is important to improve the efficiency of the system. To reduce data dimensionality of our feature set and identify prominent visual features, we apply feature selection technique, Minimum Redundancy Maximum Relevance (mRMR) on our set of boundary descriptors. A sub-optimal feature set is then computed from these visual features by applying certain evaluation criteria. Features contained in the sub-optimal set are analyzed to determine relevant features. It is seen that a small set of spatial points on the lip contour is sufficient to achieve speech recognition accuracy, otherwise obtained by using the complete set of boundary descriptors. It is also shown experimentally that lip width and corner lip segments are major visual speech articulators. Experiments also show high correlation between the upper and lower lips.

1 Introduction

Lipreading has garnered much interest in the area of human-computer interaction in manifold applications. Visual speech plays an important part in speech recognition systems where the audio signal is absent or of unacceptable quality, as maybe the case in presence of background noise [65]. It can be used to provide visual commands to hand-held devices in the presence of background noise. Visual inputs from the

P. Singh (✉)
The LNM Institute of Information Technology, Post Sumel, Jaipur, India
e-mail: prtysingh@gmail.com

V. Laxmi · M.S. Gaur
Malaviya National Institute of Technology, JLN Marg, Jaipur, India
e-mail: vlgaur@gmail.com

M.S. Gaur
e-mail: gaurms@gmail.com

© Springer International Publishing Switzerland 2016
A.I. Awad and M. Hassaballah (eds.), *Image Feature Detectors and Descriptors*,
Studies in Computational Intelligence 630, DOI 10.1007/978-3-319-28854-3_14

lip can be used for speech-to-text conversion for the benefit of hearing impaired individuals. Considering that digital media is also becoming a powerful weapon against crime, lipreading can aid visual surveillance technology of CCTV cameras and help decipher the spoken word in case the audio signal is degraded [11].

In a lipreading system, the region-of-interest (ROI) is the mouth, from where features are derived and classified to recognize speech. Visual feature extraction can be broadly classified into two categories:

- Bottom-up approach: In this approach, features consist of pixel intensities derived directly from the image of the mouth. This approach is motivated by the fact that visual cues from visible speech articulators, such as teeth, tongue, oral cavity and facial muscles, provide supplemental information to the audio signal [66]. However, since the dimensionality of data involved is high, appropriate transformation of pixel values has to be performed for feature reduction. The pixel-based approach also suffers from normalization problems caused by lighting and translation.
- Top-down approach: This approach takes a priori information of the lip shape into consideration. Features are derived from a statistical model fitted to the image. These may be shape-based features, usually derived from the contour of the lip. These include geometric features like lip height and width [50], Fourier descriptors or moments [70], templates [9, 29], Active contour or snakes [34] or Active shape models [10, 17, 40, 43]. While the top-down approach is less computationally expensive due to a limited number of features, its drawback lies in the fact that it lacks information from other visible speech articulators.

The bottom-up and top-down approaches can be combined to yield a robust feature vector [15, 41]. This technique may, however, result in the feature vector becoming quite large, requiring dimensionality reduction techniques. The obtained feature vectors may contain the pixel gray levels [22], principal components of pixel intensities [6] or coefficients obtained from transform-based compression.

To lessen the processing time of any real-time system, the number of input parameters should be as small as possible. Dealing with the large amount of data present in images of the mouth poses computational as well as storage challenges. The complete set of visual features might contain redundant information which may affect visual speech recognition adversely [4]. Reducing data dimensionality and extracting meaningful information from a large feature space is a challenging task. This can be done either by feature extraction or feature selection methods [27]. While the former transforms the features into a new feature space, feature selection retains the physical meaning of the features.

In this chapter, we use the top-down approach for extracting visual features from images of the mouth while speech is uttered. We determine a set of points, referred to as *boundary descriptors*, defining the shape of the lip. In our set of boundary descriptors, all features may not be relevant towards the target class and redundancy might also be present. Rather than exploring all possible subsets of features, which is computationally expensive, we use feature selection to compute relevant features. This assists in identifying prominent visual cues (which contribute towards increasing the performance of the lipreading system) and removes redundancy from the feature set. Our focus is to reduce the input feature set, but not at the cost of degrading the performance of the speech recognition system.

Minimum Redundancy Maximum Relevance (mRMR), a feature selection method [49] is employed to select and grade features contained in the set of boundary descriptors. We form subsets of the ranked features and classify them to attain a sub-optimal set of features. Determination of the sub-optimal set is done by taking into account the size of the feature set and various evaluation metrics. Correlation between various boundary descriptors is analyzed using classification results.

The main contributions of this chapter are as follows:

- A sub-optimal set of boundary descriptors is derived through feature selection.
- Sub-optimal feature set is analyzed to show that lip width is a vital visual cue.
- Significance of the center and corner lip sectors is shown through experiments.
- It is demonstrated that a high correlation exists between the upper and lower lips.

The chapter is structured as follows: Sect. 2 gives related literature in the area of visual speech recognition. Section 3 presents the methodology of our approach. Section 4 describes our experimental setup while Sect. 5 gives an analysis of the results. Finally, Sect. 6 concludes the chapter.

2   Related Work

Research in visual speech recognition has used various features and transformation techniques. The main focus is to curtail the number of features and provide good classification accuracy. Arsic and Thiran [3] select meaningful visual features using mutual information. Principal Component Analysis (PCA) is applied to mouth images and components with highest mutual information are used for classification. The best accuracy reported is 89.6 %. Feng and Wang [18] use Dual Tree Complex Wavelet Transform (DTCWT) to ascertain lip texture features. These are extracted by computing Canberra distances between the attributes in adjacent frames.

Jun and Hua [33] apply Discrete Cosine Transform (DCT) followed by Linear Discriminant Analysis (LDA) to extract visual features. The best accuracy obtained is 65.9 % with 90 co-efficients. In [42], feature extraction is done using Active Shape Model, Active Appearance Model and pixel intensities. Best accuracy of 44.6 % is realized with twenty principal components.

Lan et al. [38] extract two sets of visual features: low-level DCT features and shape and appearance information. Features are derived from z-score normalized features. Feature vectors are formed using LDA in conjunction with hyper-vectors. Comparison of the two feature vectors shows that the latter technique performs significantly better. In another recent work [39], the authors have shown that best results are achieved with a non-frontal view. This could be due to the fact that lip gestures are more pronounced in this view.

Faruquie et al. [17] extract the lip contour using Active Shape Model (ASM). Twelve geometric dimensions are extracted to characterize the shape of the lip. These include four heights in outer contour and eight heights in inner contour. Eight weights of eigenvectors are also considered. Seymour et al. [58] compare different transform-based feature types. They make use of the fast discrete curvelet transform for classifying isolated digits contained in the XM2VTS database. Experiments are performed using clean and corrupted data. It is reported that performance degrades when jitter or compression is introduced to the clean videos. However, DCT coefficients prove to be quite robust in case of blurred videos.

Gurban and Thiran [26] compute a visual feature vector which includes 64 low-frequency components along with their first and second temporal derivatives. Mutual information (MI) is computed and while MI between features and the class label is maximized, features are penalized for redundancy. Using these features, subsets are formed. Their performance is compared to features obtained by application of LDA. Experimental results show that maximum MI (MMI) features outperform MI and LDA features.

Huang et al. [30] propose the use of an infrared headset for the task of audio-visual speech recognition. They consider the use of an audio-visual headset which focuses on the speaker's mouth and employs infra-red illumination. For visual features, 100 DCT coefficients are taken into account from the region-of-interest. Intra-frame Linear Discriminant Analysis (LDA) is then applied followed by Maximum Likelihood Linear Transformation (MLTT). The final feature vector is of dimension 41. Experiments are performed on Large Vocabulary Continuous Speech Recognition (LVCSR) and digits in both settings, studio as well as with headset. The word error rate (WER) for visual only recognition for digit database is 47.1 % (studio setting) and 35.0 % (using headset). For LVCSR, visual-only results are not available.

Noda et al. [46] use a convolutional neural network (CNN) to extract visual features for the task of audio-visual speech recognition. The neural network is trained to recognize phonemes from input images of the mouth. A Japanese audio-visual dataset is used with speech data from six males, each recording 400 words. Average visual based phoneme recognition rates range from 43.15 to 47.91 % depending on size of input image.

3 Proposed Methodology

Visual features extracted for speech recognition can consist of pixel intensities of the region-of-interest, in this case, the mouth, [43, 52] or geometrical/shape based features defining the lip contour [17, 71]. We have used a set of points lying on the lip contour which define the movement of the lip boundary as it moves during utterance of speech. Prominent features are then extracted from these visual cues and classified for isolated word recognition. The steps involved in our methodology are as follows:

- Determination of boundary descriptors.
- Application of Minimum Redundancy Maximum Relevance method for feature selection.
- Formation of feature subsets using ranked features.
- Classification of feature sets for visual speech recognition.
- Computation of sub-optimal feature set.
- Identification of prominent features.

3.1 Extraction of Boundary Descriptors

In each image contained in the video sequence of a speech sample of a speaker, the Point Distribution Model (PDM) [25] is adopted to obtain the lip contour and six key-points are determined on it. These include the lip corners, three points on the upper lip arch and the bottom most point on the lower lip arch. We interpolate twenty points between all the key-points, resulting in 120 points $(P_1, P_2, \ldots, P_{120})$ on the lip contour, called *boundary descriptors* [25]. The $[xy]$ co-ordinates of each boundary descriptor give a set of 240 visual features, defined by feature set B:

$$B = \langle x_1 \quad y_1 \quad x_2 \quad y_2 \quad \ldots \quad x_{120} \quad y_{120} \rangle \tag{1}$$

A mean statistic model of the lip is created by taking the mean of the $[xy]$ co-ordinates of all points over the training images. This model, when placed over the input image, deforms to take the shape of the lip and stops when a change in intensity is sensed. This gives the lip contour in the input image. In each input image, the set of boundary descriptors, B, is computed, as described above. These boundary descriptors are shown in Fig. 1.

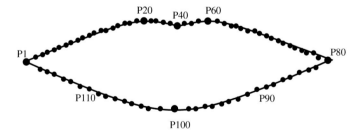

Fig. 1 Extracted *Boundary Descriptors*

3.2 Feature Selection

As the number of boundary descriptors is large, we apply feature selection on it to reduce data dimensionality. This will remove features which are redundant and select only relevant features. Feature selection method Minimum Redundancy Maximum Relevance (mRMR) [61, 62] is applied on our set of visual features, B, for ranking features according to their relevance towards the target word class. mRMR is based on the concept of mutual information. It ranks features taking into account two important criterion:

1. Maximum relevance towards the target class.
2. Minimum correlation between features, which is a cause of redundancy.

We aim to reduce the size of our input feature vector so that its usage in a real-time speech recognition system minimizes the time utilized for processing. Applying feature selection will help identify features which are irrelevant or redundant.

3.2.1 Minimum Redundancy Maximum Relevance

Minimum Redundancy Maximum Relevance (mRMR) [49] method uses mutual information between the attributes and class as well as between the attributes. It selects a subset of features such that:

- The feature set has the largest dependency on the target class, maximizing relevance.
- The features are not highly correlated, thus eliminating redundancy.

Let x and y be two variables with marginal probabilities $p(x)$ and $p(y)$ respectively. Let $p(x, y)$ be their joint probability. Then, the mutual information, I, between x and y, can be described by:

$$I(x, y) = \sum_{i,j} p(x_i, y_j) \log \frac{p(x_i, y_j)}{p(x_i)p(y_j)} \tag{2}$$

Let, $F = \{f_1, f_2 \ldots f_n\}$ represent the complete feature set of all n features. Let, S be the desired subset, $S \subset F$, containing k ($k \leq n$) features. It is desired that the feature subset should satisfy the minimal redundancy condition, $minW_I$. That is, selected features are such that they are mutually exclusive:

$$\min W_I, \quad W_I = \frac{1}{k^2} \sum_{i,j \in S} I(f_i, f_j) \tag{3}$$

Let the set of classes be represented by $C = \{C_1, C_2 \ldots C_m\}$. The relevance between target class C and feature f_i, given by I_{C,f_i}, should be maximum. Thus, the total relevance of all features in S should be maximized to satisfy the condition of maximum relevance:

$$\max V_I, \quad V_I = \frac{1}{k} \sum_{i \in S} I(C, f_i) \qquad (4)$$

Combining Eqs. 3 and 4 gives us features favouring minimum-redundancy-maximum-relevance. We have done this by computing mutual information difference (MID) criterion (Eq. 5):

$$MID = \max(V_I - W_I) \qquad (5)$$

An incremental algorithm is then applied to rank the feature set, S. Using this approach, we extract top 30 boundary descriptors from the complete set of 240 features contained in the set of boundary descriptors, B [60].

3.3 Formation of Feature Subsets

Feature subsets are constructed using the ranked 30 features. Each feature subset is denoted by B_k, where k gives the number of top features contained in the subset. Feature set F_1 contains one feature, which is the feature ranked highest by mRMR. We now add the next ranked feature to this set to form feature set F_2 (consisting of the top two features). In this manner, feature subsets $B_1, B_2, \ldots, B_{30}$ are built [64]. The process of building subsets can be seen in Fig. 2. These feature subsets are now classified to determine the sub-optimal feature subset. This is done using certain evaluation metrics, described in the following subsections.

3.4 Evaluation Metrics

Classification of the entire set of boundary descriptors and the feature subsets is evaluated using the metrics defined below:

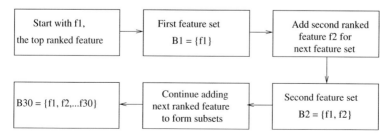

Fig. 2 Formation of feature subsets

- *Precision* (*P*): It is the ratio of correctly identified samples (True Positives) to all samples classified as existing in that class (sum of True Positives and False Positives):

$$P = \frac{TruePositives}{TruePositives + FalsePositives} \qquad (6)$$

- *F-measure* (*F*): While accuracy and word error rate are commonly used metrics, F-measure [44] is also a popular metric used in many classification problems [5, 19, 35, 69]. It can indicate the accuracy of a test [44]. It is computed as the harmonic mean of precision and recall (*R*):

$$F = \frac{2 \times Precision \times Recall}{Precision + Recall} \qquad (7)$$

where, recall (*R*) is defined by:

$$R = \frac{TruePositives}{TruePositives + FalseNegatives} \qquad (8)$$

- *ROC Area* (*A*): The overall performance of a system can be quantified by the area *A* [67] under the Receiver Operating Characteristic (ROC) curve.

3.5 Determination of Sub-optimal Feature Subset

The complete set of boundary descriptors *B* is classified using classification algorithms in WEKA [28]. We also classify the feature subsets $B_1 \ldots B_{30}$ and compare their results with the results obtained by using the complete set *B*. A set of features can be termed as optimal if it is small in size but does not degrade the recognition accuracy.

To establish a sub-optimal feature set, few terms are defined. Let, the F-measure value of the complete feature set, *B*, containing *n* features be represented by $\mathbb{F}_n$ [60, 64]. Let, $\mathbb{F}_k$ be the F-measure value of a feature subset, B_k, comprising of *k* features, where, $k < n$. The difference in F-measure value of B_k with respect to *B* is given by $\Im_k$, where:

$$\Im_k = \frac{\mathbb{F}_n - \mathbb{F}_k}{\mathbb{F}_n} \qquad (9)$$

For an enhanced recognition performance of feature subset B_k, it is desired that $\mathbb{F}_k \gg \mathbb{F}_n$ and $\Im_k < 0$. Positive values of $\Im_k$ ($\mathbb{F}_k < \mathbb{F}_n$) are indicative of degradation in the recognition rate of the feature subset. Thus, positive values are not preferable. However, if they do not exceed a certain threshold, they may be taken into consideration. $\Im_k = 0$ indicates a behaviour identical to the base vector, *B*.

For a feature subset to qualify as sub-optimal, we set thresholds for its $\Im_k$ value:

- For $\Im_k < 0$, $|\Im_k| \geq 0.1$
- For $\Im_k > 0$, $|\Im_k| \leq 0.1$

Only those feature sets which satisfy the threshold restrictions are considered. We also define η_k, the variation in feature length of feature subset as compared to the base feature vector, as:

$$\eta_k = \frac{n - k}{n} \tag{10}$$

A considerable value of η_k ($\eta_k \to 1$ for $k \to 0$) is indicative of large reduction in feature length space. Minimal length of the subset, without risking the recognition accuracy, is a desirable requirement of a sub-optimal feature subset. Establishing a sub-optimal feature vector is done in two steps:

1. Feature subsets satisfying threshold values of $\Im_k$ value are identified.
2. *Optimality Factor*, $\mathbb{O}_k$, is computed for identified feature subsets. This is defined as the product of feature size variation (η_k), precision ($\mathbb{P}$) and ROC Area ($\mathbb{A}$) of the feature subset [60, 64].

$$\mathbb{O}_k = \eta_k * \mathbb{P} * \mathbb{A} \tag{11}$$

It can be noted that the Optimality Factor, $\mathbb{O}_k$, takes into account all performance evaluators: Precision, ROC Area (maximum is desirable) and η_k, which is representative of reduction in feature length. High value of $\mathbb{O}_k$ is indicative of the fact that atleast one of the performance metric of the subset is sufficiently large for it to supersede other subsets. Feature subset with largest value of Optimality Factor is termed as the most sub-optimal feature vector.

4 Experimental Setup

We recorded an audio-visual database, influenced by some of the earlier recorded databases, AVLetters [42], OuluVS [72] and Tulips1 [45]. As much similarity as possible was maintained in terms of vocabulary, subjects and recording conditions. Additionally, recording our own database allowed us to include native speakers with different speaking habits and background, introducing more variability in the database.

Twenty subjects (ten male and ten female) spoke the English digits *zero* to *nine*. The lower half of the subject's face was recorded, in a frontal position. Since lip segmentation was not the main focus of this research, blue colour was applied to the lips of the speaker. The blue colour aids in easy extraction of the lip contour and has also been used earlier as a marker for lip detection [36, 56]. Extracted features are not affected by this.

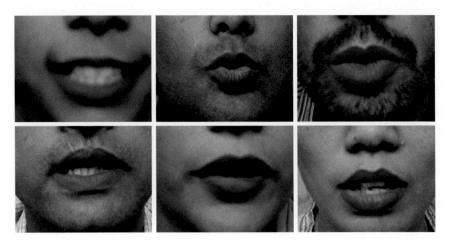

Fig. 3 Images of subjects from recorded database

To take speaker variability into consideration, two recording sessions were conducted. Three utterances of each digit per speaker were recorded in the first session and two utterances in the second session, after a gap of three months. Thus, five samples for each digit were obtained for every subject, resulting in a total of 1000 speech samples. The video sequence was frame grabbed at 30 frames per second. Some images from our database are shown in Fig. 3. The size of each image was 640×480.

From each input image, the set of boundary descriptors B are extracted. mRMR is applied on the complete set of visual features B and top 30 features are determined. Feature subsets are formed with the ranked features, as explained in Sect. 3.

While Hidden Markov Models [54] are commonly used for the task of speech recognition, we have explored the use of other popular classifiers. We have classified our feature subsets using the Random Forest (RF) and k-Nearest Neighbour (kNN) classifiers in WEKA [28]. Random Forest [8] has been used in other pattern recognition problems with excellent results [16, 47, 59, 68]. k-Nearest Neighbor [1] has also been used earlier in many classification tasks [12, 48, 57]. The 10-fold cross-validation method [37] is used.

The set of boundary descriptors B and features subsets are classified and performance of feature subsets is compared with that of base vector B using the evaluation metrics as explained in Sect. 3.5. Based on these comparisons, a sub-optimal feature subset is selected. Features contained in this set are analyzed for possible correlation between them and other features contained in B.

5   Results and Analysis

On application of mRMR on feature set B, top thirty features are obtained. Feature subsets, B_1 to B_{30}, formed with these ranked features are classified. Classification of B gives F-measure values of 0.492 (with RF) and 0.582 (with kNN) [60]. The values of the evaluation metrics of the feature subsets are shown in Fig. 4. Considering the results and feature length size, it is seen that feature vector B_{14} shows improved performance with an F-measure value of 0.501 with RF classifier and 0.613 with kNN. This is better than the F-measure values of B. The same trend is seen for Precision and ROC Area values also. Moreover, while B contains 240 features, B_{14} contains only fourteen features. Thus, B_{14} is considerably smaller in size as compared to B and also performs better than it. As discussed previously, a reduced feature set will enhance the efficiency of a visual speech recognition system, performing in a real-time scenario.

Comparing the performance of B_{14} with existing literature is not possible since the experiments have been performed on different databases, using different features and evaluation metrics. However, for the sake of completion, we take a look at the recognition performance of other lip-reading systems. Matthews et al. [42] have reported an accuracy of 26.9 % with feature length of 20 derived from an Active Shape Model. They also report an accuracy of 41.9 % with 37 features from an Active Appearance Model. Jun and Hua [33] have observed a recognition rate of 65.9 % using 90 coefficients obtained after application of DCT and LDA on their feature set. Arsic and Thiran [3] obtain an accuracy of 89.6 % using 35 most informative eigenvectors on a database of four digits only. Faruquie et al. [17] use twelve geoemtric features and eight weights of eigenvectors to obtain 31.91 % for phonetic classification.

An observation of the results obtained in literature shows that recognition results obtained using features from top-down and bottom-up approaches may vary from a low value around 20 % (in case of a large database) to about 85 % (in case of a single-speaker). Also, features consisting of transformed pixel intensities usually

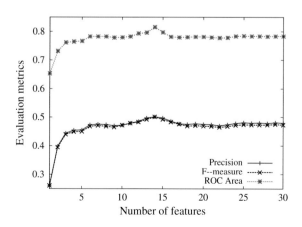

Fig. 4 Evaluation metrics of feature subsets. Results are for RF classifier

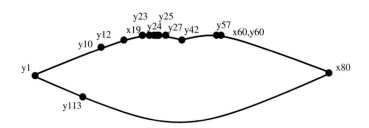

Fig. 5 Features contained in set B_{14}

deliver better results as they contain additional information contained in the speech articulators in the region around the mouth. Taking into account the reported results on different datasets and features, our near-optimal vector, B_{14} exhibits comparable performance by obtaining an F-measure value of 0.501 (using RF classifier) and 0.613 (using kNN). Moreover, it contains only fourteen boundary descriptors. A small feature set is always desirable in real-time applications. This leads to reduced computation and minimizes data handling without compromising on the recognition accuracy.

The features contained in B_{14} are shown in Fig. 5. It is observed that most of the features contained in B_{14} are y co-ordinates. According to image processing co-ordinate system, x co-ordinates will give changes in lip height while y co-ordinates will show changes in lip width. It can be inferred that most of the features contained in B_{14} are depicting changes in lip width. Thus, what is known intuitively, that human cognitive intelligence considers the changes in lip width while lip-reading, is shown here through analysis of features. It is also seen that the two lip corners, y_1 and x_{80} are in the list of prominent features for visual speech recognition.

5.1 Significance of Center and Corner Lip Contour Segments

Considering the features contained in B_{14}, we analyze the eatures contained in the center and corner areas of B_{14}. These are denoted by B_{CE} and B_{CO} respectively. The features contained in these are:

$$B_{CE} = \langle y_{57}\ y_{23}\ y_{60}\ y_{27}\ y_{24}\ y_{42}\ x_{60}\ y_{25} \rangle \tag{12}$$

$$B_{CO} = \langle y_{113}\ x_{80}\ y_1 \rangle \tag{13}$$

These feature sets are classified and their results are shown in Table 1. It can be seen from the results that the feature set containing the corner lip segment performs almost similar to the feature set containing center segment features, even though B_{CO} contains less number of features as compared to B_{CE}. The reason for this is that

Table 1 Performance of feature vectors B_{CE} and B_{CO}

Feature vector	RF			kNN		
	P	F	A	P	F	A
B_{CE}	0.445	0.440	0.753	0.231	0.231	0.572
B_{CO}	0.413	0.410	0.743	0.243	0.243	0.564
B_{14}	0.501	0.501	0.815	0.612	0.613	0.774

Also shown are the results for vector B_{14}

when lip width changes, maximum movement is seen in the corner segment while the center points do not move much. This reiterates the fact that lip corners and lip width are important visual features. However, individually, these vectors are not able to match the performance of the sub-optimal set, B_{14}.

5.2 Contribution of Upper and Lower Lips

As can be seen in Fig. 5, most of the features in B_{14} belong to the upper lip. This could be due to high correlation between upper and lower lips because of which the lower lip features were removed during the features selection process by mRMR to avoid redundancy.

To analyze this, we consider a set $B_{30-upper}$ which deletes all features lying on the upper lip (B_{upper}) from the feature set B_{30}. Similarly, B_{lower} contains boundary descriptors of the lower lip. Features contained in them are:

$$B_{30-upper} = \langle y_1 \, x_{80} \, y_{93} \, y_{96} \, y_{97} \, y_{98} \, y_{103} \, y_{113} \, x_{116} \rangle \tag{14}$$

$$B_{upper} = \langle x_1 \, y_1 \, x_2 \, y_2, \ldots, x_{80} \, y_{80} \rangle \tag{15}$$

$$B_{lower} = \langle x_{81} \, y_{81} \, x_{82} \, y_{82}, \ldots, x_{120} \, y_{120} \rangle \tag{16}$$

If high redundancy exists between upper and lower lip features, the feature vector $B_{30-upper}$ should perform almost similar to B_{14}. The classification results of these features sets are shown in Table 2. It is noted that while the performance of $B_{30-upper}$ degrades slightly as compared to B_{14}, the difference is not very significant. Thus, it can be argued that there is high correlation between upper and lower lips. The same inference can be achieved if individual results of B_{upper} and B_{lower} are seen. They are almost similar. The upper lip performs slightly better than lower lip and is able to imitate the complete lip contour (B) quite closely.

Table 2 Performance of complete feature set (B), sub-optimal vector (B_{14}), modified feature set ($B_{30} - B_{upper}$), upper lip (B_{upper}) and lower lip (B_{lower})

Feature vector	RF			kNN		
	P	F	A	P	F	A
B	0.499	0.492	0.800	0.583	0.582	0.768
B_{14}	0.501	0.501	0.815	0.612	0.613	0.774
$B_{30-upper}$	0.470	0.465	0.775	0.351	0.351	0.639
B_{upper}	0.491	0.484	0.792	0.557	0.557	0.754
B_{lower}	0.486	0.480	0.789	0.566	0.566	0.758

6 Conclusions

This chapter presents a set of boundary descriptors representing the lip shape for visual speech recognition. Feature selection technique, Minimum Redundancy Maximum Relevance, is applied to select and rank the boundary descriptors. A sub-optimal feature set, containing only fourteen features is determined by computing the Optimality Factor. It is seen that while this sub-optimal feature set has a reduced number of attributes, there is no compromise on the word classification accuracy. The features contained in the sub-optimal feature set are also analyzed. From our experiments, we infer the following results:

- Lip width is a major visual speech cue. Most of the features contained in the sub-optimal set are representative of the lip width.
- Features contained in the corner lip segment show more movement during speech as compared to features in the center lip segment.
- It is also observed that there is high correlation between the upper and lower lips.

In their future research, the authors would extend the experiments to visual recognition of continuous speech.

Acknowledgments The authors would like to thank the Department of Science & Technology, Government of India, for funding and supporting this project.

References

1. Aha, D., Kibler, D.: Instance-based learning algorithms. Mach. Learn. **6**, 37–66 (1991)
2. Aravabhumi, V.R., Chenna, R.R., Reddy, K.U.: Robust Method to Identify the Speaker using Lip Motion Features. In: International Conference on Mechanical and Electrical Technology, pp.125–129 (2010)
3. Arsic, I., Thiran, J.P.: Mutual information eigenlips for audio-visual speech recognition. In: 14th European Signal Processing Conference, pp. 1–5 (2006)
4. Bala, R., Agrawal, R.K.: Mutual information and cross entropy framework to determine relevant gene subset for cancer classification. Informatica **35**, 375–382 (2011)

5. Batista, F., Caseiro, D., Mamede, N., Trancoso, I.: Recovering punctuation marks for automatic speech recognition. Interspeech 2153–2156 (2007)
6. Bregler, C., Konig, Y.: Eigenlips for robust speech recognition. In: International Conference on Acoustics,Speech and Signal Processing, pp. 669–672 (1994)
7. Brooke, N.M., Scott, S.D.: PCA image coding schemes and visual speech intelligibility. Proc. Inst. Acoust. **16**(5), 123–129 (1994)
8. Breiman, L.: Random forests. Mach. Learn. **45**(1), 5–32 (2001)
9. Chandramohan, D., Silsbee, P.L.: A multiple deformable template approach for visual speech recognition. In: Fourth International Conference on Spoken Language Processing, pp. 50–53 (1996)
10. Cootes, T.F., Taylor, C.J., Cooper, D.H., Graham, J.: Active shape models: their training and application. Comput. Vis. Image Underst. **61**(1), 38–59 (1995)
11. Davies, A., Velastin, S.: A progress review of intelligent CCTV surveillance systems. IEEE Intelligent Data Acquisition and Advanced Computing Systems: Technology and Applications, pp. 417–423 (2005)
12. Deselaers, T., Heigold, G., Ney, H.: Speech recognition with state-based nearest neighbour classifiers. Interspeech 2093–2096 (2007)
13. Dieckmann, U., Plankensteiner, P., Schamburger, R., Froeba, B., Meller, S.: SESAM: a biometric person identification system using sensor fusion. Pattern Recogn. Lett. **18**(9), 827–833 (1997)
14. Duchnowski, P., Hunke, M., Busching, D., Meier, U., Waibel, A.: Toward movement-invariant automatic lip-reading and speech recognition. In: IEEE International Conference on Acoustics,Speech and Signal Processing, pp. 109–112 (1995)
15. Dupont, S., Luettin, J.: Audio-visual speech modeling for continuous speech recognition. IEEE Trans. Multimed. **2**(3), 141–151 (2000)
16. Fanelli, G., Gall, J., Gool, L.V.: Hough transform based mouth localization for audio visual speech recognition. In: British Machine Vision Conference (2009)
17. Faruquie, T.A., Majumdar, A., Rajput, N., Subramaniam, L.V.: Large vocabulary audio-visual speech recognition using active shape models. International Conference on Pattern Recognition, pp. 106–109 (2000)
18. Feng, X., Wang, W.: DTCWT-based dynamic texture features for visual speech recognition. In: IEEE Asia Pacific Conference on Circuits and Systems, pp. 497–500 (2008)
19. Florian, R., Ittycheriah, A., Jing, H., Zhang, T.: Named entity recognition through classifier combination. In: Seventh Conference on Natural Language Learning, pp. 168–171 (2003)
20. Frischholz, R.W., Dieckmann, U.: Bioid: a multimodal biometric identification system. IEEE Comput. **33**(2), 64–68 (2000)
21. Furui, S.: Recent advances in speaker recognition. Pattern Recogn. Lett. **18**(9), 859–872 (1997)
22. Gordan, M., Kotropoulos, C., Pitas, I.: A support vector machine-based dynamic network for visual speech recognition applications. EURASIP J. Appl. Sig. Process **11**, 1248–1259 (2002)
23. Graf, H.P., Cosatto, E., Potamianos, M.: Robust recognition of faces and facial features with a multi-modal system. IEEE Int. Conf. Syst. Man Cybern. B Cybern. 2034–2039 (1997)
24. Gudavalli, M., Raju, S.V., Babu, A.V., Kumar, D.S.: Multimodal biometrics-sources, architecture and fusion techniques: an overview. In: International Symposium on Biometrics and Security Technologies, pp. 27–34 (2012)
25. Gupta, D. Singh, P., Laxmi, V., Gaur, M.S.: Comparison of parametric visual features for speech recognition. In: IEEE International Conference on Network Communication and Computer, pp. 432–435 (2011)
26. Gurban, M., Thiran, J.-P.: Information theoretic feature extraction for audio-visual speech recognition. IEEE Trans. Signal Process. **57**(12), 4765–4776 (2009)
27. Guyon, I., Elisseeff, A.: An introduction to variable and feature selection. J. Mach. Learn. Res. **3**, 1157–1182 (2003)
28. Hall, M., Frank, E., Holmes, G., Pfahringer, B., Reutemann, P., Witten, I.H.: The WEKA data mining software: an update. SIGKDD Explor. **11**(1), 10–18 (2009)

29. Hennecke, M.E., Prasad, K.V., Venkatesh, K., David, P., Stork, D.G.: Using Deformable Templates to Infer Visual Speech Dynamics. In: 28th Annual Asilomar Conference on Signals, System and Computer, pp. 578–582 (1994)
30. Huang, J., Potamianos, G., Connell, J., Neti, C.: Audio-visual speech recognition using an infrared headset. Speech Commun. **44**, 83–96 (2004)
31. Ichino, M., Sakano, H., Komatsu, N.: Multimodal biometrics of lip movements and voice using Kernel Fisher Discriminant Analysis. In: 9th International Conference on Control, Automation, Robotics and Vision, pp. 1–6 (2006)
32. Jain, A., Ross, A., Pankanti, S.: Biometrics: a tool for information security. IEEE Trans. Inf. Forensics Secur. **1**(2), 125–143 (1997)
33. Jun, H., Hua, Z.: Research on visual speech feature extraction. In: International Conference on Computer Engineering and Technology, pp. 499–502 (2009)
34. Kass, M., Witkin, A., Terropoulos, D.: Snakes: active contour models. Int. J. Comput. Vis. **1**, 321–331 (1988)
35. Kawahara, T., Hasegawa, M.: Automatic indexing of lecture speech by extracting topic-independent discourse markers. In: IEEE International Conference on Acoustics, Speech, and Signal Processing, pp. I-1-I-4 (2002)
36. Kaynak, M.N., Zhi, Q., Cheok, A.D., Sengupta, K., Zhang, J., Ko, C.C.: Analysis of lip geometric features for audio-visual speech recognition. IEEE Trans. Syst. Man Cybern. Part A Syst. **34**(4), 564–570 (2004)
37. Kohavi, R.: A study of crossvalidation and bootstrap for accuracy estimation and model selection. In: 14th International Joint Conference on Artificial Intelligence, pp. 1137–1143 (1995)
38. Lan, Y., Theobald, B., Harvey, R., Ong, E., Bowden, R.: Improving visual features for lipreading. In: International Conference on Auditory-Visual Speech Processing, pp. 142–147 (2010)
39. Lan, Y., Theobald, B., Harvey, R.: View independent computer lip-reading. In: IEEE International Conference on Multimedia, pp. 432–437 (2012)
40. Luettin, J., Thacker, N.A., Beet, S.W.: Visual speech recognition using active shape models and hidden markov models. In: IEEE International Conference on Acoustics, Speech and Signal Processing, pp. 817–820 (1996)
41. Matthews, I.: Features for Audio-Visual Speech Recognition. Ph. D. Thesis. School of Information Systems, University of East Anglia, Norwich, United Kingdom (1998)
42. Matthews, I., Cootes, T.F., Bangham, J.A., Cox, S., Harvey, R.: Extraction of visual features for lipreading. IEEE Trans. Pattern Anal. Mach. Intell. **24**(2), 198–213 (2002)
43. Matthews, I., Potamianos, G., Neti, C., Luettin, J.: A comparison of model and transform-based visual features for audio-visual LVCSR. In: IEEE International Conference on Multimedia and Expo, pp. 825–828 (2001)
44. McCowan, I.A., Moore, D., Dines, J., Gatica-Perez, D., Flynn, M., Wellner, P., Bourlard, H.: On the use of information retrieval measures for speech recognition evaluation. Idiap-RR (2004)
45. Movellan, J.R.: Visual speech recognition with stochastic networks. In: Tesauro, G., Touretzky, D.S., Leen, T.K. (eds.) Advances in Neural Information Processing Systems. MIT Press (1995)
46. Noda, K., Yamaguchi, Y., Nakadai, K., Okuno, H.G., Ogata, T.: Audio-visual speech recognition using deep learning. J. Appl. Intell. **42**(4), 722–737 (2015)
47. Oparin, L., Gauvain, J.L.: Large-scale language modeling with random forests for mandarin Chinese speech-to-text. In: 7th International Conference on Advances in Natural Language Processing, pp. 269–280 (2010)
48. Pao, T-L., Liao, Wen-Y., Chen, Y.T.: Audio-Visual speech recognition with weighted KNN-based classification in mandarin database. In: IEEE Third International Conference on Intelligent Information Hiding and Multimedia Signal Processing, pp. 39–42 (2007)
49. Peng, H.C., Long, F., Ding, C.: Feature selection based on mutual information: criteria of max-dependency, max-relevance, and min-redundancy. IEEE Trans. Pattern Anal. Mach. Intell. **27**(8), 1226–1238 (2005)
50. Petajan, E.D.: Automatic lipreading to enhance speech recognition. In: Conferenceon computer vision and pattern recognition, pp. 40–47 (1985)

51. Potamianos, G., Graf, H.P., Cosatto, E.: An image transform approach for HMM based automatic lipreading. In: International Conference on Image Processing, pp. 173–177 (1998)
52. Potamianos, G., Verma, A., Neti, C., Iyengar, G., Basu, S.: a cascade image transform for speaker independent automatic speechreading. In: IEEE International Conference on Multimedia and Expo (II), pp. 1097–1100 (2000)
53. Potamianos, G., Neti, C., Gravier, G., Garg, A., Senior, A.W.: Recent advances in the automatic recognition of audiovisual speech. Proc. IEEE **91**(9), 1306–1326 (2003)
54. Rabiner, L.R.: A tutorial on hidden Markov models and selected applications in speech recognition. In: Waibel, A., Lee, K.-F. (eds.) Readings in Speech Recognition, pp. 267–296. Morgan Kaufmann Publishers Inc., USA (1990)
55. Ross, A., Jain, A.: Information fusion in biometrics. Pattern Recogn. Lett. Special Issue Multimodal Biometrics **24**(13), 2115–2125 (2003)
56. Saenko, K., Darrell, T., Glass, J.R.: Articulatory features for robust visual speech recognition. In: 6th International Conference on Multimodal Interfaces, pp. 152–158 (2004)
57. Saitoh, T., Hisagi, M., Konishi, R.: Analysis of Features for Efficient Japanese Vowel Recognition. IEICE Trans. Inf. Syst. **E90-D**(11), 1889–1891 (2007)
58. Seymour, R., Stewart, D., Ming, J.: Comparison of image transform-based features for visual speech recognition in clean and corrupted videos. J. Image Video Process. 14:1–14:9 (2008)
59. Su, Y., Jelinek, F., Khudanpur, S.: Large-scale random forest language models for speech recognition. Interspeech 598–601 (2007)
60. Singh, P., Laxmi, V., Gaur, M.S.: Lip peripheral motion for visual surveillance. In: 5th International Conference on Security of Information and Networks, pp. 173–177 (2012)
61. Singh, P., Laxmi, V., Gaur, M.S.: Relevant mRMR features for visual speech recognition. Int. Conf. Recent Adv. Comput. Softw. Syst. 148–153 (2012)
62. Singh, P., Laxmi, V., Gaur, M.S.: Speaker Identification using Optimal Lip Biometrics. 5th IAPR International Conference on Biometrics, pp. 472–477 (2012)
63. Singh, P., Laxmi, V., Gaur, M.S.: Visual speech as behavioural biometric. In: Kisku, D.R., Gupta, P., Sing, J.K. (eds.) Advances in Biometrics for Secure Human Authentication and Recognition. Taylor and Francis (2013)
64. Singh, P., Laxmi, V., Gaur, M.S.: Near-optimal geometric feature selection for visual speech recognition. Int. J. Pattern Recogn. Artif. Intell. **27**(8) (2013)
65. Sumby, W.H., Pollack, I.: Visual contribution to speech intelligibility in noise. J. Acoust. Soc. Am. **26**(2), 212–215 (1954)
66. Summerfield, Q., Macleod, A., McGrath, M., Brooke, M.: Lips, teeth, and the benefits of lipreading. In: Young, A.W., Ellis, H.D. (eds.) Handbook of Research on Face Processing, pp. 218–223. Elsevier Science Publishers, Amsterdam (1989)
67. Tan, P-N., Steinbach, M., Kumar, V.: Introduction to Data Mining. Addison-Wesley Longman Publishing Co., Inc. (2005)
68. Xue, J., Zhao, Y.: Random-forests-based phonetic decision trees for conversational speech recognition. In: IEEE International Conference on Acoustics, Speech and Signal Processing, pp. 4169–4172 (2008)
69. Yamahata, S., Yamaguchi, Y., Ogawa, A., Masataki, H., Yoshioka, O., Takahashi, S.: Automatic vocabulary adaptation based on semantic similarity and speech recognition confidence measure. Interspeech (2012)
70. Zekeriya, S.G., Gurbuz, S., Tufekci, Z., Patterson, E., Gowdy, J.N.: Application of affine-invariant fourier descriptors to lipreading for audio-visual speech recognition. In: IEEE International Conference on Acoustics, Speech, and Signal Processing, pp. 177–180 (2001)
71. Zhang, X., Mersereau, R.M., Clements, M., Broun, C.C.: Visual speech feature extraction for improved speech recognition. In: IEEE International Conference on Acoustics, Speech, and SignalProcessing, pp. II-1993-II-1996 (2002)
72. Zhao, G., Barnard, M., Pietikäinen, M.: Lipreading with local spatiotemporal descriptors. IEEE Trans. Multimed. **11**(7), 1254–1265 (2009)

Application of Texture Features for Classification of Primary Benign and Primary Malignant Focal Liver Lesions

Nimisha Manth, Jitendra Virmani, Vinod Kumar, Naveen Kalra
and Niranjan Khandelwal

Abstract The present work focuses on the aspect of textural variations exhibited by primary benign and primary malignant focal liver lesions. For capturing these textural variations of benign and malignant liver lesions, texture features are computed using statistical methods, signal processing based methods and transform domain methods. As an application of texture description in medical domain, an efficient CAD system for primary benign i.e., hemangioma (HEM) and primary malignant i.e., hepatocellular carcinoma (HCC) liver lesions based on texture features derived from B-Mode liver ultrasound images of Focal liver lesions has been proposed in the present study. The texture features have been computed from the inside regions of interest (IROIs) i.e., from the regions inside the lesion and one surrounding region of interest (SROI) for each lesion. Texture descriptors are computed from IROIs and SROIs using six feature extraction methods namely, FOS, GLCM, GLRLM, FPS, Gabor and Laws' features. Three texture feature vectors (TFVs) i.e., TFV1 consists of texture features computed from IROIs, TFV2 consists of texture ratio features (i.e., texture feature value computed from IROI divided by texture feature value computed from corresponding SROI) and TFV3 computed by combining TFV1 and TFV2 (IROIs texture features + texture ratio features) are subjected to classification by SVM and SSVM classifiers. It is observed that the performance of SSVM based CAD system is better than SVM based CAD system with respect to (a) overall classification accuracy (b)

N. Manth
Jaypee University of Information Technology, Solan, Himachal Pradesh, India
e-mail: manthnimisha@gmail.com

J. Virmani (✉)
Thapar University-Patiala, Patiala, Punjab, India
e-mail: jitendra.virmani@gmail.com

V. Kumar
Indian Institute of Technology-Roorkee, Roorkee, India
e-mail: vinodfee@gmail.com

N. Kalra · N. Khandelwal
Post Graduate Institute of Medical Education and Research-Chandigarh, Chandigarh, India
e-mail: navkal2004@yahoo.com

N. Khandelwal
e-mail: khandelwaln@hotmail.com

© Springer International Publishing Switzerland 2016
A.I. Awad and M. Hassaballah (eds.), *Image Feature Detectors and Descriptors*,
Studies in Computational Intelligence 630, DOI 10.1007/978-3-319-28854-3_15

385

individual class accuracy for atypical HEM class and (c) computational efficiency. The promising results obtained from the proposed SSVM based CAD system design indicates its usefulness to assist radiologists for differential diagnosis between primary benign and primary malignant liver lesions.

Keywords Texture features · Computer aided diagnostic system · Liver ultrasound images · Focal liver lesions · Primary benign lesion · Primary malignant lesion · Support vector machine classifier · Smooth support vector machine classifier

1   Introduction

The Texture descriptors have been used for efficient description of diagnostic information present in medical images. Accordingly, texture feature extraction is the first step towards designing a computer aided diagnostic system. The main idea behind feature extraction is to compute the mathematical descriptors describing the properties of a region, i.e. region of interest or ROI. These features can be extracted using different methods including statistical, signal processing based and transform domain methods. The different methods of feature extraction are depicted in Fig. 1.

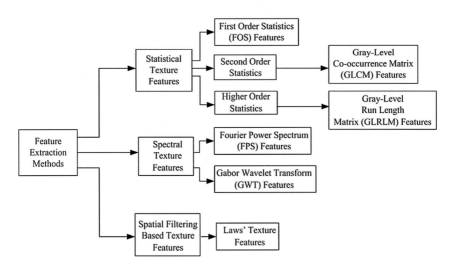

Fig. 1 Different texture feature extraction methods

1.1 Statistical Texture Features

The statistical texture features are based on spatial distribution of the gray level intensity values in image. The statistical feature extraction methods can be categorized as first order statistics (FOS), second order statistics (GLCM), and higher order statistics (GLRLM) based methods.

1.2 Spectral Texture Features

The spectral texture features can be computed by FPS method and GWT method as described below:

(*a*) *FPS Features*: For each ROI, two features i.e., angular sum and radial sum of the discrete Fourier transform (i.e., DFT) can be computed using FPS method.

(*b*) *GWT Features*: Gabor filter provides useful texture descriptors by using multi-scale features estimated at different scales and orientations. The 2D-GWT, considering N scales and M orientations, result in a group of (N × M = X) wavelets. When this group of Gabor filters family of X wavelets is convolved with a given ROI image, a set of X feature images are obtained. From these X feature images statistical features such as, mean and standard deviation are computed and are used as texture descriptors.

1.3 Spatial Filtering Based Texture Features

The Laws' texture features are spatial filtering based texture descriptors which are used to determine the texture properties of a ROI, by performing local averaging (L), edge detection (E), spot detection (S), wave detection (W), and ripple detection (R) in texture.

2 Application of Texture Features: CAD System for Differential Diagnosis Between FLLs

2.1 Background

Even though biopsy is considered as the golden standard for liver disease diagnosis, ultrasonography is commonly preferred for screening, because of its nonradioactive, inexpensive and non-invasive nature [1–5]. The sensitivity of contrast enhanced spiral computed tomography (CT), contrast enhanced Ultrasound (US), and magnetic resonance imaging (MRI) modalities for the detection and characterization of focal liver lesions (FLLs) is much higher than the conventional gray scale US. However, these imaging modalities are not generally available, costly, and produce greater operational inconvenience [1, 6–10]. Thus, a computer aided diagnostic system (CAD) for precise characterization of primary benign and primary malignant FLLs, based on conventional gray scale B-mode liver US is extremely desired to assist radiologists during routine clinical practice.

The aim of the present work is to highlight the application of texture descriptors for description of diagnostic information in the form of texture for differential diagnosis between primary and secondary malignant FLLs using B-Mode US images. According, a CAD system design for differential diagnosis between hemangioma (HEM i.e., primary benign FLL) and hepatocellular carcinoma (HCC i.e., primary malignant FLL) based on texture features derived from the B-mode liver US images of FLLs is proposed in the present study. The motivation behind considering these image classes is that the incidence of these lesions is very high in comparison to other primary benign and primary malignant lesions. Among the primary malignant lesions, hepatoblastoma (7 %), cholangiocarcinoma and cystadenocarcinoma (6 %) occur rarely therefore, the most commonly occurring primary malignant lesion i.e., HCC is considered [11–13]. Among all the primary benign lesions of liver, HEM is the most commonly occurring primary benign FLL [8, 14].

Almost 70 % of the total HEM cases encountered in day to day practice are typical HEMs, which appears as well circumscribed uniformly hyperechoic lesion [1, 7–9, 14–16]. On the other hand, atypical HEMs may be hypoechoic or even isoechoic, mimicking the sonographic appearances of HCC lesions [1, 8, 10]. The experienced participating radiologists were of the view that HCC lesions cannot be divided into typical and atypical sub-classes because there is no typical appearance for HCC. However, the participating radiologists opined that images in HCC class should include both small hepatocellular carcinomas (i.e., SHCCs, <2 cm in size) and large hepatocellular carcinomas (i.e., LHCCs, >2 cm in size). The differentiation between typical HEM and HCC is easy due to the typical sonographic appearance of the former one, but differential diagnosis between atypical HEM and HCC lesions is difficult even for the experienced radiologists [8].

In 85 % of cases, HCC develops in patient with cirrhosis [1, 6–9, 14, 15, 17–19]. In radiology, cirrhosis condition is considered as a precursor for the occurrence of HCC [1, 6–8]. Also, this feature favours the differential diagnosis of HCC with atypical HEMs. The sonographic appearances of SHCC lesions vary from hypoechoic to hyperechoic, whereas LHCC generally appears with mixed echogenecity [7, 8]. In order to design an efficient classifier, it should be ensured that the database for designing the classifier should be diversified and comprehensive i.e., it should include both typical and atypical HEMs as well as SHCC and LHCC cases. Accordingly, the dataset used in the present work includes 10 typical and 6 atypical HEM images and 13 SHCCs and 15 LHCCs images. The sample images for HEM (typical and atypical) and HCC (SHCC and LHCC) lesions are as shown in Fig. 2.

The differential diagnosis between HEM and HCC lesions with the help of conventional gray scale B-mode US is considered difficult because of various limitations including: (a) limited sensitivity for the detection of SHCCs developed on the

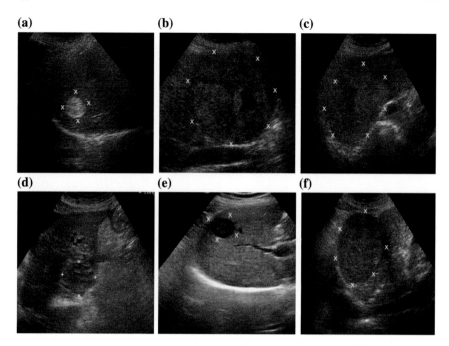

Fig. 2 Conventional gray scale B-mode liver US images. **a** typical HEM with homogeneously hyperechoic appearance; **b**, **c** atypical HEM with mixed echogenecity; **d** small HCC with mixed echogenecity; **e** small HCC with hypoechoic appearance; **f** large HCC with mixed echogenecity

cirrhotic liver which is generally coarse-textured and nodular [1, 7–9, 20], (*b*) in most of the cases, the sonographic appearances of HCC and atypical HEMs are overlapping [1, 8, 9, 20, 21], (*c*) in certain cases, it is difficult to differentiate isoechoic lesions with very slim liver to lesion contrast [8, 17]. So, it is very important to overcome these limitations by design of an efficient CAD system with representative and diversified image database containing typical and atypical cases of HEM image class and SHCC and LHCC variants of the HCC image class.

In literature, there are only few related studies on classification of FLLs. The brief description of these studies is given in Table 1. The study in [22] used regions of interest (ROIs) of size 10×10 for computing gray-level run length matrix (GLRLM) and first order statistics (FOS) features to classify normal (NOR), HEM and malignant liver lesions by using neural network (NN) and linear discriminant analysis (LDA) classifier.

The study in [23] computed texture features based on autocorrelation, gray level co-occurrence matrix (GLCM), edge frequency and Laws' mask analysis from ROI of size 10×10 for the classification of NOR, Cyst, HEM and malignant liver lesions by using neural network classifier. The research work carried out in study [18] used ROI of size 32×32 for computing GLCM, GLRLM, FPS, and Laws' texture features to classify HCC and metastatic carcinoma (MET) lesions by using support vector

Table 1 Studies on classification of FLLs using B-Mode Liver US images

Investigator	Liver image class	No. of ROIs	ROI size (pixels)	Classifiers used
Sujana et al. [22]	NOR/HEM/Malignant	113	10 × 10	NN/LDA
Poonguzhali et al. [23]	NOR/Cyst/HEM /Malignant	120	10 × 10	NN
Virmani et al. [18]	HCC/MET	174 (120 IROIs and 54 SROIs)	32 × 32	SVM
Virmani et al. [13]	NOR/Cyst/HEM/HCC/MET	491 (380 IROIs and 111 SROIs)	32 × 32	KNN/PNN/BPNN
Virmani et al. [15]	NOR/Cyst/HEM/HCC/MET	491 (380 IROIs and 111 SROIs)	32 × 32	SVM
Virmani et al. [3]	NOR/Cyst/HEM/HCC/MET	491 (380 IROIs and 111 SROIs)	32 × 32	Ensemble of NN classifiers
Yoshida et al. [16]	HEM/Malignant	193	64 × 64	NN
Present study (2015)	HEM/HCC	204 (160 IROIs and 44 SROIs)	32 × 32	SSVM

Note The HCCs evolved on cirrhotic liver only are considered as the occurrence of HCC on normal liver is rare; *Nor* Normal; *HEM* Hemangioma (primary benign lesion); *HCC* Hepatocellular carcinoma (primary malignant lesion); *MET* metastatic carcinoma (secondary malignant lesion)

machine (SVM) classifier. In studies [13, 15], five class classification between NOR, Cyst, HEM, HCC and MET classes has been carried out considering the ROI size of 32 × 32 pixels using KNN, PNN, BPNN and SVM classifiers. The study in [3] reports five class classification between NOR, Cyst, HEM, HCC and MET classes considering two stage classification approach using 11 neural network classifiers. In first stage a single five class NN was used for prediction of probability of each class and in the second stage 10 binary NN classifiers were used. Based on the first two highest probabilities predictions of the first stage five class NN, the testing instance was passed to the corresponding binary NN of the second stage. The research work reported in [3, 13, 15] computed the texture features based on FOS, GLCM, GLRLM, FPS, Gabor Wavelet Transform (GWT) and Laws' features. However, the study in [16] reports the binary classification of HCC and HEM, HCC and MET, and MET and HEM lesions by computing the single-scale and multi-scale texture features of 64 × 64 sized ROIs by using NN classifiers.

It is highlighted in other studies that minimum 800 pixels are required to compute reliable estimates of statistical features [24–26]. However, the research work reported in [22, 23] has been carried out on ROI of size 10 × 10 which yields smaller number of pixels. The studies [3, 13, 15, 18] used an ROI size of 32 × 32 and Yoshida et al. [16] used the ROI size of 64 × 64. Since, necrotic areas within the lesions should be avoided for cropping of inside ROIs (i.e., IROIs) and inhomogeneous areas include blood vessel etc. should be avoided for cropping of surrounding ROIs (i.e., SROIs), so in the present work, considering ROI size larger than 32 × 32 was not feasible. In studies [16, 22, 23] texture features of IROIs were only considered and the dataset description as to how many typical and atypical HEMs and how many SHCC and LHCC were taken into consideration is not described.

As HCC develops on cirrhotic and nodular background, therefore, the experienced radiologists carry out differential diagnosis between atypical HEM and HCC by visual analysis of texture information of regions inside and outside the lesions. Therefore, in the present work, an investigation of contribution made by texture information present in the regions inside and outside of the lesion has been carried out for characterization of HEM and HCC lesions. For the classification task, support vector machine (SVM) classifier and smooth support vector machine (SSVM) classifier have been used [20, 27, 28].

2.2 Data Collection

For the present work, 44 B-mode liver US images consisting of 16 HEM and 28 HCC liver images were collected over a time period from March 2010 to March 2014 from the patients visiting the Department of Radiodiagnosis and Imaging, Post Graduate Institute of Medical Education and Research (PGIMER), Chandigarh, India. The consent of the patients was taken prior to recording, the medical ethics board of PGIMER granted the ethical clearance to carry out this research work. The Philips ATL HDI 5000 US machine equipped with the multi-frequency transducer of 2–5

MHz range was used to record the direct digital images. The size of the images is
800×564 pixels with gray scale consisting of 256 tones, and horizontal as well as
vertical resolution is 96 dpi.

2.3 Dataset Description

In the present work, clinically acquired image database of 44 B-mode liver US images
consisting of 16 HEM images with 16 solitary HEM lesions (i.e., 10 typical HEM
lesions and 6 atypical HEM lesions) and 28 HCC images with 28 solitary HCC lesions
(i.e., 13 SHCCs lesions and 15 LHCCs lesions) have been used. The description of
the image database is given in Table 2. The final database comprising of total 160
IROIs and 44 SROIs was stored in Intel® Core™ I3-M 370, 2.40 GHz with 3 GB
RAM.

In order to design an efficient classifier, it should be ensured that the training
data includes representative cases from both the image sub-classes i.e., typical and
atypical HEMs and SHCC as well as LHCC cases. Each image class was divided
into two sets of images (training and testing set). To avoid biasing, the training ROIs
were taken from the first set while testing ROIs were taken from the other set.

The brief description of training and testing dataset is given in Table 3.

2.4 Data Collection Protocol

The following protocols were adopted for the data collection:

(*a*) The judgment regarding the representativeness and diagnostic quality of each
image was made by two participating radiologists with 15 and 25 years of experience in US imaging. (*b*) The ground truth of HEM and HCC lesions was confirmed
using liver image assessment criteria including: (*i*) visualization of sonographic

Table 2 Description: image database

Clinically acquired B-mode liver US images (44)		
Total IROIs: 160, Total SROIs: 44		
	HEM	HCC
Total images	16	28
	Typical HEM: 10	SHCC: 13
	Atypical HEM: 6	LHCC: 15
Total lesions	16	28
Total IROIs	70	90
	Typical HEM IROIs: 27	SHCC IROIs: 19
	Atypical HEM IROIs: 43	LHCC IROIs: 71
Total SROIs	16	28

Note SHCC Small Hepatocellular Carcinoma (size varies from 1.5 to 1.9 cm); *LHCC* Large Hepatocellular Carcinoma (size varies from 2.1 to 5.6 cm). *IROIs* Inside ROIs, *SROIs* Surrounding ROIs

Table 3 Description: training and testing dataset

	HEM	HCC
Description: training dataset		
Total images (26)	10	16
Total lesions	10	16
	Typical HEM lesions: 7	SHCC lesions: 7
	Atypical HEM lesions: 3	LHCC lesions: 9
Total IROIs (90)	40	50
	Typical HEM IROIs: 22	SHCC IROIs: 10
	Atypical HEM IROIs: 18	LHCC IROIs: 40
Total SROIs (26)	10	16
Description: testing dataset		
Total images (18)	6	12
Total lesions	6	12
	Typical HEM lesions: 3	SHCC lesions: 7
	Atypical HEM lesions: 3	LHCC lesions: 9
Total IROIs (90)	40	50
	Typical HEM IROIs: 22	SHCC IROIs: 10
	Atypical HEM IROIs: 18	LHCC IROIs: 40
Total SROIs (26)	10	16

appearances, imaging features of HEM and HCC based on their expertise and knowledge, (*ii*) follow up of clinical history of the patient and other associated findings, and (*iii*) imaging appearance on dynamic helical MRI/CT/biopsy and pathological examinations, which is an invasive procedure. (*c*) The difference between SHCC and LHCC was made by observing the size of the lesion in longitudinal and transverse views. The HCC lesions which are less than 2 cm in size are considered as SHCCs.

The labeling of HEM lesions as typical and atypical and HCC lesions as SHCC or LHCC lesion was done during data collection solely for the purpose of having representative data in the training set for designing the classifier.

2.5 Selection of Regions of Interest (ROIs)

In the present study, two types of ROIs i.e., inside regions of interest (IROIs) and surrounding regions of interest (SROIs) are used. The sample images for typical HEM, atypical HEM, SHCC and LHCC cases with ROIs marked are shown in Fig. 3. The cropping of ROIs from image dataset has been done according to the following protocols:

(*a*) For cropping of IROIs, maximum non-overlapping IROIs were cropped from the region well within the boundary of each lesion by avoiding the necrotic areas, if any.

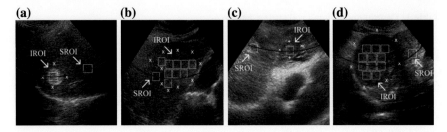

Fig. 3 Sample images with ROIs marked. **a** typical HEM; **b** atypical HEM; **c** SHCC; **d** LHCC;
Note SROI: Surrounding lesion ROI; IROI: Inside Lesion ROI; HEM: Hemangioma; SHCC: Small
hepatocellular carcinoma; LHCC: Large hepatocellular carcinoma

(*b*) For every lesion, one SROI was cropped from the surrounding liver parenchyma
approximately from the same depth as that of the center of the lesion by avoiding
the inhomogeneous structures such as blood vessels and liver ducts, etc.

2.6 Selection of ROI Size

The selection of ROI size is done carefully, considering the fact that it should provide
adequate number of pixels for computing the texture properties. The different sized
ROIs have been selected in the literature for the classification of FLLs such as,
10×10 pixels [22, 23], 32×32 pixels [3, 13, 15, 18], and 64×64 pixels [16]. In
the present study, multiple ROIs of size 32×32 pixels are manually extracted from
each lesion considering the facts given below:
(*a*) It has been shown in earlier studies that ROI size with 800 pixels or more provide
good sampling distribution for estimating reliable statistics. The ROI size of 32×32
contains 1024 pixels and therefore, the texture features computed can be considered
to be reliable estimates.
(*b*) The participating radiologists suggested avoiding larger sized ROIs because some
lesions contain necrotic areas. Therefore, radiologists opined that necrotic area inside
the lesions should be avoided during IROIs extraction. Further, the participating
radiologists were of the view that SROI for every lesion should be chosen by avoiding
inhomogeneous areas such as blood vessels and hepatic ducts etc., which is practically
not possible by considering larger sized ROIs.
(*c*) The ROIs, which are smaller in size, takes less time for feature computation in
comparison to the larger sized ROIs. Also, more number of samples are available by
considering smaller sized ROIs.

2.7 Proposed CAD System Design

In the present work, the CAD system for characterization of HEM and HCC lesions using B-Mode US images has been proposed. The block diagram of the proposed CAD system design is shown in Fig. 4.

For implementing proposed CAD system design, the dataset of 160 non-overlapping IROIs and 44 SROIs was extracted from 44 clinically acquired B-mode liver US liver images. The CAD system includes feature extraction and classification modules. In feature extraction module, texture features are computed from 160 IROIs and 44 SROIs using First Order Statistics (FOS), second order statistics which includes GLCM (Gray-Level Co-occurrence Matrix) method, higher order statistics i.e., GLRLM (Gray-Level Run Length Matrix) method, spectral features

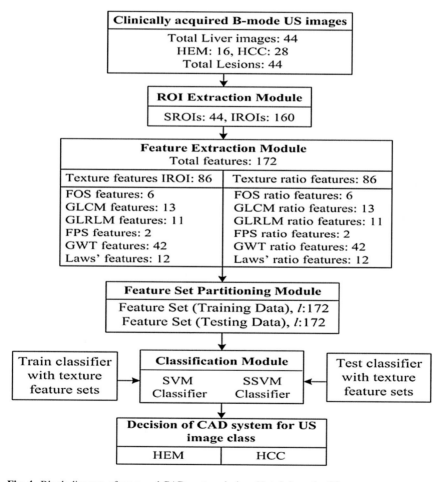

Fig. 4 Block diagram of proposed CAD system design. *Note l*: Length of feature set

such as FPS (Fourier power spectrum) and GWT (Gabor Wavelet Transform) features and spatial filtering based Laws' texture features. The texture feature set of 172 texture features containing 86 texture IROI features and 86 texture ratio features is considered for analysis. The feature set is further divided into training dataset and testing dataset.

The bifurcation of ROIs of particular class in training and testing dataset is described in Table 3.

In classification module, two different classifiers, i.e., SVM and SSVM have been used for the classification task.

2.7.1   Feature Extraction Module

The main idea behind feature extraction is to compute the mathematical descriptors describing the properties of ROI. These mathematical descriptors are further classified as shape based features and texture based features [29, 30].

The participating radiologists opined that the shape based features do not provide any significant information for differential diagnosis between HEM and HCC lesions. Accordingly, the proposed CAD system design is based on the textural features only.

The texture feature extraction methods are generally classified into statistical, spectral and spatial filtering based methods. From the exhaustive review of the related studies on classification of FLLs [3, 13, 15, 18, 22, 23], it can be observed that all these texture features are important for the classification of FLLs.

Accordingly, for the present task of classification between HEM and HCC lesions, the texture features are extracted for each ROI image using statistical, spectral and spatial filtering based methods as shown in Fig. 5.

In the present work, a total of 172 texture features (shown in Table 4) were computed by using statistical, spectral and spatial filtering based texture feature extraction

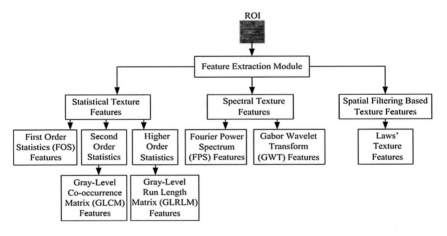

Fig. 5 Texture features computed for each ROI image

Table 4 Description of 172 texture features extracted for characterization of HEM and HCC lesions

Methods	Features description		
Statistical	FOS (F1–F6)	GLCM (F7–F19)	GLRLM (F20–F30)
	F1: average gray level	F7: angular second moment	F20: short run emphasis
	F2: standard deviation	F8: contrast	F21: long run emphasis
	F3: smoothness	F9: correlation	F22: low gray level run emphasis
	F4: third moment	F10: sum of squares variance	F23: high gray level run emphasis
	F5: uniformity	F11: inverse difference moment	F24: short run low gray level emphasis
	F6: entropy	F12: sum average	F25: short run high gray level emphasis
		F13: sum variance	F26: long run low gray level emphasis
		F14: sum entropy	F27: long run high gray level emphasis
		F15: entropy	F28: gray level non uniformity
		F16: difference variance	F29: run length non uniformity
		F17: difference entropy	F30: run percentage
		F18: information measures of correlation- 1	
		F19: information measures of correlation- 2	
Spectral	FPS (F31–F32)	GWT (F33–F74)	
	F31: angular sum	*F33–F53: mean	
	F32: radial sum	*F54–F74: standard deviation	
Spatial filtering	Laws' (F75–F86)		
	F75: LLmean	F79: LSmean	F83: SSstd
	F76: EEmean	F80: ESmean	F84: LEstd
	F77: SSmean	F81: LLstd	F85: LSstd
	F78: LEmean	F82: EEstd	F86:ESstd

Note F87–F172: 86 texture ratio features corresponding to above features (F1–F86). Features F1to F86 are computed for each IROI and SROI so as to compute another 86 texture ratio features (F87–F172) corresponding to the above features. *(F33–F74) two statistical parameters computed out of 21 feature images obtained as a result of convolving 21 Gabor filters with each ROI image. These 21 wavelet filters were computed by considering 3 scales (0, 1, 2) and 7 orientations (22.5°, 45°, 67.5°, 90°, 112.5°, 135°, 157.5°)

methods. Further, these features are applied to the CAD system with a tedious task of joining all the effective features together. The brief description of these texture features is depicted below.

Statistical Texture Features (F1–F30)

The statistical texture features are based on spatial distribution of the gray level intensity values in an image. The statistical feature extraction methods can be categorized as first order statistics (FOS), second order statistics (GLCM), and higher order statistics (GLRLM) based methods.

(a) FOS features (F1–F6): For each ROI, six textural features are computed with FOS method i.e., average gray level, smoothness, standard deviation, entropy, third moment and uniformity [2, 13, 15].

(b) GLCM features (F7–F19): For each ROI, thirteen textural features are computed with GLCM method i.e., contrast, angular second moment, inverse difference moment, correlation, sum average, variance, sum variance, difference variance, entropy, sum entropy, difference entropy, information measure of correlation-1 and information measure of correlation-2 [4, 13, 15, 31] .

(c) GLRLM features (F20–F30): For each ROI, eleven textural features are-computed with GLRLM method i.e., long run emphasis (LRE), short run emphasis (SRE), high gray level run emphasis (HGLRE), low gray level run emphasis (LGLRE), short run high gray level emphasis (SRHGLE), short run low gray level emphasis (SRLGLE), long run high gray level emphasis (LRHGLE), long run low gray level emphasis (LRLGLE), gray level non-uniformity (GLN), run length non-uniformity (RLN) and run percentage (RP) [13, 15, 32–35].

Spectral Texture Features (F31–F74)

The spectral texture features can be computed by FPS method and GWT method as described below:

(a) FPS features (F31–F32): For each ROI, two features i.e., angular sum and radial sum of the discrete Fourier transform (i.e., DFT) has been computed using FPS method [13, 15, 32].

(b) GWT features (F33–F74): Gabor filter provides useful texture descriptors by using multi-scale features estimated at different scales and orientations. The 2D-GWT, considering three scales (0, 1, 2) and seven orientations (22.5°, 45°, 67.5°, 90°, 112.5°, 135°, 157.5°), result in a group of $(7 \times 3 = 21)$ wavelets. When this group of Gabor filters family of 21 wavelets is convolved with a given ROI image, a set of 21 feature images are obtained. The real parts of Gabor filter family of twenty one feature images obtained for a sample HEM ROI image is shown in Fig. 6. From these 21 feature images, mean and standard deviation are computed as texture descriptors resulting in (21 feature images $\times$ 2 statistical parameters = 42) features for each ROI [5, 30, 36].

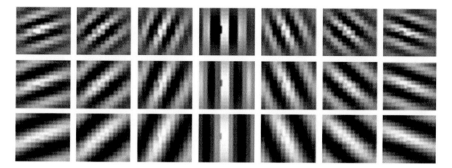

Fig. 6 The real part of Gabor filter family of 21 wavelets (feature images) obtained for a sample HEM IROI image with scales (0, 1, 2) from *top* to *bottom* and orientations (22.7°, 45°, 67.5°, 90°, 112.5°, 135°, 157.5°) from *left* to *right*

Spatial Filtering Based Texture Features (F75–F86)

The Laws' texture features are spatial filtering based texture descriptors which are used to determine the texture properties of a ROI, by performing local averaging (L), edge detection (E), spot detection (S), wave detection (W), and ripple detection (R) in texture [13, 15, 30, 37, 38]. Laws' texture features can be computed by using special 1-D filters of length 3, 5, 7, and 9 as shown in Table 5. Different filter lengths correspond to different resolutions for extraction of texture features from a ROI.

In the present work, 1-D filters of length 3, i.e., L3 = [1, 2, 1], E3 = [−1, 0, 1], and S3 = [−1, 2, −1], have been considered for analysis. A total of nine 2-D filters are generated by combining these 1-D filters as shown in Fig. 7.

The texture images (TIs) are obtained by convolving the ROI of size M × N with these 2D Laws' masks, for example

$$TI_{E3E3} = ROI \otimes E3E3 \tag{1}$$

The output texture images are processed by texture energy measurement (TEM) filters. The TEM filter performs moving average nonlinear filtering as depicted by:

$$TEI = TEM[TI(x, y)]$$
$$= \sum_{i=-7}^{7} \sum_{j=-7}^{7} |I(x+i, y+j) \tag{2}$$

Here, 15 × 15 descriptor windows are used to obtain nine texture energy images (TEIs). TEIs obtained by a pair of identical filters, for example, and are combined to obtain a rotational invariant image (90° rotational invariance) (TR).

$$TR_{E3L3} = \frac{TEI_{E3L3} + TEI_{L3E3}}{2} \tag{3}$$

Table 5 Description of Laws' masks of different lengths

Length of 1-D filter	1-D filter coefficients	No. of 2D Laws' masks	TRs obtained from identical filter pairs	Total TRs
3	L3 = [1, 2, 1] E3 = [−1, 0, 1] S3 = [1−, 2, −1]	9	3	6
5	L5 = [1, 4, 6, 4, 1] E5 = [−1, −2, 0, 2, 1] S5 = [−1, 0, 2, 0, −1] W5 = [−1, 2, 0, −2 1] R5 = [1, −4, 6, −4, 1]	25	10	15
7	L7 = [1, 6, 15, 20, 15, 6, 1] E7 = [−1 −4, −5, 0, 5, 4, 1] S7 = [−1, −2, 1, 4, 1, −2, −1]	9	3	6
9	L9 = [1, 8, 28, 56, 70, 56, 28, 8, 1] E9 = [1, 4, 4, −4, −10, −4, 4, 4, 1] S9 = [1, 0, −4, 0, 6, 0, −4, 0, 1] W9 = [1, −4, 4, −4, −10, 4, 4, −4, 1] R9 = [1, −8, 28, −56, 70, −56, 28, −8, 1]	25	10	15

Note TRs rotational invariant texture images

Fig. 7 Nine 2-D Laws' masks

L3L3	E3L3	S3L3
L3E3	E3E3	S3E3
L3S3	E3S3	S3S3

Statistics derived from these TR images provide significant texture information of ROI. Two statistics i.e., mean, and standard deviation is extracted from each TR image. Thus, twelve Laws' texture features (6 TR images × 2 statistical parameters) are computed for each ROI.

The classification of HEM and HCC lesions was initially attempted by Laws' masks of all the lengths i.e., 3, 5, 7, and 9. It was observed that the features derived by Laws' mask of length 3 resulted in better discrimination between HEM and HCC lesions. Therefore, in the present work, Laws' masks of length 3 have been considered.

For the detection and characterization of HEM and HCC initially, 3 TFVs are computed using FOS, GLCM, GLRLM, FPS, GWT and Laws' texture feature extraction methods. The description of these TFVs is given in Table 6.

Table 6 Description of TFVs

TFVs	Description	l
TFV1	TFV consisting of 86 texture features (6 FOS, 13 GLCM, 11 GLRLM, 2 FPS, 42 Gabor, 12 Laws' features)	86
TFV2	TFV consisting of 86 texture ratio features (6 FOS, 13 GLCM, 11 GLRLM, 2 FPS, 42 Gabor, 12 Laws' features)	86
TFV3	Combined TFV consisting of 86 texture features (TFV1) and 86 texture ratio features (TFV2)	172

Note TFVs Texture Feature Vectors, *l* Length of TFVs

2.7.2 Classification Module

The task of a classifier is to assign a given sample to its concerned class. In the present work, SVM and SSVM classifiers have been used for the classification task. To avoid any bias induced by unbalanced feature values the extracted features are normalized between 0 and 1, by using min-max normalization procedure.

Support Vector Machine (SVM) Classifier

The SVM classifier belongs to a class of supervised machine learning algorithms. This classifier is based on the notion of decision planes which define the decision boundary. The kernel functions are used in nonlinear mapping of training data from the input space to higher dimensional feature space. The polynomial, Gaussian radial basis function and sigmoid kernel are used in general [27, 28, 38–41].

In the present work, LibSVM library has been used for the implementation of SVM classifier [42] and the performance of the Gaussian radial basis kernel function has been examined. The optimal choice of kernel parameter and regularization parameter C is the crucial step in attaining good generalization performance. The regularization parameter C keeps low training error value and tries to maximize the margin while the kernel parameter decides the curvature of decision boundary. In present work, 10-fold cross validation is carried out on training data for each combination of (C, γ), such that $C \varepsilon \{2^{-4}, 2^{-3} \ldots 2^{15}\}$, and $\gamma \varepsilon \{2^{-12}, 2^{-11} \ldots 2^{4}\}$. The optimum values of C and γ can be obtained by this grid search procedure in the parameter space for which the training accuracy is maximum.

Smooth Support Vector Machine (SSVM) Classifier

To solve important mathematical problems related to programming, smoothing methods are extensively used. The SSVM works on the idea of smooth unconstrained optimization reformulation based on the traditional quadratic program which is associated with SVM [43, 44]. For implementing SSVM classifier, the SSVM toolbox [45] developed by Laboratory of Data Science and Machine Intelligence, Taiwan has been used. Similar to SVM implementation in case of SSVM also, 10-fold cross validation is carried out on training data for each combination of (C, γ), such that, $C \varepsilon \{2^{-4}, 2^{-3} \ldots 2^{15}\}$ and $\gamma \varepsilon \{2^{-12}, 2^{-11} \ldots 2^4\}$. The optimum values of C and γ can be obtained by this grid search procedure in the parameter space for which the training accuracy is maximum.

2.8 Results

The flow chart for design of CAD system for classification of HEM and HCC lesions is shown in Fig. 8.

For implementing the above CAD system design rigorous experiments were conducted. A brief description of these experiments is given in Table 7. The performance of the CAD system design has been compared with respect to overall classification accuracy (OCA), individual class accuracy (ICA), and the computational efficiency.

Fig. 8 Flow chart for design of CAD system for classification of HEM and HCC lesions

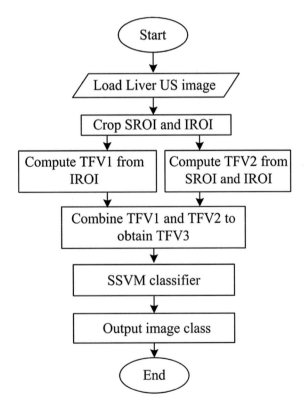

Table 7 Description of experiments carried out in the present work

Experiment no.	Description
1.	To evaluate the performance of TFV1 with SVM and SSVM classifier
2.	To evaluate the performance of TFV2 with SVM and SSVM classifier
3.	To evaluate the performance of TFV3 with SVM and SSVM classifier
4.	To evaluate the computational efficiency of SVM and SSVM classifier with TFV3

Note TFV Texture feature vector

2.8.1 Experiment 1: To Evaluate the Performance of TFV1 with SVM and SSVM Classifier

In this experiment, classification performance of TFV1 consisting of 86 IROI texture features has been evaluated using SVM and SSVM classifier. The results are reported in Table 8. It can be observed that the SVM classifier yields OCA of 52.9 % with ICA values of 90 and 75 % for HEM and HCC classes, respectively. The SSVM classifier yields OCA of 72.9 % with ICA values of 46.6 and 92.5 % for HEM and HCC classes, respectively.

2.8.2 Experiment 2: To Evaluate the Performance of TFV2 with SVM and SSVM Classifier

In this experiment, classification performance of TFV2 consisting of 86 texture ratio features has been evaluated using SVM and SSVM classifier. The results are reported in Table 9. It can be observed that the SVM classifier yields OCA of 77.1 % with ICA values of 50.0 and 97.5 % for HEM and HCC classes, respectively. The SSVM classifier yields OCA of 91.4 % with ICA values of 90 and 92.5 % for HEM and HCC classes, respectively.

Table 8 Classification performance of TFV1 with SVM and SSVM classifier

TFV (l)	Classifier used	CM			OCA (%)	ICA_{HEM} (%)	ICA_{HCC} (%)
			HEM	HCC			
TFV1 (86)	SVM	HEM	27	3	52.9	90.0	75.0
		HCC	30	10			
TFV1 (86)	SSVM	HEM	14	16	72.9	46.6	92.5
		HCC	3	37			

Note TFV1 Texture feature vector 1 (consisting of 86 IROI texture features only); *l* Length of TFV1; *CM* Confusion Matrix; *OCA* Overall classification accuracy; *ICA* Individual Class Accuracy; ICA_{HEM} ICA of HEM class; ICA_{HCC} ICA of HCC class

Table 9 Classification performance of TFV2 with SVM and SSVM classifier

TFV (*l*)	Classifier used	CM			OCA (%)	ICA_HEM (%)	ICA_HCC (%)
			HEM	HCC			
TFV2 (86)	SVM	HEM	15	5	77.1	50.0	97.5
		HCC	1	39			
TFV2 (86)	SSVM	HEM	27	3	91.4	90.0	92.5
		HCC	3	37			

Note TFV2 Texture feature vector 2 (consisting of 86 texture ratio features only); *l* Length of TFV2; *CM* Confusion Matrix; *OCA* Overall classification accuracy; *ICA* Individual Class Accuracy; ICA_{HEM} ICA of HEM class; ICA_{HCC} ICA of HCC class

Table 10 Classification performance with of TFV3 with SVM and SSVM classifier

TFV (*l*)	Classifier used	CM			OCA (%)	ICA_HEM (%)	ICA_HCC (%)
			HEM	HCC			
TFV3 (86)	SVM	HEM	28	2	92.9	93.3	92.5
		HCC	3	37			
TFV3 (86)	**SSVM**	**HEM**	**29**	**1**	**94.3**	**96.6**	**92.5**
		HCC	**3**	**37**			

Note TFV3 Texture feature vector 3 (consisting of 172 texture features, i.e., TFV1 and TFV2); *l* Length of TFV3; *CM* Confusion Matrix; *OCA* Overall classification accuracy; *ICA* Individual Class Accuracy; ICA_{HEM} ICA of HEM class; ICA_{HCC} ICA of HCC class

2.8.3 Experiment 3: To Evaluate the Performance of TFV3 with SVM and SSVM Classifier

In this experiment, the combined TFV i.e., TFV3 consisting of 172 features (86 IROI texture features + 86 texture ratio features) has been evaluated by using SVM and SSVM classifier. The results are reported in Table 10. It can be observed that the SVM classifier yields OCA of 92.9 % with ICA values of 93.3 and 92.5 % for HEM and HCC classes, respectively. The SSVM classifier yields OCA of 94.3 % with ICA values of 96.6 and 92.5 % for HEM and HCC classes, respectively.

It can be observed that the results obtained by SSVM classifier (bold) are better in comparison with SVM classifier.

2.8.4 Experiment 4: To Evaluate the Computational Efficiency of SVM and SSVM Classifier with TFV3

From the results of the exhaustive experiments carried out in the study, it can be observed that TFV3 i.e., combined TFV yields highest OCA value in comparison with the OCA value yielded by TFV1 and TFV2. Therefore, the computational efficiency of SVM and SSVM classifier with only TFV3 has been evaluated. Further, since the time taken for computing TFV3 is same for both the CAD system designs i.e., SVM based CAD system and SSVM based CAD system, the time taken for prediction of 70 cases of testing dataset of TFV3 is considered for evaluating the computational efficiency. It was observed that the proposed SSVM based CAD system design is computationally more efficient than SVM based CAD system design as the time taken for prediction was 5.1 µs for SVM classifier and 3.4 µs for SSVM classifier using MATLAB (version 7.6.0.324 R2008a) with PC configuration Intel® Core™ I3-M 370, 2.40 GHz with 3 GB RAM.

2.9 Discussion

Misclassification Analysis: Analysis of five misclassified cases out of 70 cases (i.e., 5/70) for SVM based CAD system design and four misclassified cases out of 70 cases (i.e., 4/70) for SSVM based CAD system design is given in Table 11.

Table 11 Misclassification analysis of 70 cases of testing dataset with SVM and SSVM classifier

	SVM	SSVM
Misclassification analysis of HEM cases		
Total HEM cases	30	30
Typical HEM cases	5	5
Atypical HEM cases	25	25
Correctly classified cases	28	29
Misclassified cases	2	1
ICA_{HEM}	93.3 %	96.6 %
HEM misclassified cases	2 out of 25 atypical HEM cases	1 out of 25 atypical HEM cases
$ICA_{TypicalHEM}$	100 %	100 %
$ICA_{AtypicalHEM}$	**92 %**	**96 %**
Misclassification analysis of HCC cases		
	SVM	SSVM
Total HCC cases	40	40
Small HCC cases	9	9
Large HCC cases	31	31
Correctly classified cases	37	37
Misclassified cases	3	3
ICA_{HCC}	92.5 %	92.5 %
HCC misclassified cases	3 out of 31 LHCC cases	3 out of 31 LHCC cases
ICA_{SHCC}	100 %	100 %
ICA_{LHCC}	90.3 %	90.3 %

Note ICA Individual Class Accuracy; ICA_{HEM} ICA of HEM class; $ICA_{TypicalHEM}$ ICA of Typical HEM class; $ICA_{AtypicalHEM}$ ICA of Atypical HEM class; ICA_{HCC} ICA of HCC class; ICA_{SHCC} ICA of Small HCC class; ICA_{LHCC} ICA of Large HCC class. It can be observed that the ICA value of atypical HEM class (bold) has improved with SSVM classifier

It can be observed that same ICA values of 100, 100 and 90.3 %, are obtained for typical HEM, SHCC and LHCC cases with both SVM and SSVM based CAD system designs. However, it is worth mentioning that for atypical HEM class the SSVM based CAD system design yields higher ICA value of 96 % in comparison to 92 % as yielded by SVM based CAD system design. It can also be seen from Table 10 that the proposed SSVM based CAD system design yields higher OCA of 94.3 % in comparison to 92.9 % as yielded by SVM based CAD system design.

Since, US offers limited sensitivity for detection of SHCCs and also given the fact that differential diagnosis between atypical HEM and HCC cases is considerably difficult; therefore, the improvement in ICA values for atypical HEMs and HCC cases is highly desirable. Further, it can be observed that all the SHCC cases have been correctly classified by both the CAD system designs (i.e., the ICA value for SHCC is 100 %).

Overall it can be observed that the SSVM based CAD system outperforms the SVM based CAD system with respect to the (a) OCA value, (b) ICA value for the atypical HEM cases, and (c) computational efficiency.

The participating radiologists were of the view that the results yielded by proposed SSVM based CAD system design are quite convincing keeping in view that the comprehensive and diversified database (consisting of typical and atypical HEMs as well SHCC and LHCCs cases) used in present work.

3   Conclusion

In the present work, focuses on the aspect of textural variations exhibited by primary benign and primary malignant focal liver lesions. For capturing these textural variations of benign and malignant liver lesions texture features are computed using statistical methods, signal processing based methods and transform domain methods. In the present work, an efficient CAD system has been proposed for differential diagnosis between benign and malignant focal liver lesions using texture descriptors. The following main conclusions can be drawn:

(a) The texture of the region surrounding the lesion contributes significantly towards the differential diagnosis of HEM and HCC lesions.

(b) The proposed SSVM based CAD system design is better in comparison to the SVM based CAD system design in terms of the OCA value, ICA values for atypical HEM class and computational efficiency.

The promising results yielded by proposed SSVM based CAD system design indicate its usefulness to assist radiologists for the differential diagnosis of HEM and HCC lesions during routine clinical practice.

References

1. Bates, J.: Abdominal Ultrasound How Why and When, 2nd edn, pp. 80–107. Churchill Livingstone, Oxford (2004)
2. Virmani, J., Kumar, V., Kalra, N., Khandelwal, N.: A rapid approach for prediction of liver cirrhosis based on first order statistics. In: Proceedings of IEEE International Conference on Multimedia, Signal Processing and Communication Technologies, IMPACT-2011, pp. 212–215 (2011)
3. Virmani, J., Kumar, V., Kalra, N., Khandelwal, N.: Neural network ensemble based CAD system for focal liver lesions using B-mode ultrasound. J. Digit. Imaging **27**(4), 520–537 (2014)
4. Virmani, J., Kumar, V., Kalra, N., Khandelwal, N.: SVM-based characterization of liver cirrhosis by singular value decomposition of GLCM matrix. Int. J. Artif. Intell. Soft. Comput. **4**(1), 276–296 (2013)
5. Virmani, J., Kumar, V., Kalra, N., Khandelwal, N.: Prediction of liver cirrhosis based on multiresolution texture descriptors from B-mode ultrasound. Int. J. Convergence Comput. **1**(1), 19–37 (2013)
6. Soye, J.A., Mullan, C.P., Porter, S., Beattie, H., Barltrop, A.H., Nelson, W.M.: The use of contrast-enhanced ultrasound in the characterization of focal liver lesions. Ulster Med. J. **76**(1), 22–25 (2007)
7. Colombo, M., Ronchi, G.: Focal Liver Lesions-Detection, Characterization, Ablation, pp. 167–177. Springer, Berlin (2005)
8. Harding, J., Callaway, M.: Ultrasound of focal liver lesions. Rad. Mag. **36**(424), 33–34 (2010)
9. Jeffery, R.B., Ralls, P.W.: Sonography of Abdomen. Raven, New York (1995)
10. Pen, J.H., Pelckmans, P.A., Van Maercke, Y.M., Degryse, H.R., De Schepper, A.M.: Clinical significance of focal echogenic liver lesions. Gastrointest. Radiol. **11**(1), 61–66 (1986)
11. Mitrea, D., Nedevschi, S., Lupsor, M., Socaciu, M., Badea, R.: Advanced classification methods for improving the automatic diagnosis of the hepatocellular carcinoma, based on ultrasound images. In: 2010 IEEE International Conference on Automation Quality and Testing Robotics (AQTR), vol. 2, issue 1, pp. 1–6 (2010)
12. Mitrea, D., Nedevschi, S., Lupsor, M., Socaciu, M., Badea, R.: Exploring texture-based parameters for non-invasive detection of diffuse liver diseases and liver cancer from ultrasound images. In: Proceedings of MMACTEE'06 Proceedings of the 8th WSEAS International Conference on Mathematical Methods and Computational Techniques in Electrical Engineering, pp. 259–265 (2006)
13. Mitrea, D., Nedevschi, S., Lupsor, M., Socaciu, M., Badea, R.: Improving the textural model of the hepatocellular carcinoma using dimensionality reduction methods. In: 2nd International Congress on Image and Signal Processing, 2009. CISP '09. vol. 1, issue 5, pp. 17–19 (2009)
14. Virmani, J., Kumar, V., Kalra, N., Khandelwal, N.: A comparative study of computer-aided classification systems for focal hepatic lesions from B-mode ultrasound. J. Med. Eng. Technol. **37**(4), 292–306 (2013)
15. Virmani, J., Kumar, V., Kalra, N., Khandelwal, N.: PCA-SVM based CAD system for focal liver lesions using B-mode ultrasound images. Def. Sci. J. **63**(5), 478–486 (2013)
16. Yoshida, H., Casalino, D.D., Keserci, B., Coskun, A., Ozturk, O., Savranlar, A.: Wavelet packet based texture analysis for differentiation between benign and malignant liver tumors in ultrasound images. Phys. Med. Biol. **48**, 3735–3753 (2003)
17. Tiferes, D.A., D'lppolito, G.: Liver neoplasms: imaging characterization. Radiol. Bras. **41**(2), 119–127 (2008)
18. Virmani, J., Kumar, V., Kalra, N., Khandelwal, N.: Characterization of primary and secondary malignant liver lesions from B-mode ultrasound. J. Digit. Imaging **26**(6), 1058–1070 (2013)
19. Di Martino, M., De Filippis, G., De Santis, A., Geiger, D., Del Monte, M., Lombardo, C.V., Rossi, M., Corradini, S.G., Mennini, G., Catalano, C.: Hepatocellular carcinoma in cirrhotic patients: prospective comparison of US. CT and MR imaging. Eur. Radiol. **23**(4), 887–896 (2013)

20. Virmani, J., Kumar, V., Kalra, N., Khandelwal, N.: SVM-based characterization of liver ultrasound images using wavelet packet texture descriptors. J. Digit. Imaging **26**(3), 530–543 (2013)
21. Kimura, Y., Fukada, R., Katagiri, S., Matsuda, Y.: Evaluation of hyperechoic liver tumors in MHTS. J. Med. Syst. **17**(3/4), 127–132 (1993)
22. Sujana, S., Swarnamani, S., Suresh, S.: Application of artificial neural networks for the classification of liver lesions by image texture parameters. Ultrasound Med. Biol. **22**(9), 1177–1181 (1996)
23. Poonguzhali, S., Deepalakshmi, B., Ravindran, G.: Optimal feature selection and automatic classification of abnormal masses in ultrasound liver images. In: Proceedings of IEEE International Conference on Signal Processing, Communications and Networking, ICSCN'07, pp. 503–506 (2007)
24. Kadah, Y.M., Farag, A.A., Zurada, J.M., Badawi, A.M., Youssef, A.M.: Classification algorithms for quantitative tissue characterization of diffuse liver disease from ultrasound images. IEEE Trans. Med. Imaging **15**(4), 466–478 (1996)
25. Badawi, A.M., Derbala, A.S., Youssef, A.B.M.: Fuzzy logic algorithm for quantitative tissue characterization of diffuse liver diseases from ultrasound images. Int. J. Med. Inf. **55**, 135–147 (1999)
26. Fukunaga, K.: Introduction to Statistical Pattern Recognition. Academic, New York (1990)
27. Burges, C.J.C.: A tutorial on support vector machines for pattern recognition. Data Min. Knowl. Disc. **2**(2), 1–43 (1998)
28. Guyon, I., Weston, J., Barnhill, S., Vapnik, V.: Gene selection for cancer classification using support vector machines. J. Mach. Learn. **46**(1–3), 1–39 (2002)
29. Kim, S.H., Lee, J.M., Kim, K.G., Kim, J.H., Lee, J.Y., Han, J.K., Choi, B.I.: Computer-aided image analysis of focal hepatic lesions in ultrasonography: preliminary results. Abdom. Imaging **34**(2), 183–191 (2009)
30. Rachidi, M., Marchadier, A., Gadois, C., Lespessailles, E., Chappard, C., Benhamou, C.L.: Laws' mask descriptors applied to bone texture analysis: an innovative and discrimant tool in osteoporosis. Skeletal Radiol. **37**(1), 541–548 (2008)
31. Haralick, R., Shanmugam, K., Dinstein, I.: Textural features for image classification. IEEE Trans. Syst. Man Cybern. SMC- **3**(6), 610–121 (1973)
32. Galloway, R.M.M.: Texture analysis using gray level run lengths. Comput. Graph. Image Process. **4**, 172–179 (1975)
33. Chu, A., Sehgal, C.M., Greenleag, J.F.: Use of gray level distribution of run lengths for texture analysis. Pattern Recogn. Lett. **11**, 415–420 (1990)
34. Dasarathy, B.V., Holder, E.B.: Image characterizations based on joint gray level-run length distributions. Pattern Recogn. Lett. **12**, 497–502 (1991)
35. Virmani, J., Kumar, V., Kalra, N., Khandelwal, N.: Prediction of cirrhosis based on singular value decomposition of gray level co-occurrence matrix and a neural network classifier. In: Proceedings of the IEEE International Conference on Development in E-Systems Engineering, DeSe-2011, pp. 146–151 (2011)
36. Lee, C., Chen, S.: Gabor wavelets and SVM classifier for liver disease classification from CT images. In: Proceedings of IEEE International Conference on Systems, Man and Cybernetics, pp. 548–552. IEEE, Taipei, Taiwan, San Diego, USA
37. Laws, K.I.: Rapid texture identification. SPIE Proc. Semin. Image Process. Missile Guid. **238**, 376–380 (1980)
38. Virmani, J., Kumar, V., Kalra, N., Khandelwal, N.: Prediction of cirrhosis from liver ultrasound B-mode images based on Laws' mask analysis. In: Proceedings of the IEEE International Conference on Image Information Processing, ICIIP-2011, pp. 1–5. Himachal Pradesh, India (2011)
39. Hassanein, A.E., Kim, T.H.: Breast cancer MRI diagnosis approach using support vector machine and pulse coupled neural networks. J. Appl. Logic **10**(4), 274–284 (2012)
40. Kriti., Virmani, J., Dey, N., Kumar, V.: PCA-PNN and PCA-SVM based CAD Systems for Breast Density Classification. In: Hassanien, A.E., et al. (eds.) Applications of Intelligent Optimization in Biology and Medicine, vol. 96, pp. 159–180. Springer (2015)

41. Kriti., Virmani, J.: Breast Tissue Density Classification Using Wavelet-Based Texture Descriptors. In: Proceedings of the Second International Conference on Computer and Communication Technologies (IC3T-2015), pp. 539–546 (2015)
42. Chang, C.C., Lin, C.J.: LIBSVM, a library of support vector machines. http://www.csie.ntu.edu.tw/~cjlin/libsvm. Accessed 15 Jan 2015
43. Purnami, S.W., Embong, A., Zain, J.M., Rahayu, S.P.: A new smooth support vector machine and its applications in diabetes disease diagnosis. J. Comput. Sci. **1**, 1003–1008 (2009)
44. Lee, Y.J., Mangasarian, O.L.: SSVM: a smooth support vector machine for classification. Comput. Optim. Appl. **20**(1), 5–22 (2001)
45. Lee, Y.J., Mangasarian, O.L.: SSVM toolbox. http://research.cs.wisc.edu/dmi/svm/ssvm/. Accessed 20 Feb 2015

Application of Statistical Texture Features for Breast Tissue Density Classification

Kriti, Jitendra Virmani and Shruti Thakur

Abstract It has been strongly advocated that increase in density of breast tissue is strongly correlated with the risk of developing breast cancer. Accordingly change in breast tissue density pattern is taken seriously by radiologists. In typical cases, the breast tissue density patterns can be easily classified into fatty, fatty-glandular and dense glandular classes, but the differential diagnosis between atypical breast tissue density patterns from mammographic images is a daunting challenge even for the experienced radiologists due to overlap of the appearances of the density patterns. Therefore a CAD system for the classification of the different breast tissue density patterns from mammographic images is highly desirable. Accordingly in the present work, exhaustive experiments have been carried out to evaluate the performance of statistical features using PCA-kNN, PCA-PNN, PCA-SVM and PCA-SSVM based CAD system designs for two-class and three-class breast tissue density classification using mammographic images. It is observed that for two-class breast tissue density classification, the highest classification accuracy of 94.4 % is achieved using only the first 10 principal components (PCs) derived from statistical features with the SSVM classifier. For three-class breast tissue density classification, the highest classification accuracy of 86.3 % is achieved using only the first 4 PCs with SVM classifier.

Keywords Breast tissue density classification · Statistical features · Principal component analysis (PCA) · k-nearest neighbor (kNN) classifier · Probabilistic neural network (PNN) classifier · Support vector machine (SVM) classifier · Smooth support vector machine (SSVM) classifier

Kriti
Jaypee University of Information Technology, Solan, Himachal Pradesh, India
e-mail: kriti.23gm@gmail.com

J. Virmani (✉)
Thapar University, Patiala, Punjab, India
e-mail: jitendra.virmani@gmail.com

S. Thakur
Department of Radiology, IGMC, Shimla, Himachal Pradesh, India
e-mail: tshruti878@yahoo.in

© Springer International Publishing Switzerland 2016
A.I. Awad and M. Hassaballah (eds.), *Image Feature Detectors and Descriptors*,
Studies in Computational Intelligence 630, DOI 10.1007/978-3-319-28854-3_16

1 Introduction

Cancer comes under a class of diseases that are characterized by uncontrolled growth of cells resulting in formation of tissue masses called tumors at any location in the body [1]. The malignant tumor can destroy other healthy tissues in the body and often travels to other parts of the body to form new tumors. This process of invasion and destruction of healthy tissues is called metastasis [2]. Breast cancer is the type of cancer that develops form breast cells. It is considered to be a major health problem nowadays and is the most common form of cancer found in women [3]. For the women in United Kingdom, the lifetime risk of being diagnosed with breast cancer is 1 in 8 [4]. The study in [5] reported 1.67 million new incidences of breast cancer worldwide in the year 2012. There are various risk factors associated with cancer development: (*a*) Age, (*b*) History of breast cancer, (*c*) Formation of certain lumps in the breasts (*d*) Higher breast density, (*e*) Obesity, (*f*) Alcohol consumption, (*g*) Cosmetic implants.

It has been strongly advocated by many researchers in their study that high breast density is strongly correlated with the risk of developing breast cancer [6–14]. The association between increased breast density and breast cancer risk can be explained on the basis of effects due to the hormones mitogens and mutagens. The size of the cell population in the breast and cell proliferation is affected by mitogens while the likelihood of damage to these cells is due to mutagens. Due to increased cell population, there is an increase in reactive oxygen species (ROS) production and lipid peroxidation. The products of lipid peroxidation; malondialdehyde (MDA) and isoprostanes catalyze the proliferation of cells [15].

Even though breast cancer is considered to be a fatal disease with a high mortality rate, the chances of survival are improved significantly if it can be detected at an early stage. Various imaging modalities like ultrasound, MRI, computerized tomography, etc. can be used for diagnosis of breast abnormalities but mammography is considered to be the best choice for detection due to its higher sensitivity [16–23]. Mammography is an X-ray imaging technique used to detect the breast abnormalities. Mammograms display the adipose (fatty) and fibroglandular tissues of the breast along with the present abnormalities.

On the basis of density, breast tissue can be classified into the following categories:

(a) *Fatty (F)/Dense (D) (Two-class classification)*
(b) *Fatty (F)/Fatty-glandular (FG)/Dense-glandular (DG) (Three-class classification)*
(c) *Almost entirely fatty (B-I)/Some fibro-glandular tissue (B-II)/Heterogeneously dense breast (B-III)/Extremely dense breast (B-IV) (Four-class BI-RADS classification)*

The typical fatty tissue being translucent to X-rays appears dark on a mammogram where as the dense tissues appear bright on the mammograms. The fatty-glandular breast tissue is an intermediate stage between fatty and dense tissues therefore a typical fatty-glandular breast tissue appears dark with some bright streaks on the

(a)   **(b)**   **(c)**

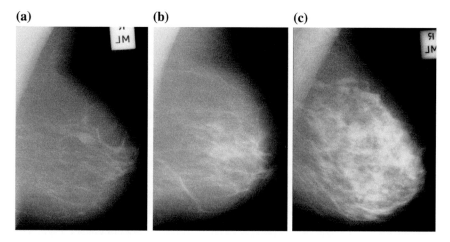

Fig. 1 Sample mammographic images depicting typical cases. **a** Typical fatty tissue 'mdb012'. **b** Typical fatty-glandular tissue 'mdb014'. **c** Typical dense-glandular tissue 'mdb108'

mammogram. The mammographic appearances of the typical breast tissues based on density are depicted in Fig. 1.

The discrimination between different density patterns by visual analysis is highly subjective and depends on the experience of the radiologist. The participating radiologist i.e. one of the co-author of this work, opined that, in case of atypical cases where there is a high overlap in appearances of the different density patterns, a clear discrimination cannot be made by visual analysis easily. The sample mammographic images depicting the atypical cases are shown in Fig. 2.

(a)   **(b)**   **(c)**

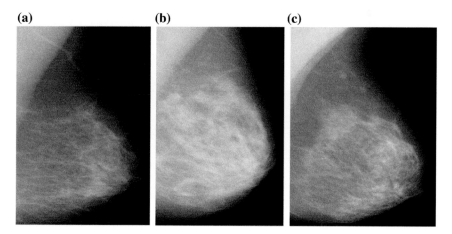

Fig. 2 Sample mammographic images depicting atypical cases **a** Atypical fatty tissue 'mdb088'. **b** Atypical fatty-glandular tissue 'mdb030'. **c** Atypical dense-glandular tissue 'mdb100'

In order to correctly identify and analyze these atypical cases various computer aided diagnostic (CAD) systems have been developed for breast tissue density classification. These proposed CAD systems can be categorized as: (*a*) CAD system designs based on segmented breast tissue versus CAD system designs based on Regions of Interest (ROIs). (*b*) CAD system designs for two class classification (fatty/dense) versus CAD system designs for three class classification (fatty/fatty-glandular/dense-glandular) versus CAD system designs for four class classification based on BI-RADS (B-I: almost entirely fatty/B-II: some fibro-glandular tissue/B-III: heterogeneously dense breast/B-IV: extremely dense breast). (*c*) CAD system designs using standard benchmark dataset (Mammographic image analysis society (MIAS), Digital database of screening mammograms (DDSM), Oxford, Nijmegen) versus CAD system designs using data collected by individual research groups. A brief description of the related studies is given in Tables 1, 2 and 3.

From the above tables, it can be observed that most of the researchers have used a subset of MIAS and DDSM databases and have worked on the segmented breast tissue. It is also observed that only a few studies report CAD systems based on ROIs

Table 1 Summary of studies carried out for two-class breast tissue density classification

Investigators	Dataset description				
	Database	No. of images	ROI size	Classifier	OCA (%)
Miller and Astley [24]	Collected by investigator	40	SBT	Bayesian	80.0
Bovis and Singh [25]	DDSM (SBMD)	377	SBT	ANN	96.7
Castella et al. [26]	Collected by investigator	352	256 × 256	LDA	90.0
Oliver et al. [27]	MIAS (SBMD)	322	SBT	Bayesian	91.0
	DDSM (SBMD)	831			84.0
Mustra et al. [28]	MIAS (SBMD)	322	512 × 384	Naïve Bayesian	91.6
	KBD-FER (Collected by investigator)	144		IB1	97.2
Sharma and Singh [29]	MIAS (SBMD)	322	200 × 200	SMO-SVM	96.4
Sharma and Singh [30]	MIAS (SBMD)	212	200 × 200	kNN	97.2
Kriti et al. [31]	MIAS (SBMD)	322	200 × 200	SVM	94.4
Virmani et al. [32]	MIAS (SBMD)	322	200 × 200	kNN	96.2

Note SBMD Standard benchmark database. *SBT* Segmented breast tissue. *OCA* Overall Classification Accuracy

Table 2 Summary of studies carried out for three-class breast tissue density classification

Investigators	Dataset description				
	Database	No. of images	ROI size	Classifier	OCA (%)
Blot and Zwiggelaar [33]	MIAS (SBMD)	265	SBT	kNN	63.0
Bosch et al. [34]	MIAS (SBMD)	322	SBT	SVM	91.3
Muhimmah and Zwiggelaar [35]	MIAS (SBMD)	321	SBT	DAG-SVM	77.5
Subashini et al. [36]	MIAS (SBMD)	43	SBT	SVM	95.4
Tzikopoulos et al. [37]	MIAS (SBMD)	322	SBT	SVM	84.4
Li [38]	MIAS (SBMD)	42	SBT	KSFD	94.4
Mustra et al. [28]	MIAS (SBMD)	322	512×384	IB1	82.0
Silva and Menotti [39]	MIAS (SBMD)	320	300×300	SVM	77.1

Note SBMD Standard benchmark database. *SBT* Segmented breast tissue. *OCA* Overall Classification Accuracy

Table 3 Summary of studies carried out for four-class breast tissue density classification

Investigators	Dataset description				
	Database	No. of images	ROI size	Classifier	OCA (%)
Karssemeijer [40]	Nijmegen (SBMD)	615	SBT	kNN	80.0
Wang et al. [41]	Collected by investigator	195	SBT	NN	71.0
Petroudi et al. [42]	Oxford (SBMD)	132	SBT	Nearest neighbor	76.0
Oliver et al. [43]	DDSM (SBMD)	300	SBT	kNN+ID3	47.0
Bosch et al. [34]	MIAS (SBMD)	322	SBT	SVM	95.4
	DDSM (SBMD)	500			84.7
Castella et al. [26]	Collected by investigator	352	256×256	LDA	83.0
Oliver et al. [27]	MIAS (SBMD)	322	SBT	Bayesian	86.0
	DDSM (SBMD)	831			77.0
Mustra et al. [28]	MIAS (SBMD)	322	512×384	IB1	79.2
	KBD-FER (collected by investigator)	144			76.4

Note SBMD Standard benchmark database. *SBT* Segmented breast tissue. *OCA* Overall Classification Accuracy

extracted from the breast [26, 28–32, 39] even though it has been shown that the ROIs extracted from the center of the breast result in highest performance as this region of the breast is densest and extraction of ROIs also eliminates an extra step of preprocessing included in obtaining the segmented breast tissue for pectoral muscle removal [44].

The chapter is organized into three sections. Section 2 presents the methodology adopted for present work, i.e. (*a*) description of the dataset on which the work has been carried out (*b*) description of texture features extracted from each ROI image and (*c*) description of the classification module. Section 3 describes the various experiments carried out in the present work for two-class and three-class breast tissue density classification using statistical texture features. Finally, Sect. 4 reports the conclusions drawn from the exhaustive experiments carried out in the present work for two-class and three-class breast tissue density classification.

2 Methodology

2.1 Dataset Description

In the present work a publicly available database, mini-MIAS has been used. This database consists of the Medio Lateral Oblique (MLO) views of both the breasts of 161 women i.e. a total of 322 mammographic images. These images are selected from the UK National Breast Screening Programme and were digitized using The Joyce-Loebl scanning microdensitometer. The images in the database are categorized into three categories as per their density namely fatty (106 images), fatty-glandular (104 images) and dense-glandular (112 images). Each image in the database is of size 1024 × 1024 pixels, with 256 gray scale tones and a horizontal and vertical resolution of 96 dpi. The database also includes location of abnormality, the radius of the circle enclosing the abnormality, its severity and nature of the tissue [45]. In the present work CAD system designs have been proposed for (*a*) two-class breast tissue density classification i.e. (fatty and dense class) and (*b*) three-class breast tissue density classification i.e. (fatty, fatty-glandular and dense-glandular classes). For implementing CAD systems for two-class breast tissue density classification, the fatty-glandular and dense-glandular classes are combined and considered as dense class resulting in 106 mammograms belonging to fatty class and 216 mammograms belonging to dense class. The description of the dataset, used for two-class and three-class CAD system designs is shown in Fig. 3.

2.2 Region of Interest (ROI) Selection

The ROI size is selected carefully considering the fact that it should provide a good population of pixels for computing texture features [44]. Different ROI sizes that have been selected in the literature for classification are 256 × 256 pixels [26], 512 × 384 pixels [28], 200 × 200 pixels [29–32] and 300 × 300 pixels [39]. Other

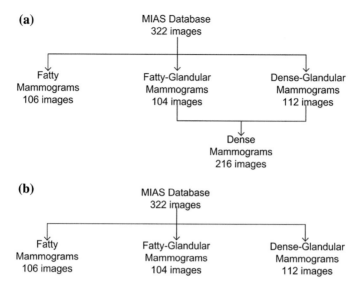

Fig. 3 Dataset description. **a** Two-class breast tissue density classification. **b** Three-class breast tissue density classification

researchers have pre-processed the mammograms by removal of the pectoral muscle and the background using segmented breast tissue for feature extraction [24, 25, 27, 33–38, 40–43]. The participating radiologist, one of the coauthors opined that for accessing the breast tissue density patterns, visual analysis of texture patterns of the center of the breast tissue is carried out during routine practice. Accordingly, for the present work, ROIs of size 200 × 200 pixels are manually extracted from each mammogram. The ROIs are selected from the center of the breast tissue as it has also been asserted by many researchers in their research that the center region of the breast tissue is the densest region and selecting ROI from this part of the breast results in highest performance of the proposed algorithms [29–32, 44]. The selection and extraction of ROI from the breast tissue is shown in Fig. 4.

The sample images of ROIs extracted from the mammographic images are shown in Fig. 5.

2.3 Experimental Workflow for Design of CAD System for Two-Class and Three-Class Breast Tissue Density Classification

With the advancement in computer technology and artificial intelligence techniques there has been a substantial increase in the opportunities for researchers to investigate the potential of CAD systems for texture analysis and tissue characterization of

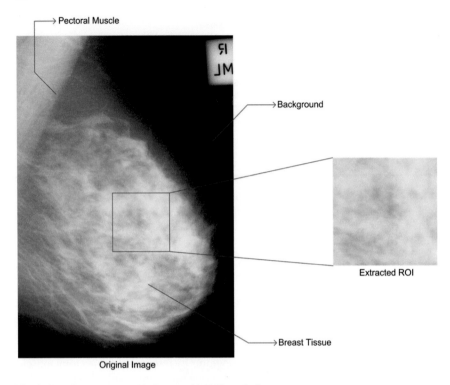

Fig. 4 Sample mammographic image with ROI marked

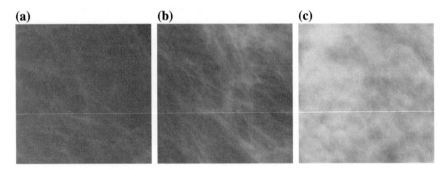

Fig. 5 Sample ROI images. **a** Fatty tissue 'mdb012', **b** Fatty-glandular tissue 'mdb014', **c** Dense-glandular tissue 'mdb108'

radiological images [46–58]. Tissue characterization refers to quantitative analysis of the tissue imaging features resulting in accurate distinction between different types of tissues. Thus, the result of tissue characterization is interpreted using numerical values. The overall aim of developing a computerized tissue characterization system is to provide additional diagnostic information about the underlying tissue which cannot be captured by visual inspection of medical images.

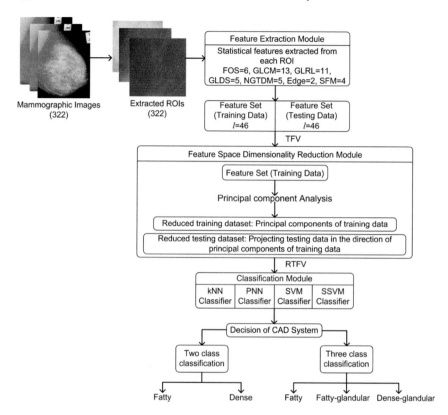

Fig. 6 Experimental workflow for design of CAD systems for two-class and three class breast tissue density classification

The CAD systems are used in the medical imaging as a second opinion tool for the radiologists to gain confidence in their diagnosis. In radiology, CAD systems improve the diagnostic accuracy for medical image interpretation helping the radiologists in detecting the lesions present in the images which might be missed by them.

In general, CAD system design consists of feature extraction module, feature space dimensionality reduction module and classification module. For implementing the proposed CAD system design for breast density classification, 322 ROIs are extracted from 322 images of the MIAS database. The block diagram of experimental workflow followed in the present work is shown in Fig. 6.

For the present CAD system design, ROIs are manually extracted from the mammograms of the MIAS database. In feature extraction module statistical features are extracted from the ROIs. In feature space dimensionality reduction module, PCA is applied to the feature set (training data) to derive its principal components (PCs). The reduced testing dataset is obtained by projecting the data points of feature set (testing data) in the direction of the PCs of feature set (training data). In feature classification module 4 classifiers i.e. k-nearest neighbor (kNN) classifier, probabilistic neural network (PNN) classifier, support vector machine (SVM) classifier and smooth support vector machine (SSVM) classifier are used for the classification task. These classifiers are trained and tested using the reduced texture feature vectors (RTFVs) i.e. set of optimal PCs obtained after applying PCA.

2.3.1 Feature Extraction Module

The feature extraction is the process used to transform the visually extractable and non-extractable features into mathematical descriptors. These descriptors are either shape-based (morphological features) or intensity distribution based (textural features). There are a variety of methods to extract the textural features including statistical, signal processing based and transform domain methods. The different methods of feature extraction are depicted in Fig. 7.

In the present work, the statistical methods are used to extract the texture features from an image based on the gray level intensities of the pixels of that image.

First Order Statistics (FOS) Features

Six features namely average gray level, standard deviation, smoothness, kurtosis and entropy are computed for each ROI [59].

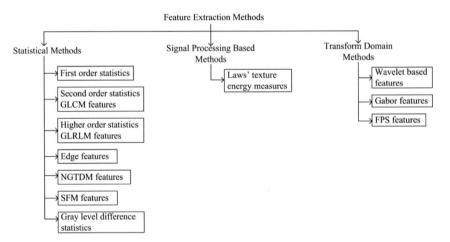

Fig. 7 Different feature extraction methods. **Note**: GLCM: Gray level co-occurrence matrix, GLRLM: Gray level run length matrix, NGTDM: Neighborhood gray tone difference matrix, SFM: Statistical feature matrix, FPS: Fourier power spectrum

Second Order Statistics-Gray Level Co-occurrence Matrix (GLCM) Features

To derive the statistical texture features from GLCM, spatial relationship between two pixels is considered. The GLCM tabulates the number of times the different combinations of pixel pairs of a specific gray level occur in an image for various directions $\theta = 0°, 45°, 90°, 135°$ and different distances d = 1, 2, 3 etc. A total of 13 GLCM features namely angular second moment (ASM), correlation, contrast, variance, inverse difference moment, sum average, sum variance, difference variance, entropy, sum entropy, difference entropy, information measures of correlation-1 and information measures of correlation-2 are computed from each ROI [60–62].

Higher Order Statistics-Gray Level Run Length Matrix (GLRLM) Features

To derive the statistical texture features from the GLRLM, spatial relationship between more than two pixels is considered. In a given direction, GLRLM measures the number of times there are runs of consecutive pixels with the same value. Total of 11 GLRLM features namely short run emphasis, long run emphasis, low gray level run emphasis, high gray level run emphasis, short run low gray level emphasis, short run high gray level emphasis, long run low gray level emphasis, long run high gray level emphasis, gray level non uniformity, run length non-uniformity and run percentage are computed from each ROI [63, 64].

Edge Features (Absolute Gradient)

The edges in an image contain more information about the texture than other parts of the image. The gradient of an image measures the spatial variation of gray levels across an image. At an edge, there is an abrupt change in gray level of the image. If the gray level variation at some point is abrupt then that point is said to have a high gradient and if the variation is smooth the point is at low gradient. Absolute gradient is used to judge whether the gray level variation in an image is smooth or abrupt. The texture features computed are absolute gradient mean and absolute gradient variance [65].

Neighborhood Gray Tone Difference Matrix (NGTDM) Features

NGTDM represents a difference in grayscale between pixels with a certain gray scale and the neighboring pixels. Features extracted from NGTDM are: coarseness, contrast, business, complexity and strength [26, 66].

Statistical Feature Matrix (SFM) Features

SFM is used to measure the statistical properties of pixels at several distances within an image. The features computed from SFM are coarseness, contrast, periodicity and roughness.

Gray Level Difference Statistics (GLDS)

These features are based on the co-occurrence of a pixel pair having a given absolute difference in gray-levels separated by a particular distance. The extracted features are: homogeneity, contrast, energy, entropy and mean [67, 68].

2.3.2 Feature Space Dimensionality Reduction Module

The texture feature vector (TFV) formed after computing the texture features in the feature extraction module may contain some redundant and correlated features which when used in the classification task can degrade the performance of the proposed CAD system. These redundant features give no extra information that proves to be helpful in discriminating the textural changes exhibited by different density patterns. Hence, to remove these redundant features and obtain the optimal attributes for the classification task, PCA is employed [69–71]. Steps used in the PCA algorithm are:

(1) Normalize each feature in dataset to zero mean and unity variance.
(2) Obtain co-variance matrix of the training dataset.
(3) Obtain Eigen values and Eigen vectors from the co-variance matrix. Eigen vectors give the directions of the PCs.
(4) Project the data points in testing dataset in the direction of the PCs of training dataset.

The obtained PCs are uncorrelated to each other and the 1st PC has the largest possible variance out of all the successive PCs. The optimal number of PCs is determined by performing repeated experiments by going through first few PCs i.e. by first considering the first two PCs, then first three PCs and so on, and evaluating the performance of the classifier for each experiment.

2.3.3 Feature Classification Module

Classification is a machine learning technique, used to predict the class membership of unknown data instances based on the training set of data containing instances whose class membership is known. In this module, different classifiers like kNN, PNN, SVM and SSVM are employed to classify the unknown testing instances of mammographic images. The extracted features are normalized in the range [0, 1] by using min-max normalization procedure to avoid any bias caused by unbalanced feature values. The different classifiers employed in the present work are described as below:

k-Nearest Neighbor (kNN) Classifier

The kNN classifier is based on the idea of estimating the class of an unknown instance from its neighbors. It tries to cluster the instances of feature vector into disjoint classes with an assumption that instances of feature vector lying close to each other in feature space represent instances belonging to the same class. The class of an unknown instance in testing dataset is selected to be the class of majority of instances among its k-nearest neighbors in the training dataset. The advantage of kNN is its ability to deal with multiple class problems and is robust to noisy data as it averages the k-nearest neighbors [71–74]. Euclidean distance is used as a distance metric. The classification performance of kNN classifier depends on the value of k. In the present work, the optimal value of k and number of PCs to be retained is determined by performing repeated experiments for the values of $k \in \{1, 2, \ldots, 9, 10\}$ and number of PCs $\in \{1, 2, \ldots, 14, 15\}$. If same accuracy is obtained for more than one value of k, smallest value of k is used to obtain the result.

Probabilistic Neural Network (PNN) Classifier

The PNN is a supervised feed-forward neural network used for estimating the probability of class membership [75–77]. The architecture of PNN has four layers: input layer, pattern layer, summation layer and output layer. Primitive values are passed to the 'n' neurons in the input unit. Values from the input unit are passed to the hidden units in the pattern layer where responses for each unit are calculated. There are 'p' number of neurons in the pattern layer, one for each class. In the pattern layer a probability density function for each class is defined based on the training dataset and optimized kernel width parameter. Values of each hidden unit are summed in the summation layer to get response in each category. Maximum response is taken from all categories in the decision layer to get the class of the unknown instance. The optimal choice of spread parameter (Sp) i.e. the kernel width parameter is critical for the classification using PNN. In the present work, the optimal values used for Sp and optimal number of PCs to design a PNN classifier are determined by performing repeated experiments for values of Sp $\in \{1, 2, \ldots, 9, 10\}$ and number of PCs $\in \{1, 2, \ldots, 14, 15\}$.

Support Vector Machine (SVM) Classifier

The SVM classifier belongs to a class of supervised machine learning algorithms. It is based on the concept of decision planes that define the decision boundary.

In SVM, kernel functions are used to map the non-linear training data from input space to a high dimensionality feature space. Some common kernels are polynomial, Gaussian radial basis function and sigmoid. In the present work, SVM classifier is implemented using LibSVM library [78] and the performance of the Gaussian Radial Basis Function kernel is investigated. The critical step for obtaining a good generalization performance is the correct choice of regularization parameter C and kernel parameter γ. The regularization parameter C tries to maximize the margin while keeping the training error low. In the present work, ten-fold cross validation is carried out on the training data, for each combination of (C, γ) such that, C $\in \{2^{-4}, 2^{-3}, \ldots, 2^{15}\}$ and $\gamma \in \{2^{-12}, 2^{-11}, \ldots, 2^{4}\}$. This grid search procedure in parameter space gives the optimum values of C and γ for which training accuracy is maximum [79–83].

Smooth Support Vector Machine (SSVM) Classifier

To solve important mathematical problems related to programming, smoothing methods are extensively used. SSVM works on the idea of smooth unconstrained optimization reformulation based on the traditional quadratic program which is associated with SVM [84, 85]. For implementing SSVM classifier, the SSVM toolbox developed by Laboratory of Data Science and Machine Intelligence, Taiwan was used [86]. Similar to SVM implementation in case of SSVM also, ten-fold cross validation is carried out on training data for each combination of (C, γ) such that, C $\in \{2^{-4}, 2^{-3}, \ldots, 2^{15}\}$ and $\gamma \in \{2^{-12}, 2^{-11}, \ldots, 2^{4}\}$. This grid search procedure in parameter space gives the optimum values of C and γ for which training accuracy is maximum.

Classifier Performance Evaluation Criteria

The performance of the CAD system for two-class and three class breast tissue density classification can be measured using overall classification accuracy (OCA) and individual class accuracy (ICA). These values can be calculated using the confusion matrix (CM).

$$\text{OCA} = \frac{\Sigma \text{ No. of correctly classified images of each class}}{\text{Total images in testing dataset}} \quad (1)$$

$$\text{ICA} = \frac{\text{No. of correctly classified images of one class}}{\text{Total no. of images in the testing dataset for that class}} \quad (2)$$

3 Results

Rigorous experimentation was carried out in the present work to characterize the mammographic images as per breast tissue density. The experiments carried out in the present work are described in Tables 4 and 5, respectively for two-class and three-class breast tissue density classification.

3.1 Experiments Carried Out for Two-Class Breast Tissue Density Classification

3.1.1 Experiment 1: To Obtain the Classification Performance of Statistical Features for Two-Class Breast Tissue Density Classification Using kNN, PNN, SVM and SSVM Classifiers

In this experiment, classification performance of TFV containing different statistical features is evaluated for two-class breast tissue density classification using different classifiers. The results of the experiment are shown in Table 6. It can be observed from the table that for statistical features, the overall classification accuracy of 92.5, 91.3, 90.6 and 92.5 % is achieved using kNN, PNN, SVM and SSVM classifiers, respectively. It can also be observed that the highest in individual class accuracy for fatty class is 83.0 % with SSVM classifier and highest individual class accuracy for dense class is 100 %, using PNN classifier. Out of total 161 testing instances, 12 instances (12/161) are misclassified in case of kNN, 14 instances (14/161) are misclassified in case of PNN, 16 instances (16/161) are misclassified in case of SVM and 12 instances (12/161) are misclassified in case of SSVM classifier.

Table 4 Description of experiments carried out for two-class breast tissue density classification

Experiment 1	To obtain the classification performance of statistical features for two-class breast tissue density classification using kNN, PNN, SVM and SSVM classifiers
Experiment 2	To obtain the classification performance of statistical features for two-class breast tissue density classification using PCA-kNN, PCA-PNN, PCA-SVM and PCA-SSVM classifiers

Table 5 Description of experiments carried out for three-class breast tissue density classification

Experiment 1	To obtain the classification performance of statistical features for three-class breast tissue density classification using kNN, PNN, SVM and SSVM classifiers
Experiment 2	To obtain the classification performance of statistical features for three-class breast tissue density classification using PCA-kNN, PCA-PNN, PCA-SVM and PCA-SSVM classifiers

Table 6 Classification performance of statistical features using kNN, PNN, SVM and SSVM classifiers for two-class breast tissue density classification

Classifier	CM			OCA (%)	ICA$_F$ (%)	ICA$_D$ (%)
		F	D			
kNN	F	43	10	92.5	81.1	98.1
	D	2	106			
PNN	F	39	14	91.3	73.5	100
	D	0	108			
SVM	F	41	12	90.6	77.3	96.2
	D	4	104			
SSVM	F	44	9	92.5	83.0	97.2
	D	3	105			

Note CM Confusion matrix, *F* Fatty class, *D* Dense class, *OCA* Overall classification accuracy, *ICA$_F$* Individual class accuracy for fatty class, *ICA$_D$* Individual class accuracy for dense class

3.1.2 Experiment 2: To Obtain the Classification Performance of Statistical Features for Two-Class Breast Tissue Density Classification Using PCA-kNN, PCA-PNN, PCA-SVM and PCA-SSVM Classifiers

In this experiment, classification performance of reduced texture feature vector (RTFV) derived by applying PCA to TFV containing different statistical features is evaluated for two-class breast tissue density classification using different classifiers. The results are shown in Table 7.

It can be observed from the table that the overall classification values of 91.9, 91.3, 93.7 and 94.4 % have been achieved using the PCA-kNN, PCA-PNN, PCA-SVM

Table 7 Classification performance of statistical features using PCA-kNN, PCA-PNN, PCA-SVM and PCA-SSVM classifiers for two-class breast tissue density classification

Classifier	*l*	CM			OCA (%)	ICA$_F$ (%)	ICA$_D$ (%)
			F	D			
kNN	6	F	43	10	91.9	81.1	97.2
		D	3	105			
PNN	4	F	39	14	91.3	73.5	100
		D	0	108			
SVM	7	F	43	10	93.7	81.1	100
		D	0	108			
SSVM	10	F	47	6	94.4	88.6	97.2
		D	3	105			

Note l No. of PCs, *CM* Confusion matrix, *F* Fatty class, *D* Dense class, *OCA* Overall classification accuracy, *ICA$_F$* Individual class accuracy for fatty class, *ICA$_D$* Individual class accuracy for dense class

and PCA-SSVM classifiers, respectively. It can also be observed that the highest individual class accuracy for fatty class is 88.6 % using PCA-SSVM classifier and that for dense class is 100 % using PCA-PNN and PCA-SVM classifiers. Out of total 161 testing instances, 13 instances (13/161) are misclassified in case of PCA-kNN, 14 instances (14/161) are misclassified in case of PCA-PNN, 10 instances (10/161) are misclassified in case of PCA-SVM and 9 instances (9/161) are misclassified in case of PCA-SSVM classifier.

From the results obtained from the above experiments, it can be observed that for two-class breast tissue density, PCA-SSVM classifier achieves highest classification accuracy of 94.4 % using first 10 PCs.

3.2 Experiments Carried Out for Three-Class Breast Tissue Density Classification

3.2.1 Experiment 1: To Obtain the Classification Performance of Statistical Features for Three-Class Breast Tissue Density Classification Using kNN, PNN, SVM and SSVM Classifiers

In this experiment, the classification performance of TFV containing different statistical features is evaluated for three-class breast tissue density classification using different classifiers. The results are shown in Table 8.

It can be observed from the table that the overall classification accuracy of 86.9, 85.0, 83.8 and 82.6 % is achieved using kNN, PNN, SVM and SSVM classifiers, respectively. The highest individual class accuracy for fatty class is 94.3 % using SVM classifier, for fatty-glandular class the highest individual class accuracy achieved is 88.4 % using SSVM classifier and for the dense-glandular class, highest individual class accuracy achieved is 96.4 % using kNN classifier. Out of total 161 testing instances, 21 instances (21/161) are misclassified in case of kNN, 24 instances (24/161) are misclassified in case of PNN, 26 instances (26/161) are misclassified in case of SVM and 28 instances (28/161) are misclassified in case of SSVM classifier.

3.2.2 Experiment 2: To Obtain the Classification Performance of Statistical Features for Three-Class Breast Tissue Density Classification Using PCA-kNN, PCA-PNN, PCA-SVM and PCA-SSVM Classifiers

In this experiment, the classification performance of RTFV derived by applying PCA to TFV containing different statistical features is evaluated for three-class breast tissue density classification using different classifiers. The results are shown in Table 9. It can be observed from the table that the overall classification of 85.0, 84.4, 86.3 and 85.0 % is achieved using PCA-kNN, PCA-PNN, PCA-SVM and PCA-SSVM

Table 8 Classification performance of statistical features using kNN, PNN, SVM and SSVM classifiers for three-class breast tissue density classification

Classifier	CM				OCA (%)	ICA_F (%)	ICA_{FG} (%)	ICA_{DG} (%)
		F	FG	DG				
kNN	F	46	2	5	86.9	86.7	76.9	96.4
	FG	2	40	10				
	DG	0	2	54				
PNN	F	41	8	4	85.0	77.3	82.6	94.6
	FG	1	43	8				
	DG	0	3	53				
SVM	F	50	3	0	83.8	94.3	67.3	89.2
	FG	12	35	5				
	DG	1	5	50				
SSVM	F	39	11	3	82.6	73.5	88.4	85.7
	FG	3	46	3				
	DG	1	7	48				

Note CM Confusion matrix, *F* Fatty class, *FG* Fatty–glandular class, *DG* Dense-glandular class, *OCA* Overall classification accuracy, ICA_F Individual class accuracy for fatty class, ICA_{FG} Individual class accuracy for fatty-glandular class, ICA_{DG} Individual class accuracy for dense-glandular class

Table 9 Classification performance of statistical features using PCA-kNN, PCA-PNN, PCA-SVM and PCA-SSVM classifiers for three-class breast tissue density classification

Classifier	*l*	CM				OCA (%)	ICA_F (%)	ICA_{FG} (%)	ICA_{DG} (%)
			F	FG	DG				
kNN	9	F	44	4	5	85.0	83.0	73.0	98.2
		FG	3	38	11				
		DG	1	0	55				
PNN	6	F	43	6	4	84.4	81.1	84.6	87.5
		FG	1	44	7				
		DG	0	7	49				
SVM	4	F	47	4	2	86.3	88.6	76.9	92.8
		FG	6	40	6				
		DG	0	4	52				
SSVM	5	F	43	9	1	85.0	81.1	84.6	89.2
		FG	4	44	4				
		DG	0	6	50				

Note l Optimal number of PCs, *CM* Confusion matrix, *F* Fatty class, *FG* Fatty–glandular class, *DG* Dense-glandular class, *OCA* Overall classification accuracy, ICA_F Individual class accuracy for fatty class, ICA_{FG} Individual class accuracy for fatty-glandular class, ICA_{DG} Individual class accuracy for dense-glandular class

classifiers, respectively. The highest individual class accuracy for fatty class is 88.6 % using PCA-SVM classifier, for fatty-glandular class the highest individual class accuracy achieved is 84.6 % using PCA-PNN and PCA-SSVM classifiers and for the dense-glandular class, highest individual class accuracy achieved is 98.2 % using PCA-kNN classifier. Out of total 161 testing instances, 24 instances (24/161) are misclassified in case of PCA-kNN, 25 instances (25/161) are misclassified in case of PCA-PNN, 22 instances (22/161) are misclassified in case of PCA-SVM and 24 instances (24/161) are misclassified in case of PCA-SSVM classifier.

For three-class breast tissue density classification, it can be observed from the above experiments that highest classification accuracy of 86.9 % is achieved using the kNN classifier, however it should also be noted that PCA-SVM classifier achieves the highest classification accuracy of 86.3 % by using only the first 4 PCs obtained by applying PCA to the TFV of statistical features. Thus CAD system design based on PCA-SVM classifier can be considered to be the best choice for three-class breast tissue density classification.

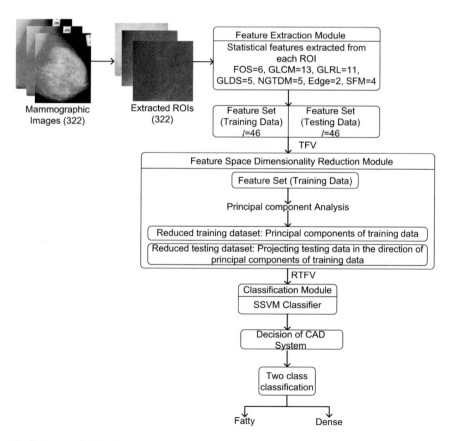

Fig. 8 Proposed SSVM based CAD system design for two-class breast tissue density classification

4   Conclusion

From the rigorous experiments carried out in the present work, it can be observed that for two-class breast tissue density, PCA-SSVM based CAD system using first 10 PCs obtained by applying PCA to the TFV derived using statistical features yields highest OCA of 94.4 % using mammographic images. It could also be observed that the PCA-SVM based CAD system using first 4 PCs obtained by applying PCA to the TFV derived using statistical features yields highest OCA of 86.3 % for three-class breast tissue density classification using mammographic images. It can be concluded that statistical features are significant to account for the textural changes exhibited by the fatty and dense breast tissues. The proposed CAD system designs derived using the above results are shown in Figs. 8 and 9 for two-class and three-class breast tissue density classification, respectively.

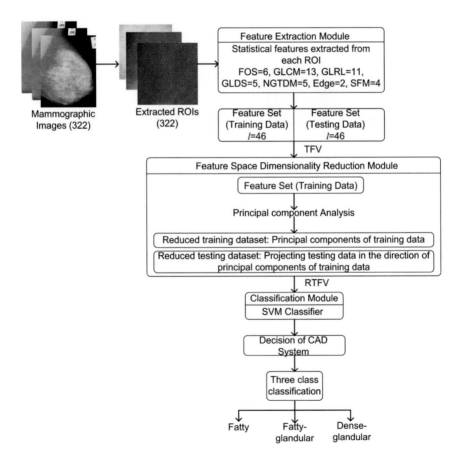

Fig. 9 Proposed CAD system design for three-class breast tissue density classification

The promising results obtained by the proposed CAD system designs indicate their usefulness to assist radiologists for characterization of breast tissue density during routine clinical practice.

References

1. Ganesan, K., Acharya, U.R., Chua, C.K., Min, L.C., Abraham, T.K., Ng, K.H.: Computer-aided breast cancer detection using mammograms: a review. IEEE Rev. Biomed. Eng. **6**, 77–98 (2013)
2. What is cancer? MNT Knowledge Center. http://www.medicalnewstoday.com/info/cancer-oncology/
3. Breast cancer awareness month in October, World Health Organisation, 2012. http://www.who.int/cancer/events/breast_cancer_month/en/
4. Cancer stats: key stats, Cancer Research UK. http://www.cancerresearchuk.org/cancer-info/cancerstats/keyfacts/
5. Globocan 2012: Estimated cancer incidence, mortality and prevalence worldwide, International Agency for Research on Cancer, 2012. http://globocan.iarc.fr/Default.aspx
6. Wolfe, J.N.: Breast patterns as an index of risk for developing breast cancer. Am. J. Roentgenol. **126**(6), 1130–1137 (1976)
7. Wolfe, J.N.: Risk for breast cancer development determined by mammographic parenchymal pattern. Cancer **37**(5), 2486–2492 (1976)
8. Boyd, N.F., Rommens, J.M., Vogt, K., Lee, V., Hopper, J.L., Yaffe, M.J., Paterson, A.D.: Mammographic breast density as an intermediate phenotype for breast cancer. Lancet Oncol. **6**(10), 798–808 (2005)
9. Boyd, N.F., Martin, L.J., Chavez, S., Gunasekara, A., Salleh, A., Melnichouk, O., Yaffe, M., Friedenreich, C., Minkin, S., Bronskill, M.: Breast Tissue composition and other risk factors for breast cancer in young women: a cross sectional study. Lancet Oncol. **10**(6), 569–580 (2009)
10. Boyd, N.F., Martin, L.J., Yaffe, M.J., Minkin, S.: Mammographic density and breast cancer risk: current understanding and future prospects. Breast Cancer Res. **13**(6), 223–235 (2011)
11. Boyd, N.F., Guo, H., Martin, L.J., Sun, L., Stone, J., Fishell, E., Jong, R.A., Hislop, G., Chiarelli, A., Minkin, S., Yaffe, M.J.: Mammographic density and the risk and detection of breast cancer. New Engl. J. Med. **356**(3), 227–236 (2007)
12. Vachon, C.M., Gils, C.H., Sellers, T.A., Ghosh, K., Pruthi, S., Brandt, K.R., Pankratz, V.S.: Mammographic density, breast cancer risk and risk prediction. Breast Cancer Res. **9**(6), 217–225 (2007)
13. Warren, R.: Hormones and mammographic breast density. Maturitas **49**(1), 67–78 (2004)
14. Al Mousa, D.S., Brennan, P.C., Ryan, E.A., Lee, W.B., Tan, J., Mello-Thomas, C.: How mammographic breast density affects radiologists' visual search patterns. Acad. Radiol. **21**(11), 1386–1393 (2014)
15. Boyd, N.F., Lockwood, G.A., Byng, J.W., Tritchler, D.L., Yaffe, M.J.: Mammographic densities and breast cancer risk. Cancer Epidemiol. Biomark. Prev. **7**(12), 1133–1144 (1998)
16. Papaevangelou, A., Chatzistergos, S., Nikita, K.S., Zografos, G.: Breast density: computerized analysis on digitized mammograms. Hellenic J. Surg. **83**(3), 133–138 (2011)
17. Colin, C., Prince, V., Valette, P.J.: Can mammographic assessments lead to consider density as a risk factor for breast cancer? Eur. J. Radiol. **82**(3), 404–411 (2013)
18. Zhou, C., Chan, H.P., Petrick, N., Helvie, M.A., Goodsitt, M.M., Sahiner, B., Hadjiiski, L.M.: Computerized image analysis: estimation of breast density on mammograms. Med. Phys. **28**, 1056–1069 (2001)
19. Heine, J.J., Carton, M.J., Scott, C.G.: An automated approach for estimation of breast density. Cancer Epidemiol. Biomark. Prev. **17**(11), 3090–3097 (2008)

20. Huo, Z., Giger, M.L., Vyborny, C.J.: Computerized analysis of multiple-mammographic views: potential usefulness of special view mammograms in computer-aided diagnosis. IEEE Trans. Med. Imaging **20**(12), 1285–1292 (2001)
21. Jagannath, H.S., Virmani, J., Kumar, V.: Morphological enhancement of microcalcifications in digital mammograms. J. Inst. Eng. (India): Ser. B. **93**(3), 163–172 (2012)
22. Yaghjyan, L., Pinney, S.M., Mahoney, M.C., Morton, A.R., Buckholz, J.: Mammographic breast density assessment: a methods study. Atlas J. Med. Biol. Sci. **1**(1), 8–14 (2011)
23. Virmani, J., Kumar, V.: Quantitative evaluation of image enhancement techniques. In: Proceedings of International Conference on Biomedical Engineering and Assistive Technology (BEATS), pp. 1–8. IEEE Press, New York (2010)
24. Miller, P., Astley, A.: Classification of Breast tissue by texture analysis. Image Vis. Comput. **10**(5), 277–282 (1992)
25. Bovis, K., Singh, S.: Classification of mammographic breast density using a combined classifier paradigm. In: 4th International Workshop on Digital Mammography, pp. 1–4 (2002)
26. Castella, C., Kinkel, K., Eckstein, M.P., Sottas, P.E., Verdun, F.R., Bochud, F.: Semiautomatic mammographic parenchymal patterns classification using multiple statistical features. Acad. Radiol. **14**(12), 1486–1499 (2007)
27. Oliver, A., Freixenet, J., Marti, R., Pont, J., Perez, E., Denton, E.R.E., Zwiggelaar, R.: A novel breast tissue density classification methodology. IEEE Trans. Inf. Technol. Biomed. **12**, 55–65 (2008)
28. Mustra, M., Grgic, M., Delac, K.: Breast density classification using multiple feature selection. Auotomatika **53**(4), 362–372 (2012)
29. Sharma, V., Singh, S.: CFS-SMO based classification of breast density using multiple texture models. Med. Biol. Eng. Comput. **52**(6), 521–529 (2014)
30. Sharma, V., Singh, S.: Automated classification of fatty and dense mammograms. J. Med. Imaging Health Inf. **5**(3), 520–526 (7) (2015)
31. Kriti., Virmani, J., Dey, N., Kumar, V.: PCA-PNN and PCA-SVM based CAD systems for breast density classification. In: Hassanien, A.E., et al. (eds.) Applications of Intelligent Optimization in Biology and Medicine. vol. 96, pp. 159–180. Springer (2015)
32. Kriti., Virmani, J.: Breast tissue density classification using wavelet-based texture descriptors. In: Proceedings of the Second International Conference on Computer and Communication Technologies (IC3T-2015), vol. 3, pp. 539–546 (2015)
33. Blot, L., Zwiggelaar, R.: Background texture extraction for the classification of mammographic parenchymal patterns. In: Proceedings of Conference on Medical Image Understanding and Analysis, pp. 145–148 (2001)
34. Bosch, A., Munoz, X., Oliver, A., Marti, J.: Modelling and classifying breast tissue density in mammograms. computer vision and pattern recognition. In: IEEE Computer Society Conference, vol. 2, pp. 1552–1558. IEEE Press, New York (2006)
35. Muhimmah, I., Zwiggelaar, R.: Mammographic density classification using multiresolution histogram information. In: Proceedings of 5th International IEEE Special Topic Conference on Information Technology in Biomedicine (ITAB), pp. 1–6. IEEE Press, New York (2006)
36. Subashini, T.S., Ramalingam, V., Palanivel, S.: Automated assessment of breast tissue density in digital mammograms. Comput. Vis. Image Underst. **114**(1), 33–43 (2010)
37. Tzikopoulos, S.D., Mavroforakis, M.E., Georgiou, H.V., Dimitropoulos, N., Theodoridis, S.: a fully automated scheme for mammographic segmentation and classification based on breast density and asymmetry. Comput. Methods Programs Biomed. **102**(1), 47–63 (2011)
38. Li, J.B.: Mammographic image based breast tissue classification with kernel self-optimized fisher discriminant for breast cancer diagnosis. J. Med. Syst. **36**(4), 2235–2244 (2012)
39. Silva, W.R., Menotti, D.: Classification of mammograms by the breast composition. In: Proceedings of the 2012 International Conference on Image Processing, Computer Vision, and Pattern Recognition, pp. 1–6 (2012)
40. Karssemeijer, N.: Automated classification of parenchymal patterns in mammograms. Phys. Med. Biol. **43**(2), 365–378 (1998)

41. Wang, X.H., Good, W.F., Chapman, B.E., Chang, Y.H., Poller, W.R., Chang, T.S., Hardesty, L.A.: Automated assessment of the composition of breast tissue revealed on tissue-thickness-corrected mammography. Am. J. Roentgenol. **180**(1), 257–262 (2003)
42. Petroudi, S., Kadir T., Brady, M.: Automatic classification of mammographic parenchymal patterns: a statistical approach. In: Proceedings of 25th Annual International Conference of IEEE on Engineering in Medicine and Biology Society, pp. 798–801. IEEE Press, New York (2003)
43. Oliver, A., Freixenet, J., Bosch, A., Raba, D., Zwiggelaar, R.: Automatic classification of breast tissue. In: Maeques, J.S., et al. (eds.) Pattern Recognition and Image Analysis. LNCS, vol. 3523, pp. 431–438. Springer, Heidelberg (2005)
44. Li, H., Giger, M.L., Huo, Z., Olopade, O.I., Lan, L., Weber, B.L., Bonta, I.: Computerized analysis of mammographic parenchymal patterns for assessing breast cancer risk: effect of ROI size and location. Med. Phys. **31**(3), 549–555 (2004)
45. Suckling, J., Parker, J., Dance, D.R., Astley, S., Hutt, I., Boggis, C.R.M., Ricketts, I., Stamatakis, E., Cerneaz, N., Kok, S.L., Taylor, P., Betal, D., Savage, J.: The mammographic image analysis society digital mammogram database. In: Gale, A.G., et al. (eds.) Digital Mammography. LNCS, vol. 169, pp. 375–378. Springer, Heidelberg (1994)
46. Tang, J., Rangayyan, R.M., Xu, J., El Naqa, I., Yang, Y.: Computer-aided detection and diagnosis of breast cancer with mammography: recent advances. IEEE Trans. Inf.Technol. Biomed. **13**(2), 236–251 (2009)
47. Tagliafico, A., Tagliafico, G., Tosto, S., Chiesa, F., Martinoli, C., Derechi, L.E., Calabrese, M.: Mammographic density estimation: comparison among BI-RADS categories, a semi-automated software and a fully automated one. The Breast **18**(1), 35–40 (2009)
48. Doi, K.: Computer-aided diagnosis in medical imaging: historical review, current status, and future potential. Comput. Med. Imaging Graph. **31**(4–5), 198–211 (2007)
49. Doi, K., MacMahon, H., Katsuragawa, S., Nishikawa, R.M., Jiang, Y.: Computer-aided diagnosis in radiology: potential and pitfalls'. Eur. J. Radiol. **31**(2), 97–109 (1997)
50. Giger, M.L., Doi, K., MacMahon, H., Nishikawa, R.M., Hoffmann, K.R., Vyborny, C.J., Schmidt, R.A., Jia, H., Abe, K., Chen, X., Kano, A., Katsuragawa, S., Yin, F.F., Alperin, N., Metz, C.E., Behlen, F.M., Sluis, D.: An intelligent workstation for computer-aided diagnosis. Radiographics **13**(3), 647–656 (1993)
51. Hui, L., Giger, M.L., Olopade, O.I., Margolis, A., Lan, L., Bonta, I.: Computerized texture analysis of mammographic parenchymal patterns of digitized mammograms. Int. Congr. Ser. **1268**, 878–881 (2004)
52. Kumar, I., Virmani, J., Bhadauria, H.S.: A review of breast density classification methods. In: Proceedings of 2nd IEEE International Conference on Computing for Sustainable Global Development. IndiaCom-2015, pp. 1960–1967. IEEE Press, New York (2015)
53. Tourassi, G.D.: Journey toward computer aided diagnosis: role of image texture analysis. Radiology **213**(2), 317–320 (1999)
54. Virmani, J., Kumar, V., Kalra, N., Khandelwal, N.: Prediction of cirrhosis from liver ultrasound B-Mode images based on laws' mask analysis. In: Proceedings of the IEEE International Conference on Image Information Processing. ICIIP-2011, pp. 1–5. IEEE Press, New York (2011)
55. Virmani, J., Kumar, V., Kalra, N., Khandelwal, N.: Neural network ensemble based CAD system for focal liver lesions from B-Mode ultrasound. J. Digit. Imaging **27**(4), 520–537 (2014)
56. Zhang, G., Wang, W., Moon, J., Pack, J.K., Jean, S.: A review of breast tissue classification in mammograms. In: Proceedings of ACM Symposium on Research in Applied Computation, pp. 232–237 (2011)
57. Chan, H.P., Doi, K., Vybrony, C.J., Schmidt, R.A., Metz, C., Lam, K.L., Ogura, T., Wu, Y., MacMahon, H.: Improvement in radiologists' detection of clustered micro-calcifications on mammograms: the potential of computer-aided diagnosis. Instigative Radiol. **25**(10), 1102–1110 (1990)

58. Virmani, J., Kumar, V., Kalra, N., Khandelwal, N.: Prediction of liver cirrhosis based on multiresolution texture descriptors from B-Mode ultrasound. Int. J. Convergence Comput. **1**(1), 19–37 (2013)

59. Virmani, J., Kumar, V., Kalra, N., Khandelwal, N.: A rapid approach for prediction of liver cirrhosis based on first order statistics. In: Proceedings of the IEEE International Conference on Multimedia. Signal Processing and Communication Technologies, pp. 212–215. IEEE Press, New York (2011)

60. Virmani, J., Kumar, V., Kalra, N., Khandelwal, N.: Prediction of cirrhosis based on singular value decomposition of gray level co-occurrence matrix and a neural network classifier. In: Proceedings of Development in E-systems Engineering (DESE-2011), pp. 146–151 (2011)

61. Vasantha, M., Subbiah Bharathi, V., Dhamodharan, R.: Medical image feature extraction, selection and classification. Int. J. Eng. Sci. Technol. **2**, 2071–2076 (2010)

62. Mohanaiah, P., Sathyanarayanam, P., Gurukumar, L.: Image texture feature extraction using GLCM approach. Int. J. Sci. Res. Publ. **3**(5), 1–5 (2013)

63. Xu, D.H., Kurani, A.S., Furst, J.D., Raicu, D.S.: Run-length encoding for volumetric texture. Heart **27**, 25–30 (2004)

64. Albregtsen, F.: Statistical texture measures computed from gray level run length matrices. Image **1**, 3–8 (1995)

65. Castellano, G., Bonilha, L., Li, L.M., Cendes, F.: Texture analysis of medical images. Clin. Radiol. **59**, 1061–1069 (2004)

66. Amadasun, M., King, R.: Textural features corresponding to textural properties. IEEE Trans. Syst. Man Cybern. **19**, 1264–1274 (1989)

67. Weszka, J.S., Dyer, C.R., Rosenfeld, A.: A comparative study of texture measures for terrain classification. IEEE Trans. Syst. Man Cybern. **6**(4), 269–285 (1976)

68. Kim, J.K., Park, H.W.: Statistical textural features for detection of microcalcifications in digitized mammograms. IEEE Trans. Med. Imaging **18**(3), 231–238 (1999)

69. Kumar, I., Bhadauria, H.S., Virmani, J., Rawat, J.: Reduction of speckle noise from medical images using principal component analysis image fusion. In: Proceedings of 9th International Conference on Industrial and Information Systems, pp. 1–6. IEEE Press, New York (2014)

70. Romano, R., Acernese, F., Canonico, R., Giordano, G., Barone, F.: A principal components algorithm for spectra normalization. Int. J. Biomed. Eng. Technol. **13**(4), 357–369 (2013)

71. Amendolia, S.R., Cossu, G., Ganadu, M.L., Galois, B., Masala, G.L., Mura, G.M.: A comparative study of k-nearest neighbor, support vector machine and multi-layer perceptron for thalassemia screening. Chemometr. Intell. Lab. Syst. **69**(1–2), 13–20 (2003)

72. Virmani, J., Kumar, V., Kalra, N., Khandelwal, N.: A comparative study of computer-aided classification systems for focal hepatic lesions from B-Mode ultrasound. J. Med. Eng. Technol. **37**(44), 292–306 (2013)

73. Yazdani, A., Ebrahimi, T., Hoffmann, U.: Classification of EEG signals using dempster shafer theory and a k-nearest neighbor classifier. In: Proceedings of 4th International IEEE EMBS Conference on Neural Engineering, pp. 327–330. IEEE Press, New York (2009)

74. Wu, Y., Ianakiev, K., Govindaraju, V.: Improved kNN classification. Pattern Recogn. **35**(10), 2311–2318 (2002)

75. Specht, D.F.: Probabilistic neural networks. Neural Netw. **1**, 109–118 (1990)

76. Specht, D.F., Romsdahl, H.: Experience with adaptive probabilistic neural network and adaptive general regression neural network. In: Proceedings of the IEEE International Conference on Neural Networks, pp. 1203–1208. IEEE Press, New York (1994)

77. Georgiou, V.L., Pavlidis, N.G., Parsopoulos, K.E., Vrahatis, M.N.: Optimizing the performance of probabilistic neural networks in a bioinformatics task. In: Proceedings of the EUNITE 2004 Conference, pp. 34–40 (2004)

78. Chang, C.C., Lin, C.J.: LIBSVM, a library of support vector machines. ACM Trans. Intell. Syst. Technol. **2**(3), 27–65 (2011)

79. Virmani, J., Kumar, V., Kalra, N., Khandelwal, N.: PCA-SVM based CAD system for focal liver lesion using B-Mode ultrasound images. Defence Sci. J. **63**(5), 478–486 (2013)

80. Virmani, J., Kumar, V., Kalra, N., Khandelwal, N.: SVM-based characterization of liver cirrhosis by singular value decomposition of GLCM matrix. Int. J. Artif. Intell. Soft Comput. **3**(3), 276–296 (2013)
81. Hassanien, A.E., Bendary, N.E., Kudelka, M., Snasel, V.: Breast cancer detection and classification using support vector machines and pulse coupled neural network. In: Proceedings of 3rd International Conference on Intelligent Human Computer Interaction (IHCI 2011), pp. 269–279 (2011)
82. Azar, A.T., El-Said, S.A.: Performance analysis of support vector machine classifiers in breast cancer mammography recognition. Neural Comput. Appl. **24**, 1163–1177 (2014)
83. Virmani, J., Kumar, V., Kalra, N., Khandelwal, N.: SVM-based characterization of liver ultrasound images using wavelet packet texture descriptors. J. Digit. Imaging **26**(3), 530–543 (2013)
84. Purnami, S.W., Embong, A., Zain, J.M., Rahayu, S.P.: A new smooth support vector machine and its applications in diabetes disease diagnosis. J. Comput. Sci. **5**(12), 1003–1008 (2009)
85. Lee, Y.J., Mangasarian, O.L.: SSVM: a smooth support vector machine for classification. Comput. Optim. Appl. **20**, 5–22 (2001)
86. Lee, Y.J., Mangasarian, O.L.: SSVM toolbox. http://research.cs.wisc.edu/dmi/svm/ssvm/

Author Index